T0324933

Graduate Texts in Mathematics 36

Linear Topological Spaces

John L. Kelley   *Isaac Namioka*

and

W. F. Donoghue, Jr.   *G. Baley Price*
Kenneth R. Lucas   *Wendy Robertson*
B. J. Pettis   *W. R. Scott*
Ebbe Thue Poulsen   *Kennan T. Smith*

Springer-Verlag   New York Heidelberg Berlin

John L. Kelley
Department of Mathematics
University of California
Berkeley, California 94720

Isaac Namioka
Department of Mathematics
University of Washington
Seattle, Washington 98195

AMS Subject Classifications
46AXX

Library of Congress Cataloging in Publication Data

Kelley, John L.
    Linear topological spaces.
    (Graduate texts in mathematics; 36)
    Reprint of the ed. published by Van Nostrand, Princeton, N.J., in series: The University series in higher mathematics.
    Bibliography: p.
    Includes index.
    1. Linear topological spaces. I. Namioka, Isaac, joint author. II. Title. III. Series.
QA322.K44   1976   514'.3   75–41498

Second corrected printing

Originally published in the University Series in Higher Mathematics (D. Van Nostrand Company); edited by M. H. Stone, L. Nirenberg and S. S. Chern.

Printed in the United States of America.

ISBN 0–387–90169–8 Springer-Verlag New York Heidelberg Berlin
ISBN 3–540–90169–8 Springer-Verlag Berlin Heidelberg New York

FOREWORD

THIS BOOK IS A STUDY OF LINEAR TOPOLOGICAL SPACES. EXPLICITLY, WE
are concerned with a linear space endowed with a topology such that scalar
multiplication and addition are continuous, and we seek invariants relative
to the class of all topological isomorphisms. Thus, from our point of view,
it is incidental that the evaluation map of a normed linear space into its
second adjoint space is an isometry; it is pertinent that this map is relatively
open. We study the geometry of a linear topological space for its own sake,
and not as an incidental to the study of mathematical objects which are
endowed with a more elaborate structure. This is not because the relation
of this theory to other notions is of no importance. On the contrary, any
discipline worthy of study must illuminate neighboring areas, and motivation for the study of a new concept may, in great part, lie in the clarification
and simplification of more familiar notions. As it turns out, the theory of
linear topological spaces provides a remarkable economy in discussion of
many classical mathematical problems, so that this theory may properly be
considered to be both a synthesis and an extension of older ideas.*

The text begins with an investigation of linear spaces (not endowed with
a topology). The structure here is simple, and complete invariants for a
space, a subspace, a linear function, and so on, are given in terms of cardinal
numbers. The geometry of convex sets is the first topic which is peculiar to
the theory of linear topological spaces. The fundamental propositions here
(the Hahn-Banach theorem, and the relation between orderings and convex
cones) yield one of the three general methods which are available for attack
on linear topological space problems.

A few remarks on methodology will clarify this assertion. Our results
depend primarily on convexity arguments, on compactness arguments (for
example, Šmulian's compactness criterion and the Banach-Alaoglu theorem),
and on category results. The chief use of scalar multiplication is made in
convexity arguments; these serve to differentiate this theory from that of

---

* I am not enough of a scholar either to affirm or deny that all mathematics is both
a synthesis and an extension of older mathematics.

topological groups. Compactness arguments—primarily applications of the Tychonoff product theorem—are important, but these follow a pattern which is routine. Category arguments are used for the most spectacular of the results of the theory. It is noteworthy that these results depend essentially on the Baire theorem for complete metric spaces and for compact spaces. There are non-trivial extensions of certain theorems (notably the Banach-Steinhaus theorem) to wider classes of spaces, but these extensions are made essentially by observing that the desired property is preserved by products, direct sums, and quotients. No form of the Baire theorem is available save for the classical cases. In this respect, the role played by completeness in the general theory is quite disappointing.

After establishing the geometric theorems on convexity we develop the elementary theory of a linear topological space in Chapter 2. With the exceptions of a few results, such as the criterion for normability, the theorems of this chapter are specializations of well-known theorems on topological groups, or even more generally, of uniform spaces. In other words, little use is made of scalar multiplication. The material is included in order that the exposition be self-contained.

A brief chapter is devoted to the fundamental category theorems. The simplicity and the power of these results justify this special treatment, although full use of the category theorems occurs later.

The fourth chapter details results on convex subsets of linear topological spaces and the closely related question of existence of continuous linear functionals, the last material being essentially a preparation for the later chapter on duality. The most powerful result of the chapter is the Krein-Milman theorem on the existence of extreme points of a compact convex set. This theorem is one of the strongest of those propositions which depend on convexity-compactness arguments, and it has far reaching consequences—for example, the existence of sufficiently many irreducible unitary representations for an arbitrary locally compact group.

The fifth chapter is devoted to a study of the duality which is the central part of the theory of linear topological spaces. The existence of a duality depends on the existence of enough continuous linear functionals—a fact which illuminates the role played by local convexity. Locally convex spaces possess a large supply of continuous linear functionals, and locally convex topologies are precisely those which may be conveniently described in terms of the adjoint space. Consequently, the duality theory, and in substance the entire theory of linear topological spaces, applies primarily to locally convex spaces. The pattern of the duality study is simple. We attempt to study a space in terms of its adjoint, and we construct part of a "dictionary" of translations of concepts defined for a space, to concepts involving the adjoint. For example, completeness of a space E is equivalent to the proposition that each hyperplane in the adjoint E^* is weak* closed whenever its intersection with every equicontinuous set A is weak* closed in A, and the topology of E is the strongest possible having E^* as the class of continuous linear functionals provided each weak* compact convex subset of E is equicontinuous. The situation is very definitely more complicated than in the case of a Banach space. Three "pleasant" properties of a space can be used to classify the type of structure. In order of increasing strength, these are: the topology for E is the strongest having E^* as adjoint (E is a Mackey space), the evaluation map of E into E^{**} is continuous (E is evaluable), and a form of the Banach-Steinhaus theorem holds for E (E is a barrelled space, or tonnelé). A complete metrizable locally convex space possesses all of these properties, but an arbitrary linear topological space may fail to possess any one of them. The class of all spaces possessing any one of these useful properties is closed under formation of direct sums, products, and quotients. However, the properties are not hereditary, in the sense that a closed subspace of a space with the property may fail to have the property. Completeness, on the other hand, is preserved by the formation of direct sums and products, and obviously is hereditary, but the quotient space derived from a complete space may fail to be complete. The situation with respect to semi-reflexiveness (the evaluation map carries E onto E^{**}) is similar. Thus there is a dichotomy, and each of the useful properties of linear topological spaces follows one of two dissimilar patterns with respect to "permanence" properties.

Another type of duality suggests itself. A subset of a linear topological space is called bounded if it is absorbed by each neighborhood of 0 (that is, sufficiently large scalar multiples of any neighborhood of 0, contain the set). We may consider dually a family $\mathscr{B}$ of sets which are to be considered as bounded, and construct the family $\mathscr{U}$ of all convex circled sets which absorb members of the family $\mathscr{B}$. The family $\mathscr{U}$ defines a topology, and this scheme sets up a duality (called an internal duality) between possible topologies for E and possible families of bounded sets. This internal duality is related in a simple fashion to the dual space theory.

The chapter on duality concludes with a discussion of metrizable spaces. As might be expected, the theory of a metrizable locally convex space is more nearly perfect than that of an arbitrary space and, in fact, most of the major propositions concerning the internal structure of the dual of a Banach space hold for the adjoint of a complete metrizable space. Countability requirements are essential for many of these results. However, the structure of the second adjoint and the relation of this space to the first adjoint is still complex, and many features appear pathological compared to the classical Banach space theory.

The Appendix is intended as a bridge between the theory of linear topological spaces and that of ordered linear spaces. The elegant theorems of Kakutani characterizing Banach lattices which are of functional type, and those which are of L^1-type, are the principal results.

A final note on the preparation of this text: By fortuitous circumstance the authors were able to spend the summer of 1953 together, and a complete manuscript was prepared. We felt that this manuscript had many faults, not the least being those inferred from the old adage that a camel is a horse which was designed by a committee. Consequently, in the interest of a more uniform style, the text was revised by two of us, I. Namioka and myself. The problem lists were revised and drastically enlarged by Wendy Robertson, who, by great good fortune, was able to join in our enterprise two years ago.

J. L. K.

Berkeley, California, 1961

Note on notation: The end of each proof is marked by the symbol

|||.

ACKNOWLEDGMENTS

WE GRATEFULLY ACKNOWLEDGE A GRANT FROM THE GENERAL RESEARCH funds of the University of Kansas which made the writing of this book possible. Several federal agencies have long been important patrons of the sciences, but the sponsorship by a university of a large-scale project in mathematics is a significant development.

Revision of the original manuscript was made possible by grants from the Office of Naval Research and from the National Science Foundation. We are grateful for this support.

We are pleased to acknowledge the assistance of Tulane University which made Professor B. J. Pettis available to the University of Kansas during the writing of this book, and of the University of California which made Professor J. L. Kelley available during the revision.

We wish to thank several colleagues who have read all or part of our manuscript and made valuable suggestions. In particular, we are indebted to Professor John W. Brace, Mr. D. O. Etter, Dr. A. H. Kruse, Professor V. L. Klee, and Professor A. Wilansky. We also wish to express our appreciation to Professor A. Robertson and Miss Eva Kallin for their help in arranging some of the problems.

Finally, Mrs. Donna Merrill typed the original manuscript and Miss Sophia Glogovac typed the revision. We extend our thanks for their expert service.

CONTENTS

x

Chapter 1

LINEAR SPACES

This chapter is devoted to the algebra and the geometry of linear spaces; no topology for the space is assumed. It is shown that a linear space is determined, to a linear isomorphism, by a single cardinal number, and that subspaces and linear functions can be described in equally simple terms. The structure theorems for linear spaces are valid for spaces over an arbitrary field; however, we are concerned only with real and complex linear spaces, and this restriction makes the notion of convexity meaningful. This notion is fundamental to the theory, and almost all of our results depend upon propositions about convex sets. In this chapter, after establishing connections between the geometry of convex sets and certain analytic objects, the basic separation theorems are proved. These theorems provide the foundation for linear analysis; their importance cannot be overemphasized.

1 LINEAR SPACES

Each linear space is characterized, to a linear isomorphism, by a cardinal number called its dimension. A subspace is characterized by its dimension and its co-dimension. After these results have been established, certain technical propositions on linear functions are proved (for example, the induced map theorem, and the theorem giving the relation between the linear functionals on a complex linear space and the functionals on its real restriction). The section ends with a number of definitions, each giving a method of constructing new linear spaces from old.

A **real (complex) linear space** (also called a **vector space** or a **linear space over the real** (respectively, **complex) field**) is a

1

non-void set E and two operations called **addition** and **scalar multiplication.**   Addition is an operation $\oplus$ which satisfies the following axioms:

(i)   For every pair of elements x and y in E, $x \oplus y$, called the **sum** of x and y, is an element of E;

(ii)   addition is commutative: $x \oplus y = y \oplus x$;

(iii)   addition is associative: $x \oplus (y \oplus z) = (x \oplus y) \oplus z$;

(iv)   there exists in E a unique element, θ, called the **origin** or the (additive) **zero element**, such that for all x in E, $x \oplus \theta = x$; and

(v)   to every x in E there corresponds a unique element, denoted by $-x$, such that $x \oplus (-x) = \theta$.

Scalar multiplication is an operation $\cdot$ which satisfies the following axioms:

(vi)   For every pair consisting of a real (complex) number a and an element x in E, $a \cdot x$, called the **product** of a and x, is an element of E;

(vii)   multiplication is distributive with respect to addition in E: $a \cdot (x \oplus y) = a \cdot x \oplus a \cdot y$;

(viii)   multiplication is distributive with respect to the addition of real (complex) numbers: $(a + b) \cdot x = a \cdot x \oplus b \cdot x$;

(ix)   multiplication is associative: $a \cdot (b \cdot x) = (ab) \cdot x$;

(x) $1 \cdot x = x$ for all x in E.

From the axioms it follows that the set E with the operation addition is an abelian group and that multiplication by a fixed scalar is an endomorphism of this group.

In the axioms, $+$ and juxtaposition denote respectively addition and multiplication of real (complex) numbers.   Because of the relations between the two kinds of addition and the two kinds of multiplication, no confusion results from the practice, to be followed henceforth, of denoting both kinds of addition by $+$ and both kinds of multiplication by juxtaposition.   Also, henceforth 0 denotes, ambiguously, either zero or the additive zero element θ of the abelian group formed by the elements of E and addition.   Furthermore, it is customary to say simply "the linear space E" without reference to the operations.   The elements of a linear space E are called **vectors.** The **scalar field** K of a real (complex) linear space is the field of real (complex) numbers, and its elements are frequently called **scalars.** The real (complex) field is itself a linear space under the convention that vector addition is ordinary addition, and that scalar multiplication

is ordinary multiplication in the field.   If it is said that a linear space
E is the real (complex) field, it will always be understood that E is a
linear space in this sense.

Two linear spaces E and F are identical if and only if $E = F$ and
also the operations of addition and scalar multiplication are the same.
In particular, the real linear space obtained from a complex linear
space by restricting the domain of scalar multiplication to the real
numbers is distinct from the latter space.   It is called the **real
restriction** of the complex linear space.   It must be emphasized
that the real restriction of a complex linear space has the same set of
elements and the same operation of addition; moreover, scalar
multiplication in the complex space and its real restriction coincide
when both are defined.   The only difference—but it is an important
difference—is that the domain of the scalar multiplication of the real
restriction is a proper subset of the domain of the original scalar
multiplication.   The real restriction of the complex field is the two-
dimensional Euclidean space.   (By definition, **real (complex)
Euclidean** n-space is the space of all n-tuples of real (complex,
respectively) numbers, with addition and scalar multiplication
defined coordinatewise.)   It may be observed that not every real
linear space is the real restriction of a complex linear space (for
example, one-dimensional real Euclidean space).

A subset A of a linear space E is (**finitely**) **linearly independent**
if and only if a finite linear combination $\sum \{a_i x_i : i = 1, \cdots, n\}$, where
$x_i \in A$ for each i and $x_i \neq x_j$ for $i \neq j$, is 0 only when each a_i is zero.
This is equivalent to requiring that each member of E which can be
written as a linear combination, with non-zero coefficients, of distinct
members of A have a unique such representation (the difference of
two distinct representations exhibits linear dependence of A).   A subset B of E is a **Hamel base** for E if and only if each non-zero element
of E is representable in a unique way as a finite linear combination
of distinct members of B, with non-zero coefficients.   A Hamel base
is necessarily linearly independent, and the next theorem shows that
any linearly independent set can be expanded to give a Hamel base.

1.1 Theorem   *Let E be a linear space.   Then:*

   (i) *Each linearly independent subset of E is contained in a maximal
   linearly independent subset.*

   (ii) *Each maximal linearly independent subset is a Hamel base, and
   conversely.*

   (iii) *Any two Hamel bases have the same cardinal number.*

PROOF   The first proposition is an immediate consequence of the maximal principle, and the elementary proof of (ii) is omitted. The proof of (iii) is made in two parts. First, suppose that there is a finite Hamel base B for E, and that A is an arbitrary linearly independent set. It will be shown that there are at most as many members in A as there are in B, by setting up a one by one replacement process. First, a member x_1 of A is a linear combination of members of B, and hence some member of B is a linear combination of x_1 and other members of B. Hence x_1 together with B with one member deleted is a Hamel base. This process is continued; at the r-th stage we observe that a member x_r of A is a linear combination of $x_1, \cdots, x_{r-1}$ and of the non-deleted members of B, that x_r is not a linear combination of $x_1, \cdots, x_{r-1}$ and that therefore one of the remaining members of B can be replaced by x_r to yield a Hamel base. This process can be continued until A is exhausted, in which case it is clear that A contains at most as many members as B, or until all members of B have been deleted. In this case there can be no remaining members of A, for every x is a linear combination of the members of A which have been selected. The proof of (iii) is then reduced to the case where each Hamel base is infinite.

Suppose that B and C are two infinite Hamel bases for E. For each member x of B let $F(x)$ be the finite subset of C such that x is a linear combination with non-zero coefficients of the members of $F(x)$. Since the finite linear combinations of members of $\bigcup \{F(x): x \in B\}$ include every member of B and therefore every member of E, $C = \bigcup \{F(x): x \in B\}$. Let $k(A)$ denote the cardinal number of A; then $k(C) = k(\bigcup \{F(x): x \in B\}) \leqq \aleph_0 \cdot k(B) = k(B)$ because B is an infinite set (see problem 1A). A similar argument shows that $k(B) \leqq k(C)$, whence $k(B) = k(C)$.|||

The **dimension** of a linear space is the cardinal number of a Hamel base for the space.

A linear space F is a **subspace (linear subspace)** of a linear space E if and only if F is a subset of E, F and E have the same scalar field K, and the operations of addition and scalar multiplication in F coincide with the corresponding operations in E. A necessary and sufficient condition that F be a subspace of E is that the set F be a nonempty subset of E, that F be closed under addition and scalar multiplication in E, and that addition and scalar multiplication in F coincide with the corresponding operations in E. If A is an arbitrary subset of E, then the set of all linear combinations of members of A is a linear subspace of E which is called the **linear extension (span, hull)** of A or the **subspace generated by** A.

It is convenient to define addition and scalar multiplication of subsets of a linear space. If A and B are subsets of a linear space E, then $A + B$ is defined to be the set of all sums $x + y$ for x in A and y in B. If x is a member of E, then the set $\{x\} + A$ is abbreviated to $x + A$, and $x + A$ is called the **translation** or **translate** of A by x. If a is a scalar, then aA denotes the set of all elements ax for x in A, and $-A$ is an abbreviation for $(-1)A$; this coincides with the set of all points $-x$ with x in A.

Using this terminology, it is clear that a non-void set F of E is a linear subspace of E if and only if $aF + bF \subset F$ for all scalars a and b. If F and G are linear subspaces, then $F + G$ is a linear subspace, but the translate $x + F$ of a linear subspace is not a linear subspace unless $x \in F$. A set of the form $x + F$, where F is a linear subspace, is called a **linear manifold** or **linear variety**, or a **flat**.

Two linear subspaces F and G of E are **complementary** if and only if each member of E can be written in one and only one way as the sum of a member of F and a member of G. Observe that if a vector x has two representations as the sum of a member of F and a member of G, then (taking the difference) the zero vector has a representation other than $0 + 0$. It follows that linear subspaces F and G are complementary if and only if $F + G = E$ and $F \cap G = \{0\}$. It is true that there is always at least one subspace G complementary to a subspace F of E, for one may choose a Hamel base B for F, adjoin a set C of vectors to get a base for E, and let G be the linear extension of C. It can be shown that, if both G and H are complementary to F in E, then the dimension of G is identical with the dimension of H (see problem 1B). The **co-dimension (deficiency, rank)** of F in E is defined to be the dimension of a subspace of E, which is complementary to F in E.

Let E and F be two linear spaces over the same scalar field, and let T be a mapping of E into F. Then T is a **linear function**[1] from E to F if and only if for all x and y in E and all scalars a and b in K, $T(ax + by) = aT(x) + bT(y)$. A linear function T is a group homomorphism of E, under addition, into F, under addition, with the additional property that $T(ax) = aT(x)$ for each scalar a and each x in E. The range of a linear function T is always a subspace of F. Notice that there exists a linear function T on E with arbitrarily

---

[1] 'Function', 'map', 'mapping', and 'transformation' are all synonymous, and they are used interchangeably throughout the text.

prescribed values on the elements of a Hamel base for E, and that every linear function T is completely determined by its values on the elements of the Hamel base. Consequently linear functions exist in some profusion.

The **null space (kernel)** of a linear function T is the set of all x such that $T(x) = 0$; that is, the null space of T is $T^{-1}[0]$. It is easy to see that T is one-to-one if and only if the null space is $\{0\}$. A one-to-one linear map of E onto F is called a **linear isomorphism** of E onto F. The inverse of a linear isomorphism is a linear isomorphism, and the composition of two linear isomorphisms is again a linear isomorphism. Consequently the class of linear spaces is divided into equivalence classes of mutually linearly isomorphic spaces. A property of one linear space which is shared by every linear isomorph is called a **linear invariant.** The dimension of a linear space is evidently a linear invariant, and, moreover, since it is easy to see that two spaces of the same dimension and over the same scalar field are linearly isomorphic, the dimension is a complete linear invariant. That is, two linear spaces over the same scalar field are isomorphic if and only if they have the same dimension.

If S is a linear map of E into a linear space G and U is a linear map of G into a space F, then the composition $U \circ S$ is a linear map of E into F. It is clear that the null space of $U \circ S$ contains that of S. There is a useful converse to this proposition.

1.2 INDUCED MAP THEOREM *Let T be a linear transformation from E into F, and let S be a linear transformation from E onto G. If the null space of T contains that of S, then there is a unique linear transformation U from G into F such that $T = U \circ S$. The function U is one-to-one if and only if the null spaces of T and S coincide.*

PROOF   If x is any element of G, and $S^{-1}[x]$ is the set of all elements y in E for which $S(y) = x$, then $S^{-1}[x]$ is a translate of $S^{-1}[0]$. Consequently (since $S^{-1}[0] \subset T^{-1}[0]$) T has a constant value, say z, on $S^{-1}[x]$. It now follows easily that $T = U \circ S$ if and only if $U(x)$ is defined to be z, and that the function U is one-to-one if and only if the null spaces coincide.|||

The scalar field is itself a linear space, if scalar multiplication is defined to be the multiplication in the field. A **linear functional** on a linear space E is a linear function with values in the scalar field. The null space N of a linear functional f which is not identically zero is of co-dimension one, as the following reasoning demonstrates. If $x \notin N$, then $f(x) \neq 0$; and if y is an arbitrary member of the linear

space E, then $f(y - [f(y)/f(x)]x) = 0$.   That is, $y - xf(y)/f(x)$ is
a member of N and hence each member of E is the sum of a member
of N and a multiple of x.   A subspace of co-dimension one can
clearly be described as a **maximal** (proper) linear subspace of E.
Moreover, if N is any maximal linear subspace of E and x is any
vector which does not belong to N, then each member y of E can be
written uniquely as a linear combination $f(y)x + z$, where $z \in N$.
The function f is a linear functional, and its null space is precisely N,
and hence each maximal subspace is precisely the null space of a linear
functional which is not identically zero.   Thus the following are
equivalent: null space of a linear functional which is not identically
zero, maximal linear subspace, and linear subspace of co-dimension
one.

It also follows from the foregoing discussion that if g is a linear
functional whose null space includes that of a linear functional f
which is not identically zero, and if $f(x) \neq 0$, then $g(y) = (g(x)/f(x))f(y)$.   That is, g is a constant multiple of f.   This result is a
special case of the following theorem.

1.3   THEOREM ON LINEAR DEPENDENCE   *A linear functional f_0 is a
linear combination of a finite set $f_1, \cdots, f_n$ of linear functionals if and
only if the null space of f_0 contains the intersection of the null spaces of
$f_1, \cdots, f_n$.*

PROOF   If f_0 is a linear combination of $f_1, \cdots, f_n$, then the null space
of f_0 obviously contains the intersection of the null spaces of $f_1, f_2, \cdots,$
f_n.   The converse is proved by induction, and the case $n = 1$ was
established in the paragraph preceding the statement of the theorem.
Suppose that the null space N_0 of f_0 contains the intersection of the
null spaces $N_1, \cdots, N_{k+1}$ of $f_1, \cdots, f_{k+1}$.   If each of the functionals
$f_0, f_1, \cdots, f_k$ is restricted to the subspace N_{k+1}, then, for x in N_{k+1},
it is true that $f_0(x) = 0$ whenever $f_1(x) = \cdots = f_k(x) = 0$.   Hence,
by the induction hypothesis, there are scalars $a_1, \cdots, a_k$ such that
$f_0(x) = \sum \{a_i f_i(x) : i = 1, \cdots, k\}$ whenever $x \in N_{k+1}$.   Consequently
$f_0 - \sum \{a_i f_i : i = 1, \cdots, k\}$ vanishes on the null space of f_{k+1}, and is
therefore a scalar multiple of f_{k+1}.|||

Up to this point the scalar field K has been either the real or
complex numbers, but in the next theorem it will be assumed that K
is complex.   Suppose then that E is a complex linear space, and that
f is a linear functional on E.   For each x in E let $r(x)$ be the real part
of $f(x)$.   It is a straightforward matter to see that r is a linear functional on the real restriction of E; the fact that $f(x) = r(x) - ir(ix)$

for every x in E is perhaps a little less expected, but it is also straight-forward.   On the other hand, if r is a linear functional on the real restriction of E and $f(x) = r(x) - ir(ix)$ for every x in E, then it is easy to see that f is linear on E.   These remarks establish the following result.

1.4   CORRESPONDENCE BETWEEN REAL AND COMPLEX LINEAR FUNC-TIONALS   *The correspondence defined by assigning to each linear functional on E its real part is a one-to-one correspondence between all the linear functionals on. E and all the linear functionals on the real restriction of E.   If f and r are paired under this correspondence, where f is a linear functional on E and r is a linear functional on the real restriction of E, then $f(x) = r(x) - ir(ix)$ for every x in E.*

This section is concluded with a few definitions on the construction of new linear spaces from old.   If F is a linear space and X is a set, then the family of all functions on X to F is a linear space, if addition and scalar multiplication are defined pointwise (that is, $(f + g)(x) = f(x) + g(x)$ and $(af)(x) = af(x)$).   Many, if not most, linear spaces which are studied are subspaces of a function space of this sort.   For example: if F is the scalar field and X is the unit interval, then each of the following families is a linear subspace of the space of all func-tions on $[0:1]$ to K: all bounded functions, all continuous functions, all n-times differentiable functions, and all Borel functions.   The family of all analytic functions on an open subset of the plane is another interesting linear space.   If X is a σ-ring of sets, then each of the families of all additive, bounded and additive, and countably additive complex functions on X is a linear space.

If E and F are linear spaces, the set of all linear functions on E to F is a subspace of the space of all functions on E to F.   If F is the scalar field, then this subspace is simply the family of all linear functionals on E.   This space is called the **algebraic dual** of E, and is denoted by E'.

The product $\times \{E: x \in X\}$ is the space of all functions on a set X to a linear space E.   More generally, if for each member t of a non-void set A there is given a linear space E_t over a fixed scalar field, then the product $\times \{E_t: t \in A\}$ is the set of all functions x on A such that $x(t) \in E_t$ for each t in A.   This product is a linear space under pointwise (coordinate-wise) addition and scalar multiplication.   The subspace $\sum \{E_t: t \in A\}$ of $\times \{E_t: t \in A\}$ consisting of all functions which are zero except at a finite number of points of A is called the **direct sum**.   The **projection** P_s of the product $\times \{E_t: t \in A\}$ onto the coordinate space E_s is defined by $P_s(x) = x(s)$. There is also a natural map I_s of the space E_s into the direct sum, defined for each x in E_s by letting $I_s(x)(t)$ be zero if $t \neq s$ and letting $I_s(x)(s) = x$. This map is called the **injection** of E_s into the direct sum $\sum \{E_t : t \in A\}$. The following simple proposition is recorded for future reference.

1.5   Projections and Injections   *The projection P_s is a linear map of the product $\times \{E_t : t \in A\}$ onto the coordinate space E_s. The injection I_s is a linear isomorphism of the coordinate space E_s onto the subspace of the direct sum $\sum \{E_t : t \in A\}$ which consists of all vectors x such that $x(t) = 0$ for $t \neq s$.*

If F is a linear subspace of a linear space E, the **quotient space (factor space, coset space, difference space)** E/F is defined as follows. The elements of E/F are sets of the form $x + F$, where x is an element of E; evidently two sets of this form are either disjoint or identical. Addition $\oplus$ and scalar multiplication $\cdot$ in E/F are defined by the following equations:

$$(x + F) \oplus (y + F) = (x + F) + (y + F)$$
$$a \cdot (x + F) = ax + F.$$

It can be verified that the class E/F with the operations thus defined is a linear space. The map Q which carries a member x of E into the member $x + F$ of E/F is called the **quotient mapping**; alternatively, Q may be described as the map which carries a point x of E into the unique member of E/F to which x belongs. It is straightforward to see that Q is a linear mapping of E onto E/F, and that F is the null space of Q. It follows that an arbitrary linear function T can be represented as the composition of a quotient map and a linear isomorphism. Explicitly, if T is a linear map of E into G and F is the null space of T, then F is also the null space of the quotient map Q of E onto E/F, and hence there is a linear isomorphism U of E/F into G such that $T = U \circ Q$ by the induced map theorem 1.2.

There is a standard construction for new linear spaces which is based on the direct sum and the quotient construction. We will begin with an example. Consider the class C of all complex functions f, each of which is defined and analytic on some neighborhood of a subset A of the complex plane. The domain of definition of a member f of the class C depends on f, and the problem is (roughly) to make a linear space of C. One possible method: define an equivalence relation, by agreeing that f is equivalent to g if and only if $f - g$ is zero on some neighborhood of A. It is then possible to define addition and scalar multiplication of equivalence classes so that a linear space results. An alternative method of defining the linear space is the following.

For each neighborhood U of A let E_U be the linear space of all analytic functions on U, and for a neighborhood V of A such that $V \subset U$ let $P_{V,U}$ be the map of E_U into E_V which carries each member of E_U into its restriction to V. Let $\mathscr{U}$ be the family of all neighborhoods of A, and let N be the subspace of $\sum \{E_U : U \in \mathscr{U}\}$ consisting of all members ϕ of this direct sum such that the sum of the non-zero values of ϕ vanishes on some neighborhood of A; equivalently, N is the set of all members ϕ of the direct sum such that for some U in $\mathscr{U}$ it is true that, if $\phi(T) \neq 0$, then $T \supset U$, and $\sum \{P_{T,U}(\phi(T)): U \subset T\} = 0$. Each member of $\sum \{E_U : U \in \mathscr{U}\}/N$ contains elements ϕ with a single non-vanishing coordinate $\phi(U) = f$, and, if g is the unique non-vanishing coordinate of another member θ of the sum, then ϕ and θ belong to the same member of $\sum \{E_U : U \in \mathscr{U}\}/N$ if and only if $f = g$ on some neighborhood of A. The quotient $\sum \{E_U : U \in \mathscr{U}\}/N$ is therefore a linear space which represents, in a reasonable way, our intuitive notion of the space of all functions analytic on a neighborhood of the set A. The advantage of using this rather complicated procedure, rather than the equivalence class procedure outlined earlier, is that there are standard ways of topologizing each E_U, the direct sum and the quotient space, so that a suitable topology for the space of functions analytic on a neighborhood of A is more or less self-evident.

The notion of inductive limit of spaces is a formalization of the process described in the preceding. An **inductive system (direct system)** consists of the following: an index set A, directed by a partial ordering $\geqq$; a linear space E_t for each t in A; and for every pair of indices t and s, with $t \geqq s$, a canonical linear map Q_{ts} of E_s into E_t such that: $Q_{ts} \circ Q_{sr} = Q_{tr}$, for all r, s, and t such that $t \geqq s \geqq r$, and Q_{it} is the identity map of E_t for all t. The **kernel** of an inductive system is the subspace N of the direct sum $\sum \{E_t : t \in A\}$ consisting of those f for which there is an index t in A such that $s \leqq t$ whenever $f(s) \neq 0$ and such that $\sum \{Q_{ts}(f(s)): s \leqq t\}$ is the zero of E_t. The **inductive limit,** lim ind $\{E_t : t \in A\}$, is defined to be the quotient space $(\sum \{E_t : t \in A\})/N$. The term "inductive limit" is justified by the fact that if B is a cofinal subset of A, then lim ind $\{E_t : t \in B\}$ is linearly isomorphic in a natural way to lim ind $\{E_t : t \in A\}$ (see problem 1I).

There is a construction which is dual, in a certain sense, to that of the inductive limit. A **projective system** (**inverse system**) of linear spaces consists of the following: an index set A directed by a partial ordering $\geq$: a linear space E_t for each t in A; for each pair of indices s and t with $t \geq s$, a canonical linear transformation P_{st} from E_t into E_s with the property that if $t \geq s \geq r$, then $P_{rt} = P_{rs} \circ P_{st}$, and the property that P_{ss} is the identity transformation for all s in A. The spaces E_t are to be thought of intuitively as approximations to a limit space, the accuracy of the approximation increasing as the indices increase. The canonical maps P_{st} relate the various approximations. The **projective limit** (**inverse limit**) of the system is the subspace of the product $\mathsf{X} \{E_t : t \in A\}$ which consists of all x such that for every pair of indices t and s with $t \geq s$, $P_{st}(x(t)) = x(s)$. The projective limit of the system is denoted by lim proj $\{E_t : t \in A\}$.

A simple example of a projective limit is the following. Suppose that A is an index set for each of whose members t there is defined a linear space E_t in such a way that the intersection of each pair of the spaces contains a third. Then A can be directed by agreeing that $t \geq s$ if and only if $E_t \subset E_s$. The resulting system is a projective system if each P_{st} is taken to be the identity transformation. There is clearly an algebraic isomorphism between the projective limit of the system and the intersection of the spaces E_t. The union of a family of subspaces which is directed by $\supset$ is isomorphic, in a dual fashion, to an inductive limit.

PROBLEMS

A CARDINAL NUMBERS

If $\mathscr{F}$ is the family of all finite subsets of an infinite set A, then $k(\mathscr{F}) = k(A)$. What is the cardinal number of the family of countable subsets of A?

B QUOTIENTS AND SUBSPACES

(a) If F is a subspace of a linear space E, and G is any subspace of E complementary to F, then G is isomorphic to E/F and hence dim $G =$ dim (E/F).

(b) If F and G are subspaces of a linear space E, then dim $(F + G) +$ dim $(F \cap G) =$ dim $F +$ dim G.

C DIRECT SUMS AND PRODUCTS

(a) Any linear space E is isomorphic to the direct sum $\sum \{K_t : t \in A\}$ where K_t is the field K for each t and A is a Hamel base for E.

(b) Let E be the product space $\mathsf{X}\{K_n : 1, 2, \cdots\}$ where K_n is the field K for every n. For each a in K let $x(a)$ be the element $(a, a^2, \cdots, a^n, \cdots)$, and let $Q = \{x(a) : 0 < a < 1\}$. Then Q is a linearly independent set of cardinal $2^{\aleph_0}$. Hence any space which is the product of infinitely many non-trivial factor spaces necessarily has dimension at least $2^{\aleph_0}$; thus there are linear spaces which are not isomorphic to any product of copies of K.

(c) If E is a direct sum $\sum \{E_t : t \in A\}$ of linear spaces, then the algebraic dual E' of E is isomorphic to the product $\mathsf{X}\{E_t' : t \in A\}$. For any linear space E, E' is isomorphic to $\mathsf{X}\{K_t : t \in A\}$ where K_t is the field K for each t and A is a Hamel base for E. In particular, if E is finite dimensional then E' and E are isomorphic. (Cf. 14.7.)

D   SPACE OF BOUNDED FUNCTIONS

Let $B(S)$ be the linear space of all bounded scalar valued functions on an infinite set S and let $a_1, a_2, \cdots \in S$. For each $\alpha \neq 0$ such that $|\alpha| \leq 1$ define u_α by $u_\alpha(x) = \alpha^n$ for $x = a_n$ and $u_\alpha(x) = 0$ otherwise. The family $\{u_\alpha\}$ is linearly independent. If s is the cardinal of S and if c is the cardinal of the scalars, then the dimension of $B(S)$ is c^s.

E   EXTENSION OF LINEAR FUNCTIONALS

If F is a subspace of a linear space E and f is a linear functional on F, then there is a linear functional $\bar{f}$ on E which coincides with f on F.

F   NULL SPACES AND RANGES

Let T be a linear mapping of one linear space E into another linear space F and let N be the null space of T. Then E/N is isomorphic to the subspace $T[E]$ of F.

G   ALGEBRAIC ADJOINT OF A LINEAR MAPPING

If T is a linear mapping of a linear space E into another linear space F, then the mapping T' that assigns to each element g of F' the element $g \circ T$ of E' is linear on F' to E'; it is called the algebraic adjoint of T. The null space of T' consists of those elements of F' which vanish on $T[E]$, and $T'[F']$ consists of those elements of E' which vanish on the null space of T.

H   SET FUNCTIONS

Let $\mathscr{A}$ be a family of sets in a linear space E, and assume that for each subset A of E there exists a smallest member $\phi(A)$ of $\mathscr{A}$ containing A. Let $A, A_1, \cdots, A_n$, and A_t for t in B be arbitrary subsets of E. Then $\phi(\bigcup \{A_t : t \in B\}) = \phi(\bigcup \{\phi(A_t) : t \in B\})$. If $\mathscr{A}$ is closed under scalar multiplication, then, $\phi(aA) = a\phi(A)$. If $\mathscr{A}$ is closed under translation, then $\phi(\sum \{A_i : i = 1, 2, \cdots, n\}) \supset \sum \{\phi(A_i) : i = 1, 2, \cdots, n\}$. If $\mathscr{A}$ is closed under addition, then $\phi(\sum \{A_i : i = 1, 2, \cdots, n\}) \subset \sum \{\phi(A_i) : i = 1, 2, \cdots, n\}$.

I   INDUCTIVE LIMITS (see 16C, 17G, 19A, 22C)

If B is a cofinal subset of A, then $\lim \text{ind} \{E_t : t \in B\}$ is isomorphic to $\lim \text{ind} \{E_t : t \in A\}$. Write $F = \sum \{E_t : t \in A\}$, $G = \sum \{E_t : t \in B\}$, and let M and N be the respective kernels. Let $\mathscr{J}$ be the natural injection of G

into F, and let Q_M and Q_N be the quotient maps. Put $I = Q_M \circ \mathcal{J} \circ Q_N^{-1}$. Then I is an isomorphism of G/N onto F/M. (First, $\mathcal{J}(N) \subset M$, and this shows that I is well-defined: $I(x)$ is independent of the member of x chosen in $Q_N^{-1}(x)$. Next, $\mathcal{J}(y) \in M$ implies $y \in N$, and hence I is one-one. Finally, if $z \in F/M$, choose $w \in Q_M^{-1}(z)$. Take t so large that $w(s) \neq 0$ implies $s \leq t$ and so that $t \in B$; define $y \in G$ by taking $y(s)$ zero for all $s \in B \sim \{t\}$ and $y(t)$ equal to the sum in E_t of the images of all the values of w. Then if $x = Q_N(y)$, $I(x) = z$.)

2 CONVEXITY AND ORDER

This section begins with a few elementary propositions about convex sets and circled sets. The two principal theorems of the section establish a correspondence between certain convex sets and Minkowski functionals (subadditive, non-negatively homogeneous functionals), and a correspondence between cones and partial orderings.

The **line segment** joining two points x and y of a linear space is the set of all points of the form $ax + by$ with a and b non-negative real numbers such that $a + b = 1$, or equivalently, the set of all points $ax + (1 - a)y$ with a real and $0 \leq a \leq 1$. This set is denoted by $[x:y]$. The open line segment joining x and y, denoted by $(x:y)$, is the set of all points of the form $ax + (1 - a)y$ with a real and $0 < a < 1$. The set $(x:y)$ with the point x adjoined is denoted by $[x:y)$, and the set $(x:y)$ with the point y adjoined is denoted by $(x:y]$.

A set in a linear space E is **convex** if, whenever it contains x and y, it also contains $[x:y]$. Clearly any subspace is convex, and so is any translate of a convex set; a simple computation shows also that any finite linear combination of convex sets is again convex. From the definition it is obvious that the intersection of the members of any family of convex sets is convex, and the union of the members of a family of convex sets which is directed by $\supset$ (the union of any two members of the family is contained in some third member) is convex. Since E is convex, the family of all convex sets containing any particular set A in E is non-void, and the intersection of the members of this family is the smallest convex set containing A. This intersection is the **convex extension** (**hull, envelope, cover**) of A and will be denoted by $\langle A \rangle$. It is easy to verify that the set of all finite linear combinations $\sum \{a_i x_i : i = 1, \cdots, n\}$, where n is any positive integer, each x_i is in A, each a_i is real and non-negative, and $\sum \{a_i : i = 1, \cdots, n\} = 1$, is a convex set containing A. On the other hand, it follows by a finite induction that, if A is convex, it contains all such combinations. It is then clear that $\langle A \rangle$ is the set of all such combinations. The following theorem states these and a few other simple facts.

2.1 COMPUTATION RULES FOR CONVEX SETS   *Let E be a linear space.*
Then:

- (i) *For any subset A of E the smallest convex set $\langle A \rangle$ which contains A consists of all finite linear combinations of the form $\sum \{a_i x_i : i = 1, \cdots, n\}$, where $x_i \in A$, a_i is real and non-negative, and the sum of the coefficients a_i is one.*
- (ii) *If A and B are non-void subsets of E and a and b are scalars, then $\langle aA + bB \rangle = a \langle A \rangle + b \langle B \rangle$.*
- (iii) *The intersection of convex sets is convex.*
- (iv) *If a family of convex sets is directed by $\supset$, then the union of the members of the family is convex.*
- (v) *If A and B are non-void subsets of E, then $\langle A \cup B \rangle = \bigcup \{[x:y]: x \in \langle A \rangle, y \in \langle B \rangle\}$.*

The union of all line segments with one end point in a set A and the other in a set B is sometimes called the **join** of A and B. The last proposition in the preceding list then implies that the join of two non-void convex sets is convex, and is the convex extension of the union.

A set A is **circled** if and only if $aA \subset A$ whenever $|a| \leq 1$. A circled set A has the property that $aA = A$ whenever $|a| = 1$. In particular, a circled set A is always **symmetric**, in the sense that $A = -A$. It is easy to see that A is circled and convex if and only if A contains $ax + by$ whenever x and y are in A and $|a| + |b| \leq 1$. In a real linear space, a set is convex and circled if and only if it is convex and symmetric. The smallest linear subspace containing a non-void convex circled set A is simply the set $\bigcup \{nA: n = 1, 2, \cdots\}$. The smallest circled set containing a set A is called the **circled extension** of A and is denoted by (A). Clearly $(A) = \bigcup \{aA: |a| \leq 1\}$. The convex extension of a circled set is circled. The smallest convex circled set containing A is $\langle (A) \rangle$, and this is precisely the set of all linear combinations $\sum \{a_i x_i : i = 1, \cdots, n\}$ where $x_i \in A$ and $\sum \{|a_i|: i = 1, \cdots, n\} \leq 1$.

A subset A of a linear space is called **radial at a point** x if and only if A contains a line segment through x in each direction. More precisely stated, A is radial at x if and only if for each vector y different from x there is z, $z \neq x$, such that $[x:z] \subset [x:y] \cap A$. The **radial kernel** of a set A is the set of all points at which A is radial. It should be observed that, even in two-dimensional real Euclidean space, a set may be radial at a single point (see problem 2H). The radial kernel of the radial kernel of A is usually quite different from the radial kernel of A.   However, if A is convex, then the situation is simpler.

2.2   Theorem   *Let A be a convex subset of a linear space E.   If x is a member of the radial kernel of A and $y \in A$, then the half-open interval $[x:y)$ is contained in the radial kernel of A.*

Consequently the radial kernel of a convex set is convex and is its own radial kernel.

Proof   Let $v = ax + (1 - a)y$, with $0 < a \leqq 1$.   For any fixed z in E there is a positive number r such that $x + bz \in A$ whenever $0 < b < r$.   Since A is convex, $a(x + bz) + (1 - a)y = v + abz \in A$ for all b such that $0 < b < r$.   Hence $v + cz \in A$ if $0 < c < ar$, and it follows that v is in the radial kernel of A.|||

It is worth observing that if f is a linear map of E onto a vector space F, and if a subset A of E is radial at x, then the set $f[A]$ is radial at $f(x)$.   In particular, if f is a linear functional which is not identically zero and A is radial at x, then $f[A]$ is radial at $f(x)$; and if A is convex and radial at each of its points, then $f[A]$ is open.

Certain convex sets which are radial at 0 can be described by means of real valued functions.   If U is a set which is radial at 0, the **Minkowski functional** for U is defined to be the real valued function p defined on E by $p(x) = \inf\{a : \frac{1}{a} x \in U, a > 0\}$.   A simple computation shows that $p(ax) = ap(x)$ whenever $a \geqq 0$; that is, p is **non-negatively homogeneous**.   If U is convex, then p is **subadditive**; that is, $p(x + y) \leqq p(x) + p(y)$.   If U is circled, then $p(ax) = |a| p(x)$; that is, p is **absolutely homogeneous**.   A non-negative functional on E which is absolutely homogeneous and subadditive is called a **pseudo-norm**.

2.3   Convex Sets and Minkowski Functionals

   (i) *If U is a convex set which is radial at 0, and if p is the Minkowski functional for U, then $\{x : p(x) < 1\}$ is the radial kernel of U and $U \subset \{x : p(x) \leqq 1\}$.*

   (ii) *If p is a non-negative, non-negatively homogeneous subadditive functional on E, and if $V = \{x : p(x) \leqq 1\}$, then V is a convex set which is radial at 0, and p is the Minkowski functional for V. Moreover, V is circled if and only if p is a pseudo-norm.*

The proof is a straightforward verification and is omitted.   If p is a pseudo-norm, then the set of all vectors x such that $p(x) = 0$ is a linear subspace.   In case this subspace consists simply of 0, the pseudo-norm is called a norm. Formally, a **norm** is an absolutely homogeneous, subadditive, non-negative functional p such that $p(x) = 0$ only if $x = 0$. When only one norm p is being considered, the notation $\|x\|$ is frequently used for $p(x)$.

The final theorem of this section relates the notion of partial ordering of a linear space E to a geometrical object in E. Suppose that $\geqq$ is a **partial ordering** of a real linear space E; that is, $\geqq$ is a relation such that $x \geqq x$ for each vector x of E and such that $x \geqq z$ whenever $x \geqq y$ and $y \geqq z$. Let us suppose further that the ordering is **translation invariant**, in the sense that $x + z \geqq y + z$ whenever $x \geqq y$. In this case the ordering is determined by the set P of all vectors x such that $x \geqq 0$, for then $x \geqq y$ if and only if $x - y \geqq 0$. Conversely, if P is any subset of E such that $0 \in P$ and $P + P \subset P$, then a partial ordering which is translation invariant is defined by agreeing that $x \geqq y$ if and only if $x - y \in P$; moreover, P is precisely the set of vectors which are greater than or equal to 0 relative to this ordering. Finally suppose that $\geqq$ is a translation invariant partial ordering which also has the property that if $x \geqq y$, then $ax \geqq ay$ for all non-negative scalars a. Then the set P of non-negative vectors (that is, $\{x : x \geqq 0\}$) has the properties: $P + P \subset P$ and $aP \subset P$ for each non-negative scalar a. It is evident that, conversely, a set P with these two properties is precisely the set of all non-negative vectors relative to the ordering: $x \geqq y$ if and only if $x - y \in P$.

It remains to formalize the discussion of the preceding paragraph. A **vector ordering** of E is a partial ordering $\geqq$ such that $x + z \geqq y + z$ whenever $x \geqq y$ and $ax \geqq ay$ whenever $x \geqq y$ and a is a non-negative scalar. A **cone** in E is a non-void subset P such that $P + P \subset P$ and $aP \subset P$ whenever a is a non-negative scalar. If $\geqq$ is a vector ordering, then $\{x : x \geqq 0\}$ is a cone, and this cone is called the **positive cone** of the ordering. If P is a cone in E, then the **corresponding** order is defined by setting $x \geqq y$ if and only if $x - y \in P$. The discussion above can be summarized, in this terminology, as follows.

2.4   Cones and Orderings   *Each vector ordering is precisely the ordering corresponding to the cone of non-negative vectors. Each cone P is the positive cone of the vector ordering corresponding to P.*

The section is concluded with a few remarks about cones. A cone is always convex, and in fact a non-void set P is a cone if and only if it is convex and $aP \subset P$ for each non-negative scalar a. Each linear subspace is a cone, and if P is a cone in a real linear space, then

$P - P$ is a linear subspace.   If P is an arbitrary cone in a real linear space, then it is not generally true that $x = y$ if $x \geqq y$ and $y \geqq x$. This condition is equivalent to the requirement that $P \cap (-P) = \{0\}$.   The set $P \cap (-P)$ is always a real linear subspace.   If A is an arbitrary non-void subset of E, then there is a smallest cone which contains A; this cone is **generated** by A, and is called the **conical extension** of A.

PROBLEMS

A MIDPOINT CONVEXITY

A subset A of a linear space E is *midpoint convex* iff $\frac{1}{2}(x + y)$ is in A whenever x and y are in A.

(a) The following conditions are equivalent:

  (i) A is midpoint convex;

  (ii) $A + A = 2A$;

  (iii) $a_1 A + a_2 A + \cdots + a_n A = A$ whenever $a_1, a_2, \cdots, a_n$ is   any finite collection of non-negative dyadic rationals whose sum is 1.

(b) A midpoint convex set of real numbers is convex if it is either open or closed.

B DISJOINT CONVEX SETS

(a)   (Kakutani's lemma)   If A and B are disjoint convex sets and x is a point not in their union, then either $\langle A \cup \{x\}\rangle$ and B are disjoint or else $\langle B \cup \{x\}\rangle$ and A are disjoint.

(b)   (Stone's theorem)   If A and B are disjoint convex sets in a linear space E, there are disjoint convex sets C and D such that $A \subset C, B \subset D$, and $E = C \cup D$.

C MINKOWSKI FUNCTIONALS

(a) Let U be a convex set radial at 0 and assume that $U = \bigcap \{U_t : t \in A\}$ where each U_t is convex.   If p and p_t are Minkowski functionals for U and U_t, then $p(x) = \sup \{p_t(x) : t \in A\}$.

(b) If U and V are radial at 0 and if p and q are the Minkowski functionals for U and V respectively, then $\inf \{p(v) + q(v - x) : v \in V\}$ is the Minkowski functional for the join of U and V.

(c) Let p be the Minkowski functional for a convex set U radial at 0. Then $\{x : p(x) > 1\}$ is the radial kernel of the complement of U.

D CONVEX EXTENSIONS OF SUBSETS OF FINITE DIMENSIONAL SPACES

(a) A real linear space E has finite dimension $\leqq n$ if and only if for each subset A of E and for each point x of $\langle A \rangle$, there is a subset $\{x_1, \cdots, x_{n+1}\}$ of A such that x belongs to the convex extension of $\{x_1, \cdots, x_{n+1}\}$.

(b) Each point of the convex extension of a connected subset A of Euclidean n-space is in the convex extension of a subset $\{x_1, \cdots, x_n\}$ of A.

E CONVEX FUNCTIONALS

For a real-valued function p on a linear space E with $p(0) = 0$, any two of the following statements imply the third:

(i) p is non-negatively homogeneous;

(ii) p is subadditive;

(iii) p is convex, that is, $p(\sum \{a_i x_i : 1 \leq i \leq n\}) \leq \sum \{a_i p(x_i) : 1 \leq i \leq n\}$ for any finite combination with each a_i non-negative and $\sum \{a_i : 1 \leq i \leq n\} = 1$.

F FAMILIES OF CONES

The intersection of an arbitrary family of cones is again a cone; if the family is directed by $\subset$, then the union of the members is also a cone. Any finite linear combination of cones is a cone. If, for any set A, $C(A)$ temporarily denotes the conical extension of A, then $C(A) = \bigcup \{a\langle A \rangle : a \geq 0\}$; for any two sets A and B and scalars a and b, $C(aA + bB) \subset aC(A) + bC(B)$; and $C(\bigcup \{A_t : t \in B\}) = C(\bigcup \{C(A_t) : t \in B\})$ for any family $\{A_t : t \in B\}$ of sets.

G VECTOR ORDERINGS OF R^2

There are ten essentially different vector orderings of R^2.

H RADIAL SETS

In R^2 there exists a set which is radial at only one point.

I $Z - A$ DICTIONARY ORDERING

Let B be a Hamel base for an infinite dimensional real linear space E and let B be well ordered without the maximum element. Define an ordering of E by: $x \geq 0$ iff $x = 0$ or, on writing x as a linear combination of members of B, its last (relative to the ordering of B) non-zero coefficient is positive. Then E is linearly ordered by $\geq$, and $\geq$ is a vector ordering. Moreover, for each x in E there is y in E such that $y \geq 0$ and $x + ay \geq 0$ for every positive number a.

J HELLY'S THEOREM

In R^n let $\{A_i : i = 1, \cdots, r\}$, for $r > n + 1$, be convex sets such that for each k, $\bigcap \{A_i : i \neq k\}$ is non-void. Then $\bigcap \{A_i : i = 1, \cdots, r\}$ is also non-void. (Let $x_k \in \bigcap \{A_i : i \neq k\}$. It is possible to choose r numbers α_i, not all zero, such that $\sum \{\alpha_i : 1 \leq i \leq r\} = 0$ and $\sum \{\alpha_i x_i : 1 \leq i \leq r\} = 0$. Separate the terms having $\alpha_i \geq 0$ and those for which $\alpha_i < 0$.)

Let $\mathscr{C}$ be a family of compact (see 4) convex subsets of R^n such that every $n + 1$ sets in $\mathscr{C}$ have a non-void intersection. Then the intersection of all the members of $\mathscr{C}$ is non-void.

3 SEPARATION AND EXTENSION THEOREMS

This section contains the fundamental separation and extension theorems. These theorems are essential to the study of duality. Sufficient conditions are given for the separation of convex sets by a hyperplane, and for the extension of a linear functional from a subspace to the entire space, preserving positivity or preserving a bound.

There are a number of theorems on separation of convex sets and extension of linear functionals which, although superficially different, are more or less mutually equivalent.   The presentation given here begins with a theorem about cones and derives the other results as corollaries, but this arrangement is primarily a matter of taste.   Most of the theorems of the section concern real linear spaces, and it will be assumed that the linear space under discussion is real unless explicitly stated otherwise.

If f is a non-identically-zero linear functional on a real linear space E and t is any real number, each of the sets $\{x : f(x) \geq t\}$ and $\{x : f(x) \leq t\}$ is a **half-space**, and the pair are **complementary half-spaces**.   Their intersection, the set $f^{-1}[t]$, is a linear manifold.   It is the translate of the null space of f by any vector x such that $f(x) = t$. The null space of f is evidently a maximal proper subspace of E; each translate of a maximal proper subspace of a linear space is called a **hyperplane**.   Each hyperplane is a linear manifold, and is evidently contained in no other linear manifold except the space E itself.   Each hyperplane is the intersection of two complementary half-spaces.

A **proper** cone in E is a cone which is non-empty and is not identical with E.   Recall (theorem 2.2) that the radial kernel of a convex set is convex, and that the half-open line segment $(x : y]$ joining a point x of a convex set to a point y of its radial kernel is always contained in the radial kernel.

A half-space which is a cone is called conical.   Evidently P is a conical half-space if and only if for some linear functional f, not identically zero, $P = \{x : f(x) \geq 0\}$.

3.1   Lemma   *Let E be a real linear space.   A proper cone P in E is a half-space if and only if it has a non-void radial kernel P_0 and the union of the sets P and $-P_0$ is the entire space.*

proof   If P is a half-space, then clearly the radial kernel P_0 of P is non-void and $E = P \cup (-P_0)$.   Conversely, suppose P_0 satisfies the latter conditions.   The sets $-P_0$ and P are disjoint, for if $-x \in P_0$ and $x \in P$, then 0 belongs to the half-open line segment $[-x : x)$ and is consequently in the radial kernel of P; but then P is not a proper cone.   It follows that P_0, $-P_0$, and $P \cap (-P)$ are mutually disjoint, and that the union of these three sets is the entire space.

The intersection $F = P \cap (-P)$ is a linear subspace, and the proof will be completed by showing that F is of co-dimension one. If this is shown, then it will follow that there is a non-identically-zero linear functional f whose null space is F, f will map P_0 onto a convex subset of the non-zero real numbers, and P must then be identical with one of the conical half-spaces $\{x: f(x) \leq 0\}$ or $\{x: f(x) \geq 0\}$. The proof then reduces to showing that, if x is a fixed point of P_0 and y an arbitrary point which is distinct from x, then y is a linear combination of x and some member of F. If $y \in -P_0$, then the line segment $[x:y]$ must intersect F, for each of P_0 and $-P_0$ is radial at every one of its points, and the non-void disjoint open sets $\{t: tx + (1 - t)y \in P_0\}$ and $\{t: tx + (1 - t)y \in -P_0\}$ cannot cover the unit interval. Hence for each member y of $-P_0$ there is a number a such that $y + ax \in F$. Finally, if $y \in P_0$, then $-y \in -P_0$ and $-y + ax \in F$ for some scalar a; and if y belongs to neither P_0 nor $-P_0$, then $y \in F.|||$

3.2 THEOREM *In a real linear space E each proper cone which is radial at some point is contained in some conical half-space.*

PROOF   Let C be a proper cone which is radial at a point x. Then the vector $-x$ does not belong to C, for in this case 0 would be in the radial kernel of C and C would not be a proper cone. The cone C is contained in a cone P which is maximal with respect to the property of not containing $-x$, in view of the maximal principle. It will be shown that P is a conical half-space.

Suppose $y \notin P$. The set of all points $ay + p$, for $a \geq 0$ and p in P, is a cone which properly contains P, and the maximality property of P implies that $ay + p = -x$, for some a and some p. Clearly $a \neq 0$, and since $(-a/2)y$ belongs to the line segment $(p:x]$, it follows that P is radial at $(-a/2)y$ and hence at $-y$. Hence, if $y \notin P$, then $-y$ is a member of the radial kernel P_0 of P, and the preceding lemma shows that P is a conical half-space.$|||$

The principal theorem on the extension of positive functionals is easily deduced from the preceding. As a preliminary, notice that if P is a cone and f is a non-identically-zero linear functional which is non-negative on P, then $f(x) > 0$ for each point x at which P is radial, for $f[P]$ is radial at $f(x)$.

3.3 EXTENSION OF POSITIVE FUNCTIONALS *Let P be a cone in a real linear space E, and let F be a linear subspace which intersects the radial kernel of P. Then each linear functional f on F which is non-negative on $P \cap F$ can be extended to a linear functional on E which is non-negative on P.*

PROOF   If f is identically zero, the extension is clearly possible. If not, let x be a fixed point of F which belongs to the radial kernel of P, and observe that $f(x) > 0$.   If $A = \{y : f(y) \geq 0\}$, then $P + A$ is evidently a cone; and since $f(-x) < 0$, it follows that $-x \notin P + A$ and consequently $P + A$ is a proper cone which is radial at x.   Hence by theorem 3.2 there is a conical half-space containing $P + A$; that is, there is a non-identically-zero functional g which is non-negative on $P + A$.   If $f(y) = 0$, then both y and $-y$ belong to A and hence $g(y) = 0$.   Consequently $g(y) = af(y)$ for some scalar a and for all y in F.   Finally, both $f(x)$ and $g(x)$ are positive; therefore $a > 0$, and $(1/a)g$ is the required extension of f.|||

The following theorem is the first, historically, of the theorems of the section; it is probably the most immediately applicable form of the extension-separation principle.

3.4   Hahn-Banach Theorem   *Let E be a real linear space, let F be a subspace, let p be a subadditive non-negatively homogeneous functional on E, and let f be a linear functional on F such that $f(x) \leq p(x)$ for all x in F.   Then there is a linear functional f^- on E, an extension of f, such that $f^-(x) \leq p(x)$ for all x in E.*

PROOF   Consider the linear space $E \times K$, where K is the real field, and let P be the set of all vectors (x,t) which are "above" the graph of p; explicitly, $P = \{(x,t) : t \geq p(x)\}$.   It is easy to see that P is a cone and that P is radial at $(0,1)$.   For (x,t) in $F \times K$ let $g(x,t) = t - f(x)$.   Then g is non-negative on $P \cap (F \times K)$, the extension theorem 3.3 can be applied, and there is consequently an extension g^- of g which is non-negative on P.   Then for $t \geq p(x)$ it is true that $g^-(x,t) = g^-(x,0) + g^-(0,t) = g^-(x,0) + t \geq 0$, and in particular, $g^-(x,0) + p(x) \geq 0$ for all x.   It follows that $f^-(x) = -g^-(x,0)$ is the required extension of f.|||

There is a geometric form of the preceding theorem which is frequently useful.   The connection between the preceding and the geometric statement is based on this fact: if A is a convex subset of a real linear space and A is radial at 0, then a linear functional is less than or equal to one on A if and only if $f(x) \leq p(x)$ for all x, where p is the Minkowski functional for A.   (Recall that $p(x) = \inf \{t : t > 0$ *and* $(x/t) \in A\}$, so that $f(x) \leq p(x)$ for all x if and only if $f(x) \leq t$ for all positive t such that $(x/t) \in A$; that is, $f(x/t) \leq 1$ for x/t in A.) Theorem 3.4 can then be rewritten in the following form.

3.5   Corollary   *Let E be a real linear space, F a subspace, A a convex subset of E which is radial at 0, and let f be a linear functional on F which is at most one on $F \cap A$.   Then there is a linear functional f^- on E, an extension of f, such that $f^-(x) \leq 1$ for all x in A.*

The preceding result implies a theorem concerning the extension of linear functionals on complex linear spaces. To obtain the following, notice that if f is a linear functional, then the supremum of $|f(x)|$ for x in a convex circled set is the same as the supremum of the real part of $f(x)$. The following proposition is a consequence of this remark and of theorem 1.4.

3.6 Corollary   *Let E be a real or complex linear space, F a linear subspace, A a convex circled set which is radial at 0, and let f be a linear functional on F such that $|f(x)| \leqq 1$ for x in $A \cap F$. Then there is a linear functional f^- on E, an extension of f, such that $|f^-(x)| \leqq 1$ on A.*

The Minkowski functional of a convex circled set which is radial at 0 is a pseudo-norm. The preceding result can then be restated as follows.

3.7 Corollary   *Let E be a real or complex linear space, F a linear subspace, p a pseudo-norm, and f a linear functional on F such that $|f(x)| \leqq p(x)$ for x in F. Then there is a linear functional f^- on E, an extension of f, such that $|f^-(x)| \leqq p(x)$ for all x.*

Two subsets, A and B, of a real linear space can be **separated** if there are complementary half-spaces which contain A and B respectively. A linear functional f is said to **separate** A and B if and only if f is not identically zero and $\sup \{f(x): x \in A\} \leqq \inf \{f(x): x \in B\}$. Clearly f separates A and B if and only if $-f$ separates B and A. The linear functional f **strongly separates** A and B if the inequality above is strong; that is, if $\sup \{f(x): x \in A\} < \inf \{f(x): x \in B\}$. A linear functional g on a complex linear space is said to **separate** (**strongly separate**) two sets if and only if the real part of g separates (strongly separates, respectively) the sets.

The problem of separating (or strongly separating) two sets can always be reduced to the problem of separating (strongly) a point and a set. Explicitly, a linear functional f separates A and B if and only if f separates $\{0\}$ and $B - A$, and f separates A and B strongly if and only if f strongly separates $\{0\}$ and $B - A$.

3.8 Separation Theorem   *Let A and B be non-void convex subsets of a complex or real linear space E and suppose that A is radial at some point. Then there is a linear functional f separating A and B if and only if B is disjoint from the radial kernel of A.*

proof   There is no loss in generality in assuming E is a real linear space. If f separates A and B and A_0 is the radial kernel of A, then

$f[A_0]$ is an open subset of the space of all real numbers and is disjoint from $f[B]$, and consequently A_0 is disjoint from B. On the other hand, suppose that A_0 is disjoint from B. In view of theorem 3.2 there will be a linear functional f separating A and B provided the cone consisting of all non-negative multiples of $B - A$ is a proper subset of the entire space E. Assuming that $\bigcup \{t(B - A) : t \geqq 0\} = E$, choose a member x of A_0 and a member y of B, and then choose u in A, v in B, and $t \geqq 0$ so that $-(y - x) = t(v - u)$. Then $x + tu = y + tv$ and hence $[1/(1 + t)]x + [t/(1 + t)]u = [1/(1 + t)]y + [t/(1 + t)]v$. That is, the line segment $[x:u)$ intersects the segment $[y:v)$. This is a contradiction, for $[x:u) \subset A_0$, and A_0 was supposed to be disjoint from B.|||

3.9   THEOREM ON STRONG SEPARATION   *Two convex non-void subsets, A and B, of a complex or real linear space E can be strongly separated by a linear functional if and only if there is a convex set U which is radial at 0 such that $(A + U) \cap B$ is void.*

PROOF   It may be assumed that E is a real linear space. If f strongly separates A and B, and $\inf \{f(y) : y \in B\} - \sup \{f(x) : x \in A\} = d > 0$, then the inverse under f of the open interval $(-d/2 : d/2)$ is the required set U. To prove the converse, assume that U is convex and radial at 0 and that $(A + U) \cap B$ is void. Then $0 \notin B - A - U$, and $B - A - U$ has a non-void radial kernel. The preceding separation theorem applies, and there is a non-identically-zero linear functional f such that f is non-negative on $B - A - U$. Then $f(y) - f(x) \geqq f(z)$ for y in B, x in A, and z in U. The set $f[U]$ is a neighborhood of 0 in the space of real numbers, and it follows that $\inf \{f(y) : y \in B\} - \sup \{f(x) : x \in A\} > 0$.|||

PROBLEMS

A   SEPARATION OF A LINEAR MANIFOLD FROM A CONE

If F is a linear manifold in a real linear space E and P is a cone in E, the radial kernel of which is non-void and disjoint from F, then there exists a linear functional f on E and a non-positive constant a such that $f(x) = a$ for every x in F, $f(x) \geqq 0$ for each x in P, and $f(x) > 0$ for each x in the radial kernel of P.

B   ALTERNATIVE PROOF OF LEMMA 3.1

Suppose P is a proper cone in a real linear space E, the radial kernel P_0 of P is non-void, and $P \cup (-P_0) = E$. Let $F = P \cap (-P)$, and consider the quotient space E/F. There is in E/F an ordering that is induced by the ordering in E generated by P; this ordering in E/F is linear and

Archimedean, and hence E/F has dimension 1, so that F is the null space of a linear functional that does not vanish identically.

C EXTENSION OF THEOREM 3.2

In theorem 3.2 the hypothesis can be weakened by assuming that the given cone is radial at some point with respect to its linear extension.

D EXAMPLE

Not every proper cone is contained in a half-space. (See 2I.)

E GENERALIZED HAHN-BANACH THEOREM

Let A be a convex set in a real linear space E and suppose that p is a convex functional on A (that is, $p(\sum \{a_i x_i: 1 \leq i \leq n\}) \leq \sum \{a_i p(x_i): 1 \leq i \leq n\}$ whenever $x_i \in A$, $a_i \geq 0$, and $\sum \{a_i: 1 \leq i \leq n\} = 1$). Let F be a subspace of E and f a linear functional of F such that $f \leq p$ on $A \cap F$. Then if $A \subset F - C$ where C is the conical extension of A, there is a linear functional f^- on E such that f^- is an extension of f and $f^- \leq p$ on A.

This proposition obviously implies a dual statement, in which p is a concave functional and $p \leq f$ on $A \cap F$ by hypothesis; then if $A \subset F - C$ there is a linear functional f^- which extends f and which dominates p on A.

F GENERALIZED HAHN-BANACH THEOREM (VARIANT)

Let A be a circled convex subset of a (real or complex) linear space E, and let F be a subspace of E. Let f be a linear functional of F and let p be an absolutely convex functional on A (that is, $p(\sum \{a_i x_i: 1 \leq i \leq n\}) \leq \sum \{|a_i| p(x_i): 1 \leq i \leq n\}$ whenever $x_i \in A$ and $\sum \{|a_i|: 1 \leq i \leq n\} = 1$). If $|f(x)| \leq p(x)$ whenever x is in $A \cap F$, and if $A \subset F - C$, where C is the conical extension of A, then there exists a linear functional f^- on E which is an extension of f and is such that $|f^-(x)| \leq p(x)$ for every x in A.

NOTE: The condition that A be contained in $F - C$ used in problems 3E and 3F is satisfied in case $A = E$ or, more generally, in case C is radial at some point in F. This, combined with the remark that any constant function is convex and concave, and if non-negative is also absolutely convex, yields a variety of corollaries to the above two theorems, including propositions 3.3 through 3.7 of the text.

G EXAMPLE ON NON-SEPARATION

In R^3 consider the closed cone defined by the equations $x \geq 0$, $y \geq 0$, $z^2 \leq xy$, and the line having the equations $x = 0$ and $z = 1$. There is no plane in R^3 which contains the line and has the cone lying on one side of it; yet the line is disjoint from the cone.

H EXTENSION OF INVARIANT LINEAR FUNCTIONALS

Let E be a linear space, F a subspace of E, f a linear functional on F, and p a non-negatively homogeneous, non-negative, subadditive real functional on E such that $f \leq p$. In addition, let $\mathscr{L}_1$ be a family of linear operators in E.

(a) For each x in E let $g(x) = \inf\{p(x + \sum\{R_i(y_i): 1 \leq i \leq k\}): y_i \in E,$ $R_i \in \mathscr{L}_1,$ *k a positive integer*$\}$. Then g is a non-negatively homogeneous, non-negative, subadditive functional on E and $g \leq p$.

(b) The inequality $f(x) \leq g(x)$ holds for each x in F if and only if there is an extension $\bar{f}$ of f such that $\bar{f} \leq p$ on E and $\bar{f} \circ R = 0$ for each R in $\mathscr{L}_1$. (Apply the Hahn-Banach theorem to f and g.)

For present purposes a family $\mathscr{L}$ of linear operations in E is admissible if $S[F] \subset F$ and $f \circ S = f$ for every S in $\mathscr{L}$; a linear extension $\bar{f}$ of f is $\mathscr{L}$-admissible if $\bar{f} \leq p$ on E and $\bar{f} \circ S = \bar{f}$ for every S in $\mathscr{L}$.

(c) Suppose $\mathscr{L}$ is an admissible family of linear operators; then the following are equivalent:

  (i) there exists a linear extension $\bar{f}$ of f that is $\mathscr{L}$-admissible,
 (ii) for each finite subset $\mathscr{L}'$ of $\mathscr{L}$ there exists a linear $\mathscr{L}'$-admissible extension of f,
(iii) $f(x) \leq p(x + \sum\{S_i(y_i) - y_i: 1 \leq i \leq k\})$ for x in F, y_i in E, and S_i in $\mathscr{L}$.

Chapter 2

LINEAR TOPOLOGICAL SPACES

This chapter is largely preliminary in nature; it consists of a brief review of some of the terminology and the elementary theorems of general topology, an examination of the new concept "linear topological space" in terms of more familiar notions, and a comparison of this new concept with the mathematical objects of which it is an abstraction. After an introductory section on topology, we consider linear topological spaces, subspaces, quotient spaces, product spaces, and linear functions. With the exception of a few simple propositions relating to circled sets, these theorems are specializations of familiar results on topological groups (in other words, little use is made of scalar multiplication).

In the third section we compare an arbitrary linear topological space with its antecedents: normed spaces and metrizable spaces. This is an example of a classical procedure. Each new mathematical construct is compared with the examples from which it derives (in this case, normed spaces, metrizable spaces, and products of these) in order to determine the extent to which the central ideas of the examples have been isolated. As frequently happens, this comparison is made by representing the new object in terms of the old, and embedding theorems for arbitrary linear topological spaces in products of metrizable (or in some cases, products of normable spaces) result. It is noteworthy that two new notions, that of bounded set and that of locally convex topology, arise. These notions lie deeper than the theorems of this chapter indicate. They signal the beginning of the theory of linear topological spaces, in distinction to the theory of topological groups (in other words, we are beginning to utilize scalar multiplication).

26

The last two sections, on completeness and on function spaces respectively, are essentially fragments of elementary analysis (that is, general topology). With a few exceptions, the results listed there are specializations of theorems concerning topological groups or, even more generally, uniform spaces.

4   TOPOLOGICAL SPACES

This section reviews[1] a number of topological definitions and results, the most noteworthy being the Tychonoff product theorem.

A **topology** for a set X is a family $\mathscr{T}$ of subsets of X, to be called **open sets**, such that the void (empty) set, the set X itself, the union of arbitrarily many open sets, and the intersection of finitely many open sets, are open. If several topologies for X are being considered, the members of $\mathscr{T}$ will be called $\mathscr{T}$**-open**, or **open relative to** $\mathscr{T}$, and a similar convention will hold for other topological objects to be defined. The set X with the topology $\mathscr{T}$ is a **topological space**; it is denoted by $(X,\mathscr{T})$. A subset A of X is **closed** if and only if the complement $X \sim A$ of A is open. The intersection of all closed sets containing a set A is a closed set called the **closure** of A; the closure of A is denoted by $\bar{A}$ or A^-. A subset B of A whose closure contains A is **dense** in A. The union of all open sets contained in A is an open set called the **interior** of A and denoted by A^i. A set U is a neighborhood of a point x if and only if x belongs to the interior of U. The family $\mathscr{U}_x$ of neighborhoods of x is the **neighborhood system** of x, and a subfamily $\mathscr{B}$ of $\mathscr{U}_x$ such that each member of $\mathscr{U}_x$ contains a member of $\mathscr{B}$ is a **base for the neighborhood system** of x. A topological space is **regular** if and only if for each point x, the family of closed neighborhoods of x is a base for the neighborhood system of x. Thus, X is regular if and only if for each x in X and each neighborhood U of x there is a closed neighborhood V of x such that $V \subset U$. A topological space X is **separable** if and only if there exists a countable subset A of X which is dense in X.

If X and Y are topological spaces, then Y is a **topological subspace** of X if and only if $Y \subset X$ and the sets which are open in Y are precisely the intersections of Y with open subsets of X. Equivalently, a subset of Y is closed in Y if and only if it is the intersection of Y and a closed subset of X. If Z is a subset of X, then Z may be given a topology such that it becomes a topological subspace of X. The **relative (induced)** topology on Z, or the **relativization** of the

---

[1] For more complete treatment of the topics in this section, see Kelley [5].

topology of X to Z, is the family of all subsets U of Z such that for some open subset V of X, $U = Z \cap V$. Clearly Z with the relative topology is a topological subspace of X. A subset of Z which is open with respect to the relative topology is **open in Z**.

If $(X, \mathcal{T})$ and $(Y, \mathcal{U})$ are topological spaces, and if f is a map of X into Y, then f is **continuous at a point** x of X if and only if the inverse image of each $\mathcal{U}$-neighborhood of $f(x)$ is a $\mathcal{T}$-neighborhood of x. The function f is **continuous** if and only if f is continuous at each point of X, or, equivalently, if and only if the inverse image of each $\mathcal{U}$-open set of Y is $\mathcal{T}$-open. If A is a subset of X, then the function f is **continuous on A** if and only if the restriction of f to A is continuous as a function on the topological subspace A. The function f is **open (interior)** if and only if the image under f of each open set is open. The function f is **relatively open** if and only if the image of each open set is open in $f[X]$, that is, open in the relative topology for $f[X]$. Finally, the function f is **topological,** or a **homeomorphism**, if and only if it is continuous, one-to-one (no two points have the same image) and the inverse map, f^{-1}, is a continuous map of $f[X]$ onto X; equivalently, f is topological if and only if it is one-to-one, continuous, and relatively open.

A generalization of sequential convergence is necessary; Moore-Smith convergence will be employed. A relation $\geq$ **directs** a set A if it is transitive on A (if $\alpha \geq \beta$ and $\beta \geq \gamma$, then $\alpha \geq \gamma$, for α, β, and γ in A), reflexive on A ($\alpha \geq \alpha$ for α in A), and has the property: for α in A and β in A there is γ in A such that $\gamma \geq \alpha$ and $\gamma \geq \beta$. A **net** is a pair $\{x, \geq\}$ such that x is a function and $\geq$ directs the domain of x. More generally, $\{x_\alpha, \alpha \in A, \geq\}$, or simply $\{x_\alpha, \alpha \in A\}$, is called a net if x is a function whose domain contains A and $\geq$ directs A. A net $\{x_\alpha, \alpha \in A\}$ is **eventually in** a set B if and only if, for some α in A, $x_\beta \in B$ whenever $\beta \geq \alpha$. A net in a topological space **converges** to a point (the point is a **limit** of the net) if and only if the net is eventually in each neighborhood of the point. The fact that the net $\{x_\alpha, \alpha \in A\}$ converges to y in X will be denoted by $x_\alpha \to y$ or $y = \lim x_\alpha$. A point is a **cluster point** of a net if and only if the net is not eventually in the complement of any neighborhood of the point (in the usual terminology, the net is **frequently** or **repeatedly** in each neighborhood of the point). A net $\{y_\beta, \beta \in B, \geq\}$ is a **subnet** of $\{x_\alpha, \alpha \in A, \geq\}$ if and only if there exists a function n on B to A such that n is order preserving, $y_\beta = x_{n(\beta)}$ for all β in B, and for each α in A there is a β in B such that $n(\beta) \geq \alpha$. If a net converges to a point z, each subnet also converges to z; a point z is a limit of a subnet if and only if z is a cluster point of the net.   Note that a subnet of a sequence (a **sequence** is a function on the set of positive integers with the usual ordering) need not be a sequence, although a subsequence of a sequence is a subnet of this sequence.   A function f is continuous at a point x' of the domain space if and only if for each net $\{x_\alpha, \ \alpha \in A\}$ converging to x', the net $\{f(x_\alpha), \ \alpha \in A\}$ converges to $f(x')$ in the range space.

A topological space is **Hausdorff (separated, T$_2$)** if and only if for distinct points x and y there are disjoint neighborhoods U of x and V of y.   Equivalently, a space is Hausdorff if and only if no net in the space converges to more than one point.   A topological space is **compact (bicompact)** if and only if it has the Borel-Lebesgue property: each cover by open sets has a finite subcover; equivalently, a topological space is compact if and only if each net has a cluster point.   A space is **countably compact** if and only if each countable cover by open sets has a finite subcover.   A space is countably compact if and only if each sequence has a cluster point.   A space is **sequentially compact (quasi-compact)** if and only if each sequence has a subsequence which converges to a point in the space.   Although countable compactness is implied by the two other kinds of compactness, it implies neither of these, nor does either of these imply the other.   A space is **locally compact** if and only if each point has a compact neighborhood.   Each locally compact Hausdorff space is regular.

A **pseudo-metric (semi-metric, écart)** on a set X is a non-negative function d, defined for each pair of points of X, such that $d(x,y) = d(y,x), d(x,x) = 0$, and $d(x,z) \leqq d(x,y) + d(y,z)$ for all x, y, and z in X.   A **metric** on X is a pseudo-metric d such that if $d(x,y) = 0$, then $x = y$.   A set X with a pseudo-metric (metric) is called a **pseudo-metric** (respectively, **metric**) **space**.   In a pseudo-metric space the **open** (respectively, **closed**) **sphere of radius r about a point** x is the set of all points y such that $d(x,y) < r$ (respectively, $d(x,y) \leqq r$).   Each pseudo-metric (metric) d on X generates a topology for X in the following fashion.   A set U in X is open relative to the **pseudo-metric** (respectively, **metric**) **topology** for x if and only if for each point x in U there is a positive number r such that the open sphere of radius r about x is contained in U.   A topological space X is **pseudo-metrizable** (respectively, **metrizable**) if and only if there is a pseudo-metric (metric) such that the pseudo-metric (metric) topology is the topology of X.   If X is pseudo-metrizable, there are, in general, many pseudo-metrics on X such that the pseudo-metric topology is the topology of X; consequently, the pseudo-metric topology does not determine the pseudo-metric. Clearly, a pseudo-metrizable topological space is Hausdorff if and only if the space is metrizable.

If X is the set of real or complex numbers and $d(x,y) = |x - y|$ for x and y in X, then d is a metric on X. Unless specifically stated otherwise, it will be assumed that the field of complex numbers and the field of real numbers are assigned the topology of the metric d. It is called the **usual topology** for the complex (respectively, real) numbers.

The comparison of topologies is of interest. If $\mathscr{T}$ and $\mathscr{U}$ are two topologies for a set X such that each $\mathscr{U}$-open set is $\mathscr{T}$-open, then the topology $\mathscr{T}$ is **stronger (larger, finer)** than $\mathscr{U}$, and $\mathscr{U}$ is **weaker (smaller, coarser)** than $\mathscr{T}$. Equivalently, $\mathscr{T}$ is stronger than $\mathscr{U}$ if and only if $\mathscr{T} \supset \mathscr{U}$. More precise expressions would be "$\mathscr{T}$ is at least as strong as $\mathscr{U}$" and "$\mathscr{U}$ is at least as weak as $\mathscr{T}$," but one usually foregoes such grammatical exactness. (Some authors give the terms "stronger" and "weaker" a significance exactly the reverse of that defined here, but the definitions given are those which are most common in the literature of linear topological spaces.) If a net converges relative to a topology, it also converges relative to weaker topologies, and a set compact relative to a topology is compact relative to weaker topologies. Topologies for a set are in general not comparable, but the collection of all topologies for a set X is a lattice under the ordering $\subset$; the strongest topology is the **discrete topology**, for which all sets are open, and the weakest topology is the **trivial (indiscrete) topology**, for which only X and the void set are open.

There is a natural way to topologize the cartesian product $\bigtimes \{X_t : t \in A\}$ of topological spaces X_t. Recall that for each t in A, the projection P_t of the product into the t-th coordinate space X_t is defined by $P_t(x) = x(t)$ for each x in the product, and that $P_t(x)$ is the t-th coordinate of x. For convenience, a cylinder will be defined to be a subset of the product of the form $P_t^{-1}[U]$, where U is open in X_t; that is, a set C is a cylinder if and only if for some t in A and some open subset U of X_t, C is the set of all x in the product such that $x(t) \in U$. The **product topology** for $\bigtimes \{X_t : t \in A\}$ is the family of all arbitrary unions of finite intersections of cylinders; the product topology is sometimes called the **product of the topologies** for X_t. A base for the neighborhood system of a point x, relative to the product topology, is the family of all finite intersections of cylinders containing x; that is, the family of all sets of the following form: the set of all y such that $y(t) \in U_t$ for each t in a *finite* set of indices, where U_t is an open neighborhood of $x(t)$ in X_t. A net $\{x_\alpha, \alpha \in B, \geqq\}$ in the product converges to a point x' of the product if and only if for each t in A the net $\{x_\alpha(t), \alpha \in B, \geqq\}$ converges to $x'(t)$. For this reason the product topology is often called the **topology of coordinatewise convergence.** It is easy to verify that the projection of the product into each coordinate space is open and continuous, and that the product topology is the weakest topology such that each projection is continuous. A map into the product is continuous if and only if the map followed by the projection P_t is continuous for each t in A. Finally, for a fixed point θ in the product, and for each t in A, the map I_t, defined as follows, is a homeomorphism of X_t into the product: for each s in A, and each point z in X_t, $I_t(z)(s)$, the s-th coordinate of $I_t(z)$, is either z or $\theta(s)$, depending on whether s is or is not equal to t. Thus a product (of non-void factors) contains topological copies of the factors.

4.1 TYCHONOFF THEOREM *The product of compact spaces is, with the product topology, a compact space.*

A particular case of a product is that in which all coordinate spaces are identical. Then the product $\bigtimes \{X : t \in A\}$ is simply the set of all functions on A to X. In this case the product topology is frequently called the **topology of pointwise convergence** (the **simple topology,** or the **topology of simple convergence**).

The following corollary to the Tychonoff theorem is stated for reference.

4.2 COROLLARY *The set of all functions on a set A to a compact space X is compact relative to the topology of pointwise convergence.*

There is a useful extension of the method whereby the product topology is defined. Suppose F is a class of functions, each member f of F being on a set X to a topological space Y_f. In general there will be many topologies for X which will make each member of F continuous—the discrete topology always has this property. Among these topologies there is a weakest, namely, the topology having for a subbase the family of all inverses under arbitrary members f of F of open subsets of Y_f. This topology is called the **projective topology** for X (or the F-**projective topology**). The projective topology can be described alternatively by specifying that a subset U of X is open relative to the projective topology if and only if U is the inverse under P of an open subset of $\bigtimes \{Y_f : f \in F\}$, where P is the map which sends a point x of X into the point with f-th coordinate $f(x)$ (that is, $P(x)(f) = f(x)$). An arbitrary function g on a topological space Z into X is continuous relative to the projective topology for X if and only if the composition $f \circ g$ is a continuous map of Z into Y_f for each f in F. The product topology for a cartesian product is itself the F-projective topology, where F is the class of all projections into coordinate spaces.

There is a procedure which is dual to the foregoing. Suppose G is a family of functions, each member g of G being on a topological space X_g to a fixed set Y. Then there is a strongest topology for Y which makes each member of G continuous, namely the family of all subsets U of Y such that $g^{-1}[U]$ is open in X_g for each g in G. This topology is called the **induced topology** for Y (the G-**induced topology**). A map h of Y into an arbitrary topological space Z is continuous relative to the induced topology if and only if $h \circ g$ is a continuous map of X_g into Z for each g in G. Finally, if q is an arbitrary continuous open map of a topological space X onto a topological space Z, then Z necessarily has the $\{q\}$-induced topology.

PROBLEMS

A   COMPACT AND LOCALLY COMPACT SPACES

(a) If $(X, \mathcal{T})$ is compact and $(X, \mathcal{U})$ is Hausdorff, and if $\mathcal{U}$ is weaker than $\mathcal{T}$, then $\mathcal{U} = \mathcal{T}$.

A compact subset of a regular space has a compact closure.

(b) Let X be a locally compact Hausdorff space. If A is a compact subset of X and U is an open set containing A, then there is a compact set B with $A \subset B^i \subset B \subset U$.

If also X is the union of a sequence of compact sets, then there is a sequence $\{B_n : n = 1, 2, \cdots\}$ of compact sets whose union is X such that $B_n \subset B_{n+1}{}^i$ for each n.

B   SEPARABILITY

(a) A compact metric space is separable.

(b) If $k(A) \leq 2^{\aleph_0}$ and if, for each $t \in A$, X_t is a separable space, then $\bigtimes \{X_t : t \in A\}$ is separable.

C   COMPLETE METRIC SPACES

A sequence $\{x_n : n = 1, 2, \cdots\}$ in a metric space is *Cauchy* iff $d(x_m, x_n) \to 0$ as $m, n \to \infty$; a metric space is *complete* iff every Cauchy sequence is convergent.

A compact metric space is complete; the converse is not true.

Completeness is not a topological property: there are metrics d_1 and d_2 on

$I = \{x: 0 < x < 1\}$, both giving the usual topology, such that (I,d_1) is complete but (I,d_2) is not.

D   HAUSDORFF METRIC ON A SPACE OF SUBSETS

Let E be a metric space, with metric d, and let $\mathscr{E}$ be the set of closed non-void subsets of E.   For each $x \in E$ and each $A \in \mathscr{E}$ let $d(x,A) = \inf$ $\{d(x,y): y \in A\}$, and for $A \in \mathscr{E}$ and $B \in \mathscr{E}$ let $d(A,B) = \max \{\sup \{d(a,B): a \in A\},\ \sup \{d(A,b): b \in B\}\}$.   The function $d(A,B)$ is a metric on $\mathscr{E}$, except that it may take infinite values.   (Infinite values may always be avoided by using an equivalent bounded metric d' (e.g., $d'(x,y) = \min \{d(x,y), 1\}$ or $d'(x,y) = d(x,y)/(1 + d(x,y))$).   This is called the *Hausdorff metric* on the space of closed non-void sets.

When E is complete, $\mathscr{E}$ is also complete.   (Suggestion for proof: Suppose that $\{A_n: n = 1, 2, \cdots\}$ is a Cauchy sequence in $\mathscr{E}$; let A be the set of limits of all Cauchy sequences $\{x_n\}$ in E with $x_n \in A_n$ for each n.   Show $d(A,A_n) \to 0$.)

E   CONTRACTION MAPPING

Let (X,d) be a complete metric space and f a *contraction mapping* of X into itself: that is, $d(f(x), f(y)) \leqq rd(x,y)$ for some constant $r < 1$ and all x and y in X.   Then there is a unique point z in X with $f(z) = z$.

It is an irresistible temptation to describe here the elegant classical application of this result, which proves Picard's theorem on the existence of solutions of a differential equation.   Consider the differential equation $dy/dx = \phi(x,y)$.   Suppose that there is a neighborhood U of the point (a,b) in which ϕ is continuous, and a constant l such that $|\phi(x,y_1) - \phi(x,y_2)| \leqq l|y_1 - y_2|$ whenever $(x,y_1) \in U$ and $(x,y_2) \in U$.   Then there is one and only one solution $y = y(x)$ of the differential equation with $y(a) = b$.

(First, let $m = \sup \{|\phi(x,y)| : (x,y) \in U\}$.   Choose h and k so that $\{(x,y): |x - a| \leqq h,\ |y - b| \leqq k\}$ lies in U and so that $hl < 1$ and $hm \leqq k$.   Next, let X be the set of continuous functions on the interval $[a - h: a + h]$, to the interval $[b - k: b + k]$, with the metric $d(y,z) = \sup \{|y(x) - z(x)| : |x - a| \leqq h\}$.   Now define a mapping f of X into itself by $f(y)(x) = b + \int_a^x \phi(t, y(t))dt$.   Then f is a contraction mapping, and the unique fixed point is the required solution.)

5   LINEAR TOPOLOGICAL SPACES, LINEAR FUNCTIONS, QUOTIENTS, AND PRODUCTS

It is shown that the topology of a linear topological space can be localized, in the sense that the topology is entirely determined by the family of neighborhoods of 0.   The notion of continuity of a linear function can also be localized; a linear function is continuous if and only if it is continuous at 0, and this is the case if and only if the function is uniformly continuous.   A theorem on the extension of linear functions is proved.   The quotient topology is defined, and a topological form of the induced map theorem proved.   Elementary propositions on product spaces are demonstrated.

A **linear topological** space is a linear space E with a topology such that addition and scalar multiplication are each continuous simultaneously in both variables; more precisely, such that each of the following maps is continuous: (a) the map of the product, $E \times E$, with the product topology, into E, which is given by $(x,y) \rightarrow x + y$ for x and y in E; (b) the map of the product, $K \times E$, of the scalar field and E, with the product topology, into E, which is given by $(a,x) \rightarrow ax$ for a in K and x in E. The topology of a linear topological space is called a **vector topology**. Thus a topology for a linear space is a vector topology if and only if addition and scalar multiplication are continuous, and a linear topological space is a linear space with a vector topology.

Suppose E is a linear topological space. Because of the continuity of addition, translation by a member x of E, where the translation by x of a vector y is $x + y$, is continuous. Since translation by x has a continuous inverse, namely translation by $-x$, it is a homeomorphism. Similarly, multiplication by a non-zero scalar is a homeomorphism. It follows that, if a set A is open (closed), so are $x + A$, and aA, for each x in E and each non-zero scalar a. Hence, a set U is a neighborhood of a point x if and only if $-x + U$ is a neighborhood of 0; in other words, the neighborhood system at x is simply the family of translates by x of members of the neighborhood system at 0. A base for the neighborhood system of 0 is called a **local base** or a **local base for the topology** (or a **system of nuclei**). The topology is completely determined by any local base $\mathscr{U}$, for a set V is open if and only if for each x in V there is a U in $\mathscr{U}$ such that $x + U \subset V$. It is, therefore, of some interest to describe those families of subsets of a linear space E which are a local base for some vector topology.

5.1 THEOREM ON LOCAL BASES *Let E be a linear topological space, and let $\mathscr{U}$ be a local base. Then*

(i) *for U and V in $\mathscr{U}$ there is a W in $\mathscr{U}$ such that $W \subset U \cap V$;*

(ii) *for U in $\mathscr{U}$ there is a member V of $\mathscr{U}$ such that $V + V \subset U$;*

(iii) *for U in $\mathscr{U}$ there is a member V of $\mathscr{U}$ such that $aV \subset U$ for each scalar a with $|a| \leq 1$;*

(iv) *for x in E and U in $\mathscr{U}$ there is a scalar a such that $x \in aU$; and*

(v) *for U in $\mathscr{U}$ there is a member V of $\mathscr{U}$ and a circled set W such that $V \subset W \subset U$.*

(vi) *If E is a Hausdorff space, then $\bigcap \{U : U \in \mathscr{U}\} = \{0\}$.*

Conversely, let E be a linear space and $\mathscr{U}$ a non-void family of subsets which satisfy (i) through (iv), and let $\mathscr{T}$ be the family of all sets W such *that, for each x in W, there is U in $\mathcal{U}$ with $x + U \subset W$.   Then $\mathcal{T}$ is a vector topology for E, and $\mathcal{U}$ is a local base for this topology. If, further, (vi) holds, then $\mathcal{T}$ is a Hausdorff topology.*

The proof of this proposition is straightforward and will be omitted.

If $\mathcal{U}$ and $\mathcal{V}$ are families satisfying the conditions (i) to (iv) of theorem 5.1, and if each member of $\mathcal{U}$ contains a member of $\mathcal{V}$, then the topology determined by $\mathcal{V}$ is stronger than the topology determined by $\mathcal{U}$.   The converse of this proposition is also true.   If each member of $\mathcal{U}$ contains a member of $\mathcal{V}$, and if each member of $\mathcal{V}$ contains a member of $\mathcal{U}$, then $\mathcal{U}$ and $\mathcal{V}$ determine the same topology. In this case the families $\mathcal{U}$ and $\mathcal{V}$ are said to be **equivalent**.

Part (iii) of the preceding theorem implies that for each linear topological space the family of circled neighborhoods of 0 is a local base.

There are several elementary propositions about linear topological spaces which will be used frequently, sometimes without explicit mention.   The following facts are among those most commonly used; the proofs are given as samples of computations in linear topological spaces.   Recall that A^i is the interior of A and A^- is the closure of A.

5.2   ELEMENTARY   COMPUTATIONS   *Let E be a linear topological space, and let $\mathcal{U}$ be the family of neighborhoods of 0.   Then*

   (i) *the closure of $x + A$ is $x + A^-$ for each vector x and each subset A of E; the closure of aA is aA^- for each non-zero scalar a and each set A;*
   (ii) *if A and B are subsets of E, then $A^- + B^- \subset (A + B)^-$;*
   (iii) *$A + U$ is open if U is open, and hence $A + B^i \subset (A + B)^i$;*
   (iv) *$C + D$ is compact if C and D are compact;*
   (v) *the closure A^- of the set A is $\bigcap \{A + V : V \in \mathcal{U}\}$;*
   (vi) *if C is compact and U open, and if $C \subset U$, then there is V in $\mathcal{U}$ such that $C + V \subset U$;*
   (vii) *$C + F$ is closed if C is compact and F is closed;*
   (viii) *the closure of a subspace is a subspace;*
   (ix) *the closure of a circled set is circled;*
   (x) *if the interior of a circled set contains 0, it is also circled;*
   (xi) *the family of all closed circled neighborhoods of 0 is a local base;*
   (xii) *each convex neighborhood of 0 contains the closure of a convex circled neighborhood of 0; and*
   (xiii) *the closure of a convex set is convex.*

PROOF   (i) This is a direct consequence of the fact that translation and multiplication by non-zero scalars are homeomorphisms.

(ii) If $x \in A^-$ and $y \in B^-$, then $(x,y) \in (A \times B)^-$ in the product; and since addition is continuous, $x + y \in (A + B)^-$.

(iii) Observe that $A + U = \bigcup \{x + U : x \in A\}$, and $x + U$ is open for every x.

(iv) This follows from the fact that $C + D$ is the image of the compact set $C \times D$ under the continuous map, addition.

(v) A point x belongs to A^- if and only if for each neighborhood V of 0, $x + V$ intersects A; that is, if and only if $x \in A - V$. Hence $A^- = \bigcap \{A - V : V \in \mathcal{U}\}$, but V is a neighborhood of 0 if and only if $-V$ is a neighborhood of 0. The required result follows.

(vi) Suppose that for each neighborhood V in the neighborhood system $\mathcal{U}$ of 0 there is an element x_V in C such that $x_V + V$ intersects the complement of U. Consider the net $\{x_V, V \in \mathcal{U}, \subset\}$. Since C is compact, this net has a cluster point x_0 in C. Thus every neighborhood of x_0 intersects the complement of U. This situation is impossible since x_0 is an element in the open set U.

(vii) If x does not belong to $C + F$, then the compact set $x - C$ is disjoint from F. Hence, by (vi), there is a neighborhood V of 0 such that $V + x - C$ is disjoint from F. Consequently, $x + V$ is disjoint from $C + F$, and x does not belong to the closure of $C + F$. Hence, $C + F$ is closed.

(viii) Suppose that F is a subspace of E, and that b and c are scalars. Then $bF^- + cF^- = (bF)^- + (cF)^- \subset (bF + cF)^- \subset F^-$, in view of (i) and (ii), and F^- is therefore a subspace.

(ix) If A is circled, $x \in A^-$, and $0 < |a| \leq 1$, then $ax \in aA^- = (aA)^- \subset A^-$ and A^- is therefore circled.

(x) If B is the interior of a circled set A, then $aB \subset aA \subset A$ whenever $0 < a \leq 1$, and, since aB is open, $aB \subset B$.

(xi) For each neighborhood V of 0 there is a circled neighborhood U such that $U + U \subset V$. Then U^- is circled, by (ix), and is contained in $U + U$, by (v).

(xii) If U is a convex neighborhood of 0 and $V = \bigcap \{aU : |a| = 1\}$, then V is a circled neighborhood of 0 because U contains a circled neighborhood of 0. Clearly V is a convex subset of U. The set $(1/2)V$ is a convex circled neighborhood of 0, and its closure is contained in $(1/2)V + (1/2)V$. The latter set is a subset of V because V is convex.

(xiii) The proof is similar to that of (viii) and is omitted.|||

A particular consequence of part (xi) of the foregoing theorem is that the family of closed neighborhoods of a point is a base for the neighborhood system of the point.   In other words, the topology is regular.

The notion of continuity of a linear function can be localized, as might have been expected in view of the localization of the topology of a linear topological space.   Recall that a function T on a topological space E to a topological space F is continuous at a point x if for each neighborhood W of $T(x)$ there is a neighborhood of x whose image under T is a subset of W.   If T is linear, if E is a linear topological space with local base $\mathcal{U}$, and if F is a linear topological space with local base $\mathcal{V}$, then T is continuous at x if and only if for each V in $\mathcal{V}$ there is U in $\mathcal{U}$ such that $T[U + x] \subset V + T(x)$; that is, if and only if $T[U] \subset V$.   Thus a linear function is continuous at some point x of its domain if and only if it is continuous at 0, and this is the case if and only if it is continuous at every point.   This proves the following proposition.

5.3   Localization of Continuity   *A linear function on a linear topological space to a linear topological space is continuous at some point of its domain if and only if it is continuous at every point of its domain.*

If f is a linear functional, then continuity can be deduced from premises which are even weaker than continuity at some point.

5.4   Continuity of Linear Functionals   *If f is a linear functional on the linear topological space E and f is not identically zero, then the following conditions on f are equivalent:*

(i) *f is continuous;*
(ii) *the null space of f is closed;*
(iii) *the null space of f is not dense in E;*
(iv) *f is bounded on some neighborhood of 0;*
(v) *the image under f of some non-void open set is a proper subset of the scalar field; and*
(vi) *the function r, where $r(x)$ is the real part of $f(x)$ for x in E, is continuous.*

PROOF   If f is continuous, the null space of f, being the inverse image of a closed set, is closed; and if the null space of f is closed, then, since f is not identically zero, the null space of f is not dense in E.   Hence, (i) implies (ii), and (ii) implies (iii).   If the null space N of f is not dense in E, there is a point x of E and a circled neighborhood U of 0 such that $x + U$ is disjoint from N.   Then f must be bounded on U, for otherwise, since U is circled and radial, $f[U]$ is the entire scalar field, and for a suitably chosen u in U, $f(x + u) = 0$. This conclusion contradicts the fact that $x + U$ is disjoint from N. Consequently, (iii) implies (iv). If f is bounded by M on a neighborhood U of 0, then $f^{-1}[\{t: |t| \leq e\}]$ contains $(e/M)U$ and hence f is continuous at 0. Thus (iv) implies (i). It is clear that condition (v) is implied by continuity of f. Conversely, if for some scalar a the set $f^{-1}[a]$ is disjoint from some open set, then $f^{-1}[a]$ is not dense in E and hence, by a translation argument, $f^{-1}[0]$ is not dense in E. Consequently f is continuous by (iii). Finally (vi) is obvious in view of the definition of r and the fact that $f(x) = r(x) - ir(ix)$ for all x.|||

Continuous linear functions have an important extension property. This property is a direct consequence of the fact that a continuous linear function T is automatically **uniformly continuous**, in the following sense: for each neighborhood V of 0 in the range space there is a neighborhood U of 0 in the domain such that for every x and y it is true that $T(x) - T(y) \in V$ whenever $x - y \in U$. (Intuitively, two points are close together if their difference is near 0, so that there is a "uniform" notion of nearness.) The next proposition is a form of a well-known principle applying to arbitrary uniformly continuous functions.

The **graph** of a function T with domain D is the subset of $E \times F$ consisting of all pairs $(x, T(x))$ for x in D. (Many authors define a function T to be what is here called the graph of T.) Let E and H be linear spaces with the same scalar field K. Then a subset G of $E \times H$ is the graph of a linear transformation whose domain and range are linear subspaces of E and H respectively if and only if G is a linear subspace of $E \times H$ such that $(0, y) \in G$ implies $y = 0$.

In the next theorem the following notation is used: if T is a function, G_T is the graph of T.

5.5   Extension by Continuity   *If T is a continuous linear function on a subspace F of a linear topological space E to a Hausdorff linear topological space H, then the closure $\overline{G}_T$ of G_T in the product $E \times H$ is the graph of a continuous linear extension of T.*

proof   It must be shown first that $\overline{G}_T$ is the graph of a function; that is, if $(0, y) \in \overline{G}_T$, then $y = 0$. If $(0, y) \in \overline{G}_T$, then there is a net $\{(x_\alpha, y_\alpha), \alpha \in A\}$ in G_T such that $(x_\alpha, y_\alpha) \to (0, y)$. Then $x_\alpha \to 0$ and $y_\alpha = T(x_\alpha) \to y$, and $y = 0$ because H is Hausdorff and T is continuous. Since the closure of a subspace is a subspace, $\overline{G}_T$ is the graph of a certain linear transformation $\overline{T}$, which will be shown to be continuous. For this purpose it is sufficient to prove the continuity of $\bar{T}$ at 0, and this will follow if it is shown that $\bar{T}[U] \subset V$ whenever V is a closed neighborhood of 0 in H and U is an open neighborhood of 0 in E such that $T[U] \subset V$. If x is a point of U which belongs to the domain of $\bar{T}$, then there is x_α in the domain of T such that $x_\alpha \to x$ and $T(x_\alpha) \to \bar{T}(x)$. Since U is open, x_α is eventually in U; hence $T(x_\alpha)$ is eventually in V, and $\bar{T}(x) \in V$ because V is closed.|||

5.6 REMARK It is always true that the domain $D_{\bar{T}}$ of the extension $\bar{T}$ of a continuous linear T is contained in the closure $\bar{D}_T$ of the domain of T. Under certain circumstances $D_{\bar{T}} = \bar{D}_T$. Using the definitions of "Cauchy net" and "complete" which are given in the first paragraph of Section 7, it can be shown that $D_{\bar{T}} = \bar{D}_T$ if H is complete. For, if $x \in \bar{D}_T$, there exists a net $\{x_\alpha, \alpha \in A\}$ in D_T such that $x_\alpha \to x$. Then $\{x_\alpha, \alpha \in A\}$ is a Cauchy net, and therefore $\{T(x_\alpha), \alpha \in A\}$ is a Cauchy net. When H is complete, there exists a z in H such that $T(x_\alpha) \to z$; hence, $(x,z) \in \bar{G}_T = G_{\bar{T}}$ and $x \in D_{\bar{T}}$. From this fact it follows that $\bar{D}_T \subset D_{\bar{T}}$, and hence that $\bar{D}_T = D_{\bar{T}}$.

If F is a linear subspace of a linear topological space E, then F itself, with the relativized topology of E, is a linear topological space. The space F is called a **linear topological subspace** of E if and only if F is both a linear subspace and a topological subspace of E.

If F is a linear subspace of a linear topological space E, the **linear topological quotient space** is the quotient space E/F with the topology such that a set U in E/F is open if and only if $Q^{-1}[U]$ is open in E, where Q is the quotient map; that is, $Q(x) = x + F$. This topology for E/F is the **quotient topology**, and this topology is a vector topology. Unless there is a statement to the contrary, the quotient space always has the quotient topology.

5.7 THEOREM *Let F be a linear topological subspace of a linear topological space E, let E/F be the quotient space, and let Q be the quotient map. Then the map Q is linear, continuous, and open, and the linear topological space E/F is Hausdorff if and only if F is closed. Moreover, a function T on E/F is continuous (open) if and only if the composition $T \circ Q$ is continuous (open).*

PROOF It is not difficult to see that E/F is a Hausdorff space if and only if F is closed (see part (v) of theorem 5.2 and part (vi) of 5.1). If U is open in E/F, then by the definition of the quotient topology $Q^{-1}[U]$ is open in E and Q is continuous. To show that $Q[V]$ is open in E/F when V is open in E, it is sufficient to show that

$Q^{-1}[Q[V]] = V + F$ is open in E. But the sum of the open set V and any set is open. If T is a map of E/F into G and W is a subset of G, then $T^{-1}[W]$ is open in E/F if and only if $Q^{-1}[T^{-1}[W]] = (T \circ Q)^{-1}[W]$ is open in E; this fact proves that T is continuous if and only if $T \circ Q$ is continuous. In a similar manner it follows, from the fact that Q is both open and continuous, that T is open if and only if $T \circ Q$ is open.|||

The following is a convenient form of the induced map theorem for linear topological spaces; it follows immediately from theorem 5.7. A **topological isomorphism** is a linear isomorphism which is also a homeomorphism.

5.8 INDUCED MAP THEOREM *Let T be a linear function on a linear topological space E to a linear topological space G, let F be the null space of T, and let S be the induced map of E/F into G. Then S is continuous (open) if and only if T is continuous (open). The map T is continuous and open if and only if S is a topological isomorphism from E/F onto G.*

If, for each member t of an index set A, E_t is a linear topological space, then the product $\times \{E_t : t \in A\}$ is a linear space and may be topologized by the product topology. Recall that for t in A, P_t is the projection of the product onto the coordinate space E_t, and I_t is the injection of E_t into the product (see Section 1). The proof of the following theorem requires only a direct application of the definitions and the preceding results.

5.9 THEOREM *The product $\times \{E_t : t \in A\}$ of the linear topological spaces E_t is a linear space and the product topology is a vector topology. The projections P_t are continuous open linear functions, and the injections I_t are topological isomorphisms.*

There is a lemma on products which is frequently useful. A **projection (idempotent operator)** on a linear space E is a linear function T on E to E such that $T \circ T = T$. If T is a projection, then for each vector x the vector $x - T(x)$ belongs to the null space of T, and hence each vector x is the sum of the member $T(x)$ of the range of T and the member $x - T(x)$ of the null space. Moreover, if x belongs to the range of T and also to the null space of T, then $x = T(y)$ for some y, and $0 = T(x) = T \circ T(y) = T(y) = x$. Consequently E is the direct sum of the range of T and the null space of T. In case E is a linear topological space and T is continuous, this decomposition of E is a decomposition into a topological product, in a sense made precise by the following proposition.

5.10   Lemma on Projections *Let E be a linear topological space, let T be a continuous projection on E, and let N and R be respectively the null space and the range of T. Then E is topologically isomorphic to $R \times N$ under the map which carries a vector x into $(T(x), x - T(x))$.*

The straightforward proof of this lemma is omitted.

The lemma above can be applied to show that an arbitrary linear topological space differs but little from a Hausdorff space. We begin by observing that because a linear topological space is necessarily regular, it is Hausdorff if and only if $\{0\}$ is closed. Let E be an arbitrary linear topological space and let F be $\{0\}^-$; then F is a linear subspace of E, and there is a complementary subspace G. The relative topology for F is the trivial topology (only F and the void set are open), and G, with the relative topology, is Hausdorff because $\{0\} = G \cap \{0\}^-$ is closed in G. Finally, the linear function whose null space is G, and which projects E onto F, is continuous, because every linear function to F is continuous. It follows from the lemma above that E is topologically isomorphic to $F \times G$.

5.11   Theorem on Non-Hausdorff Spaces *Let E be a linear topological space, let F be the closure of $\{0\}$, and let G be an arbitrary subspace of E which is complementary to F. Then the relative topology for G is Hausdorff, and E is topologically isomorphic to $G \times F$, where F has the trivial topology.*

PROBLEMS

A   Exercises

(1) Any hyperplane in a linear topological space is either dense or closed.

(2) If A is a closed set of scalars not containing 0 and B is a closed subset of a linear topological space not containing 0, then $\bigcup \{aB : a \in A\}$ is closed.

(3) If $\mathscr{T}$ is a topology for a linear space E such that the mappings $(x,y) \to x + y$ and $(a,x) \to ax$ are continuous in each variable separately, and continuous at $(0,0)$ in $E \times E$ and at $(0,0)$ in $K \times E$ respectively, then $\mathscr{T}$ is a vector topology.

B   Natural, Non-Vector Topologies

In no linear space containing more than one point is the discrete topology a vector topology.

Let Y be a set with the topology induced by a family G of mappings, each member g of G mapping a topological space X_g into Y (see the end of Section 4).   Then the relative topology for $Y \sim \bigcup \{g(X_g): g \in G\}$ is discrete.   Consequently if Y is a linear space, each X_g a linear topological space and the members of G linear, the G-induced topology on Y is not a vector topology if $Y \neq \bigcup \{g(X_g): g \in G\}$.

In particular, if Y is the direct sum $\sum \{X_g: g \in G\}$ and each g is the injection of X_g into Y, the G-induced topology is a vector topology only if every X_g except one is zero dimensional.

C   PROJECTIVE TOPOLOGY

Let X be a linear space and F a class of linear functions, each member f of F mapping X into a linear topological space Y_f.   Then the projective topology for X, the weakest making each member of F continuous (see Section 4) is a vector topology for X.   (Cf. 16D.)

D   ATTEMPT AT A STRONGEST VECTOR TOPOLOGY

If E is an infinite dimensional linear space, the family of all circled sets radial at 0 is not a local base for a vector topology for E.

(Let $\{e_i: i = 1, 2, \cdots\}$ be a linearly independent set, $A_n = \sum \{a_i e_i: |a_i| \leq n^{-1}, i = 1, \cdots, n\}$ and $A = \bigcup \{A_n: n = 1, 2, \cdots\}$.   If B is a subspace complementary to that generated by A, then there is no circled set C radial at zero with $C + C \subset A + B$.)

E   STRONGEST VECTOR TOPOLOGY I (see 14E; also 6C, 6I)

On any linear space there exists a strongest vector topology, the upper bound of all vector topologies.

F   BOX TOPOLOGY

Let $\{E_\alpha: \alpha \in A\}$ be an infinite family of non-zero Hausdorff linear topological spaces and let $F = \times \{E_\alpha: \alpha \in A\}$.   Then the topology which has for a local base all sets of the form $\times \{V_\alpha: \alpha \in A\}$, where V_α belongs to a local base in E_α, is not a vector topology on F.   With this topology, F is not connected; the component of 0 is the direct sum $\sum \{E_\alpha: \alpha \in A\}$.

G   ALGEBRAIC CLOSURE OF CONVEX SETS I (see 14F)

If A is a convex subset of a linear topological space and $0 \in A$, then $\bigcap \{rA: r > 1\} \subset A^-$.   If the origin is an interior point of A, then $\bigcap \{rA: r > 1\} = A^-$.   (Use the continuity of scalar multiplication and 5.2(v).)

H   LINEARLY CLOSED CONVEX SETS I (see 14G; also 18H)

A convex subset of R^n or C^n is closed if its intersection with every straight line is closed.

A convex subset of a linear topological space with a non-void interior has the same property.   (Use the previous problem.)

I   LOCALLY CONVEX SETS

A subset A of a linear topological space is *locally convex* iff for each $x \in A$ there is a neighborhood V of x such that $V \cap A$ is convex.

A closed, connected, locally convex subset A of a Hausdorff linear topological space is convex.

((i) If $x \in A$ and $y \in A$, then there are points $z_i (1 \leqq i \leqq n)$ of A with $x = z_1$, $y = z_n$ and $[z_i : z_{i+1}] \subset A$ for $1 \leqq i < n$.

(ii) If $x \in A$ and $y \in A$ and if $z \in A$ with $[x : z] \subset A$ and $[z : y] \subset A$, then $[x : y] \subset A$.)

6 NORMABILITY, METRIZABILITY, AND EMBEDDING; LOCAL CONVEXITY

This section gives necessary and sufficient conditions that a linear topological space be normable, or metrizable, or topologically isomorphic to a subspace of a product of normable or metrizable spaces. The solution of these problems requires two new notions—that of a bounded set in a linear topological space and that of a locally convex linear topological space. A few very elementary propositions about these notions are proved.

The simplest sort of linear topological space is that in which the topology is defined by means of a norm. Recall that a norm p for a linear space E is a subadditive ($p(x + y) \leqq p(x) + p(y)$), absolutely homogeneous ($p(tx) = |t| p(x)$), non-negative functional on E, such that $p(x) = 0$ if and only if $x = 0$. If p is a norm for E, then the **metric associated with** p is defined by $d(x,y) = p(x - y)$. There is no difficulty in seeing that d is actually a metric—the triangle inequality is a direct consequence of subadditivity of p. The metric topology $\mathscr{T}$ associated with d is defined by calling a set U $\mathscr{T}$-open if and only if for each member x of U there is a positive number e such that the open sphere of radius e about x (that is, $\{y : d(x,y) < e\}$) is contained-in U. It is easy to verify that the linear space E, with the topology $\mathscr{T}$, is a linear topological Hausdorff space, and $\mathscr{T}$ is called the **norm** topology. An arbitrary linear topological space is called **normable** if and only if there is a norm whose topology is that of the space. Of course there may be many norms which give the same topology for E. Two norms, p and q, for E have the same norm topology if and only if they assign the same neighborhoods to 0, since a local base determines the topology of a linear topological space. Evidently p and q have the same norm topology if and only if $q(x) \leqq ap(x) \leqq bq(x)$ for some real numbers a and b, and for all x. Such norms are said to be **equivalent.**

A pseudo-norm p has all of the properties of a norm except that p may vanish for non-zero vectors. If p is a pseudo-norm for E and $d(x,y) = p(x - y)$, then d is a pseudo-metric for E, the linear space E with the pseudo-metric topology $\mathscr{T}$ is a linear topological space (not necessarily Hausdorff) and $\mathcal{T}$ is called the **pseudo-norm topology**. A linear topological space is **pseudo-normable** if and only if there is a pseudo-norm whose topology is that of the space, and two pseudo-norms are **equivalent** if and only if they have the same topologies.

The characterization of pseudo-normable linear topological spaces depends on the observation: If p is a pseudo-norm for E and V is the open unit sphere about 0 (that is, $V = \{x : p(x) < 1\}$), then V is a convex, circled neighborhood of 0, and the family of all multiples aV of V, with $a > 0$, is a base for the neighborhood system of 0. Rephrasing this last requirement, V has the property that if U is a neighborhood of 0, then for some real number a it is true that $V \subset aU$. A subset A of a linear topological space is called **bounded** if and only if for each neighborhood U of 0 there is a real number a such that $A \subset aU$. It is then clear that in each pseudo-normable linear topological space there is a convex bounded neighborhood of 0. The converse of this proposition is the following theorem.

6.1 NORMABILITY THEOREM   *A linear topological space is pseudo-normable if and only if there is a bounded convex neighborhood of 0. It is normable if and only if it is pseudo-normable and Hausdorff.*

PROOF   If there is a bounded convex neighborhood of 0 in a linear topological space E, then there is a bounded circled convex neighborhood U of 0, because each convex neighborhood of 0 contains a convex circled neighborhood of 0. Let p be the Minkowski functional of U; explicitly, $p(x) = \inf \{t : t > 0 \text{ and } x \in tU\}$. Then p is a pseudo-norm, and the open unit p-sphere about 0 contains $(1/2)U$ and is contained in U. If G is an arbitrary neighborhood of 0, then for some positive number a it is true that $aU \subset G$. Hence the positive multiples of U are a local base for the topology of E, and since this family is also a local base for the pseudo-norm topology, the two topologies coincide and E is pseudo-normable.

The converse of this result has already been observed, and it is evident that a space is normable if and only if it is Hausdorff and pseudo-normable.|||

A subspace F of a normable space E is also normable because a norm for E restricted to F is a norm for F. If F is a closed subspace of a pseudo-normable space, then E/F is normable; and even if F is not closed, E/F is pseudo-normable. Moreover, a finite product of normable spaces is normable. However, an infinite product of non-trivial normable spaces is never normable.

We digress from the principal topic of the section to discuss briefly the notion of boundedness.   This notion is highly important in the sequel, and the elementary facts about boundedness will be used frequently without explicit mention.   It is easy to see that a subset A of a pseudo-normed space (E,p) is bounded if and only if it is of finite diameter (that is, sup $\{p(x - y): x, y \in A\}$ is finite).   However, a set may be of finite diameter relative to a pseudo-metric and still fail to be bounded (if d is a pseudo-metric, then $e(x,y) =$ min $[1,d(x,y)]$ defines a pseudo-metric such that the diameter of the entire space is at most one).   A linear function is called **bounded** if and only if it carries each bounded set into a bounded set.   Several elementary facts about bounded sets and bounded functions are summarized in the following proposition, whose proof is omitted.

6.2   Elementary Facts on Boundedness

   (i) *If A and B are bounded subsets of a linear topological space and a is a scalar, then each of the sets aA, A^{-}, $A + B$, and $A \cup B$ is bounded.*

   (ii) *A sufficient condition that a linear function be continuous is that the image of some neighborhood of 0 be bounded.   This condition is also necessary if the range space is pseudo-normable.*

   (iii) *Each continuous linear function is bounded.   If the domain space is pseudo-normable, then a linear function is bounded if and only if it is continuous.*

It is possible to characterize those linear topological spaces which are topologically isomorphic to subspaces of a product of normable spaces, or to a product of pseudo-normable spaces.   The key to this characterization is the following observation.   If E_t is a linear space with pseudo-norm p_t, for each member t of an index set A, and $E = \bigtimes \{E_t: t \in A\}$ is the product, then a subbase for the neighborhoods of 0 in E is the family of all sets of the form $\{x: p_t(x) < a\}$, where a is an arbitrary positive number and t is an arbitrary member of A.   Each of these sets is convex, and since the family of finite intersections of these is a local base for E, it follows that the family of convex neighborhoods of 0 is a local base.   A linear topological space is called **locally convex**, and its topology is called a **locally convex topology**, if and only if the family of convex neighborhoods of 0 is a local base.   Clearly each subspace of a locally convex space is locally convex, and the preceding discussion may be summarized as follows: each subspace of a product of pseudo-normable linear topological spaces is locally convex.   The converse of this proposition is also true; it will be demonstrated after the proof of a preliminary lemma. Recall that a map R of E into F is called relatively open if the image of each open set U is open in $R[E]$, and that R is a topological map of E into F if R is one-to-one, continuous, and relatively open.

6.3 Embedding Lemma *For each t in an index set A let R_t be a linear map of a linear topological space E into a linear topological space F_t. Let R be the map of E into the product $F = \bigtimes \{F_t : t \in A\}$ which is defined (coordinatewise) by $R(x)_t = R_t(x)$ for x in E and t in A. Then R is continuous if each R_t is continuous, and R is relatively open if it is true that for each neighborhood U of 0 in E there is t in A and a neighborhood V of 0 in F_t such that $R_t^{-1}[V] \subset U$. The map R is a topological isomorphism of E onto a subspace of F if it is continuous, relatively open, and one-to-one.*

Proof A linear map into a product is continuous if and only if the map followed by projection into each coordinate space is continuous, and, in view of the definition of R, R followed by projection into F_t is simply R_t. Consequently R is continuous if each R_t is continuous. To prove the assertion concerning relative openness, observe that if U is a neighborhood of 0 in E and V is an open neighborhood of 0 in F_t such that $R_t^{-1}[V] \subset U$, then $R[U]$ contains $\{y : y \in F \text{ and } y_t \in V\} \cap R[E]$; this set is open in $R[E]$, and the lemma follows.|||

6.4 Embedding of Locally Convex Spaces *A linear topological space E is locally convex if and only if it is topologically isomorphic to a subspace of a product of pseudo-normable spaces.*

Proof It has already been observed that each subspace of a product of pseudo-normable spaces is locally convex. Suppose that E is locally convex. Each convex neighborhood V of 0 contains a convex circled neighborhood of 0 (namely $\bigcap \{aV : |a| = 1\}$), and hence the family $\mathscr{U}$ of convex circled neighborhoods of 0 is a local base. For each U in $\mathscr{U}$ let p_U be the Minkowski functional of U; then p_U is a pseudo-norm. Let F_U be E with the p_U pseudo-norm topology, and let R_U be the identity map. Then the embedding lemma 6.3 applies, and the map of E into $\bigtimes \{F_U : U \in \mathscr{U}\}$ is a linear isomorphism onto a subspace of this product.|||

If a linear topological space E is locally convex and Hausdorff, then the foregoing theorem can be sharpened to state that E is topologically isomorphic to a subspace of a product of normed spaces (problem 6D). A topological isomorphism of a space E onto a subspace of a space F is sometimes called an **embedding** of E in F. The result above can then be stated as: each locally convex linear topological space can be embedded in a product of pseudo-normable spaces.

The class of locally convex spaces has many important properties other than that of the preceding theorem, and, in fact, almost all of the theory of linear topological spaces concerns locally convex spaces. It is evident that each subspace of a locally convex space is locally convex, and it is not hard to see that quotients and products of locally convex spaces are of the same sort. There are, for locally convex spaces, certain simplifications which can be made in the specification of the topology. The propositions having to do with this simplification are given here for later reference; the straightforward proofs are omitted.

6.5 Local Bases for Locally Convex Spaces   *In each locally convex linear topological space E there is a local base $\mathscr{U}$ such that*

    (i) *the sets in $\mathscr{U}$ are convex circled sets each of which is radial at 0;*

    (ii) *the intersection of two sets in $\mathscr{U}$ contains a set in $\mathscr{U}$;*

    (iii) *for each U in $\mathscr{U}$ and each scalar $a \neq 0$, $aU \in \mathscr{U}$; and*

    (iv) *each set in $\mathscr{U}$ is closed.*

Conversely, if E is a linear space, then any non-void family $\mathscr{U}$ of subsets which satisfies (i), (ii), and (iii) is a local base of a unique locally convex topology for E.

Notice that the family of convex neighborhoods of 0 in any linear topological space is a local base for some locally convex topology $\mathscr{T}$, in view of the foregoing proposition. Clearly $\mathscr{T}$ is the strongest locally convex topology which is weaker than the original topology of the space.

6.6 Continuous Pseudo-norms   *Let E be a linear space with a locally convex topology $\mathscr{T}$, and let P be the family of all continuous pseudo-norms. Then*

    (i) *the family of p-unit spheres about 0, for p in P, is a local base for $\mathscr{T}$; that is, a set U is a $\mathscr{T}$-neighborhood of 0 if and only if there is a member p of P and a positive number r such that $\{x : p(x) < r\} \subset U$;*

    (ii) *a net $\{x_\alpha, \alpha \in D\}$ in E converges to a point x if and only if $p(x_\alpha - x)$ converges to zero for each p in P;*

(iii) *a linear mapping T of a linear topological space F into E is continuous if and only if the composition p ∘ T is continuous for each p in P; and*

(iv) *a subset of E is bounded if and only if it is of finite p-diameter for each p in P.*

There are, of course, many linear topological spaces which are not locally convex. The theory of such spaces is meager; however, it is possible to describe in a natural way those spaces which are metrizable. A consequence of this description will be the fact that an arbitrary linear topological space can be embedded in a product of pseudo-metrizable spaces.

Notice that if p is a pseudo-norm for E and d is the associated pseudo-metric, then d is **invariant** in the sense that $d(x + z, y + z) = d(x, y)$ for all x, y, and z in E. Moreover, d is **absolutely homogeneous** in the sense that $d(tx,ty) = |t|\,d(x,y)$ for all x and y in E and for every scalar t. Conversely, if d is an invariant, absolutely homogeneous pseudo-metric, then d is the pseudo-metric associated with the pseudo-norm p, where $p(x) = d(0,x)$. It follows that if a pseudo-metric is to be constructed for a linear topological space which is not locally convex, then one cannot hope to obtain both invariance and absolute homogeneity.

6.7 METRIZATION THEOREM   *A linear topological space E is pseudo-metrizable if and only if there is a countable local base for the topology.*

If E is pseudo-metrizable, there is a pseudo-metric d for E, whose topology is that of E, such that d is invariant, and such that each sphere about 0 is circled.

PROOF   It is clear that the topology of a pseudo-metrizable space has a countable local base. Conversely, assume that $\{U_n\}$ is a countable local base for the topology. By theorem 5.1, it may be supposed that each U_n is circled, that $U_0 = E$, and that $U_n + U_n + U_n \subset U_{n-1}$ for each integer n, $n \geq 1$. Define $g(x)$ to be 2^{-n} if $x \in U_{n-1} \sim U_n$, and to be 0 if x belongs to every U_n. The proof proceeds by treating $g(x - y)$ as a first approximation to the desired pseudo-metric, which is then constructed by a chaining argument. Observe that for each positive number a the set $\{x : g(x) < a\}$ is circled, since this set is one of the sets U_n.

For each x and y in E, define $d(x,y)$ to be the infimum, over all finite subsets $x_0 = x, x_1, \cdots, x_{n+1} = y$, of the sum $\sum \{g(x_i - x_{i+1}) :$

$i = 0, \cdots, n\}$.   It is not difficult to see that d is an invariant pseudo-metric, and that each sphere about 0 is circled.   To prove that the pseudo-metric topology is the topology of E, it is sufficient to show, in view of the definition of g, that $d(x,y) \leq g(x - y) \leq 2d(x,y)$, for all x and y in E.   The first of these inequalities is clear, and the second will be proved by induction on the number n of links in the chain. For each n it must be shown that $g(x_0 - x_{n+1}) \leq 2 \sum \{g(x_1 - x_{i+1}): i = 0, \cdots, n\}$.   For convenience, call the number $\sum \{g(x_i - x_{i+1}): i = r, \cdots, s\}$ the length of the chain from r to $s + 1$.   Let a be the length of the chain from 0 to $n + 1$; one can suppose that $a < 2^{-2}$ (otherwise the assertion is trivial).   Assume that the required inequality is proved for integers smaller than n.   Clearly one may suppose $n \geq 1$.   Break the chain $0, 1, \cdots, n + 1$ into three parts, from 0 to k, k to $k + 1$, and $k + 1$ to $n + 1$ (here k may be equal to 0 or to n), in such a way that the lengths of the first and third parts are at most $a/2$ each.   By the induction hypothesis, $g(x_0 - x_k)$ is at most $2(a/2) = a$, and $g(x_{k+1} - x_{n+1})$ is at most a.   If m is the integer ≥ 2 such that $2^{-m} \leq a < 2^{-m+1}$, then $x_0 - x_k$, $x_k - x_{k+1}$, and $x_{k+1} - x_{n+1}$ all belong to U_{m-1}; hence, $x_0 - x_{n+1} \in U_{m-2}$, and $g(x_0 - x_{n+1}) \leq 2^{-m+1} \leq 2a$.|||

Since a space is metrizable if and only if it is pseudo-metrizable and Hausdorff, a linear topological space is metrizable if and only if it is Hausdorff and there is a countable local base.   Each metrizable linear topological space can be assigned an invariant metric such that the spheres about 0 are circled.

6.8   COROLLARY   *Let E be a linear space, and let $\{V_n\}$ be a sequence of circled sets which are radial at 0 and such that $V_n + V_n \subset V_{n-1}$ for $n \geq 2$.   Then there exists an invariant pseudo-metric d on E such that E with the pseudo-metric topology of d is a linear topological space in which $\{V_n\}$ is a local base.*

PROOF   It follows from theorem 5.1 that the family $\{V_n\}$ is a local base for some topology on E with which E is a linear topological space. The conclusion of the theorem is now a consequence of the metrization theorem.|||

The metrization theorem may be applied to show that an arbitrary linear topological space can be embedded in a product of pseudo-metrizable spaces, in much the same way that the existence of a "large" collection of continuous pseudo-norms was used to show that a locally convex space is embeddable in a product of pseudo-normed spaces.   The two essential facts are given in the foregoing corollary 6.8, and the embedding lemma 6.3. The straightforward proof of the following proposition is omitted.

6.9 EMBEDDING THEOREM FOR LINEAR TOPOLOGICAL SPACES   *Each linear topological space can be embedded in a product of pseudo-metrizable linear topological spaces.*

It is shown in the problems at the end of this section that the continuous image of a metrizable linear topological space may fail to be metrizable, although the image under a continuous open map of a metrizable space is always metrizable. The same statement holds with "metrizable" replaced by "pseudo-metrizable". Subspaces and quotient spaces of pseudo-metrizable spaces are always of the same type. The question of whether the product of pseudo-metrizable spaces is pseudo-metrizable is of some interest, and it can be answered easily. In order to avoid complications in the statement of the theorem, a linear topological space E is called **non-trivial** (not indiscrete) if and only if there are open sets other than E and the void set.

6.10 PRODUCTS OF PSEUDO-METRIZABLE SPACES   *The product $\times \{E_t : t \in A\}$ of non-trivial linear topological spaces is pseudo-metrizable if and only if each coordinate space E_t is pseudo-metrizable and A is countable.*

PROOF   Observe that, if the product is pseudo-metrizable, then, since each E_t is isomorphic to a subspace of the product, each coordinate space is pseudo-metrizable. Suppose that A is uncountable. Each neighborhood of 0 in the product contains a finite intersection of neighborhoods of the form $\{x : x(t) \in U\}$, where U is a neighborhood of 0 in a coordinate space E_t. Hence, if N is the intersection of countably many neighborhoods of 0 in the product, then there is a countable set B of indices and a neighborhood U_t of 0 in E_t for each t in B such that N contains $\{x : x(t) \in U_t \text{ for all } t \text{ in } B\}$. If $s \in A \sim B$, and U_s is a neighborhood of 0 in E_s which is different from E_s, then N is not contained in the neighborhood $\{x : x(s) \in U_s\}$ of 0. Therefore, the product cannot have a countable local base, and is not pseudo-metrizable.

Finally, suppose that the index set A is the set of positive integers, and that, for each n in A, d_n is a pseudo-metric for E_n. By replacing each d_n by $\min[1, d_n]$, it can be assumed that d_n is everywhere less than or equal to one. It is then easy to verify that $d(x,y) = \sum \{2^{-n} d_n(x(n), y(n)) : n \in A\}$ is a pseudo-metric for the product and has the required properties.|||

PROBLEMS

A EXERCISES

(1) A family $\mathcal{U}$ of subsets of a linear space E is a local base for a locally convex vector topology for E if the following conditions are satisfied.

(i) Each $U \in \mathcal{U}$ is midpoint convex $(U + U \subset 2U)$, circled and radial at 0;

(ii) the intersection of any two members of $\mathcal{U}$ contains a member of $\mathcal{U}$;

(iii) for each $U \in \mathcal{U}$ there is some $V \in \mathcal{U}$ with $2V \subset U$.

(2) If E is a real linear space and $\mathcal{T}$ is a topology for E such that the family of all convex neighborhoods of 0 is a base for the neighborhood system of 0, and if $x + y$ and ax are continuous in each variable separately, then $\mathcal{T}$ is a vector topology.

(3) A linear topological space E is pseudo-metrizable if and only if there exists a nested local base (that is, a base linearly ordered by inclusion).

(4) If E and F are real linear topological spaces with F Hausdorff and T is an additive bounded mapping of E into F, then T is linear.

(5) Let E be a separable pseudo-metrizable linear topological space of dimension greater than one. Then E contains a countable dense set A no three points of which lie on any straight line. The complement of A is radial at each of its points, but has void interior. (Let $\{x_n\}$ be dense, and put $S_n = \{x : x \in E, d(x, x_n) < 1/n\}$. Let $y_1 = x_1$; choose y_2 in $S_2 \sim \{x_1\}$; choose y_3 in S_3 but not on the line determined by y_1 and y_2 and so on. Let $A = \{y_n : n = 1, 2, \cdots\}$.)

(6) The following four conditions on a subset A of a linear topological space are equivalent:

(i) A is bounded;

(ii) given any neighborhood U of 0 there is some $e > 0$ such that $aA \subset U$ for all a with $|a| < e$;

(iii) every countable subset of A is bounded;

(iv) if $\{x_n\}$ is any sequence in A, then $\lim x_n/n = 0$.

B MAPPINGS IN PSEUDO-NORMED SPACES I (see 8B)

Let E and F be pseudo-normed spaces, with pseudo-norms p and q, respectively. A linear mapping T of E into F is continuous if and only if there is some constant k with $q(T(x)) \leq kp(x)$ for all x in E. The smallest such constant, $r(T)$ say, is equal to $\sup \{q(T(x)) : p(x) \leq 1\}$, and $q(T(x)) \leq r(T)p(x)$ for all x in E.

The function r is a pseudo-norm on the space of all continuous linear mappings of E into F, and is a norm if and only if q is a norm provided $\dim E > 0$. In particular, the set of all continuous linear functionals on a pseudo-normed space E is a normed space, called the *adjoint* (conjugate, dual) of E and denoted by E^*; the norm in E^* is given by $\|f\| = \sup \{|f(x)| : \|x\| \leq 1\}$. Then $\|x\| = \sup \{|f(x)| : \|f\| \leq 1\}$. (Use 3.7.)

C TOPOLOGIES DETERMINED BY PSEUDO-METRICS

(a) Let E be a pseudo-metrizable linear topological space. If d is an invariant pseudo-metric such that each sphere about 0 is circled (c.f. theorem 6.7) and $q(x) = d(x,0)$, then q has the following properties:

(i) $q(x) \geqq 0$ for all x in E and $q(0) = 0$;

(ii) $q(x + y) \leqq q(x) + q(y)$ for all x,y in E;

(iii) $q(ax) \leqq q(x)$ whenever $|a| \leqq 1$; and

(iv) $\lim_{n \to \infty} q(x/n) = 0$ for all x in E.

Conversely, if q is a functional on a linear space E satisfying (i) through (iv), then the function d given by $d(x,y) = q(x - y)$ is a pseudo-metric for E defining a vector topology.

(b) Given any family $\{q_\alpha : \alpha \in A\}$ of functions on a linear space E each satisfying (i) through (iv), the finite intersections of the sets of the type $\{x : q_\alpha(x) < e\}$, with $\alpha \in A$ and $e > 0$, form a local base for a vector topology on E, the topology determined by the family $\{q_\alpha\}$. Conversely, every linear topological space has a topology which can be determined in this way.

(c) The strongest vector topology on a linear space is determined by the family of all functions q satisfying (i) through (iv).

D   PRODUCTS AND NORMED SPACES

(a) A locally convex Hausdorff space is topologically isomorphic to a subspace of a product of normed spaces. (For each continuous pseudo-norm p, $E/p^{-1}(0)$ is a normed space.)

(b) A product of linear topological spaces is pseudo-normable if and only if a finite number of the factor spaces are pseudo-normable and the rest have the trivial topology.

E   POSITIVE LINEAR FUNCTIONALS

Let S be any set and $B(S)$ the space of bounded real-valued functions on S, with $\|x\| = \sup \{|x(s)| : s \in S\}$ for each x in $B(S)$. If ϕ is a real-valued linear functional on $B(S)$ such that $\phi(x) \geqq 0$ whenever x is non-negative on S, then ϕ is continuous on $B(S)$ and $\phi(1) = \sup \{|\phi(x)| : x \in B(S), \|x\| \leqq 1\}$. The same statement holds if S has a topology and $B(S)$ is replaced by the space of all continuous bounded real-valued functions.

F   LOCALLY CONVEX, METRIZABLE, NON-NORMABLE SPACES

Let S be a topological space containing a countable sequence $\{S_n\}$ of compact proper subsets such that $S_1 \subset S_2 \subset \cdots$ and $\bigcup \{S_n : n = 1, 2, \cdots\} = S$. Let E be any linear family of scalar-valued functions continuous on S. For each n and each x in E let $\|x\|_n = \sup \{|x(s)| : s \in S_n\}$, and let $U_n = \{x : x \in E, \|x\|_n \leqq 1/n\}$. Then $\{U_n : n = 1, 2, \cdots\}$ is a local base for a locally convex metrizable vector topology for E. If for each pair of positive integers m and n and each point s_0 in $S \sim S_n$ there is some x in U_n such that $|x(s_0)| \geqq m$, then this topology is not pseudo-normable.

For particular cases, take S to be R^n and E to be all the scalar-valued continuous functions, or S to be an open set in R^2 and E to be all complex functions analytic in S. (Cf. 8I, 20H.)

G   TOPOLOGY OF POINTWISE CONVERGENCE

Let S be the interval $[0,1]$ and E the set of continuous real valued functions on S. For each finite subset T of S, and each $e > 0$ let $U_{T,e} = \{x: |x(s)| < e \text{ for all } s \in T\}$. The sets $\{U_{T,e}: e > 0, T \subset S, T \text{ finite}\}$ form a local base for a locally convex Hausdorff topology on E, called the topology of pointwise convergence. With this topology, E is not metrizable. (Suppose that $\{U_{T_i, e_i}\}$ is a countable local base, and choose $s \in S$ with $s \notin \bigcup \{T_i: i = 1, 2, \cdots\}$. Show that no U_{T_i, e_i} is contained in $U_{(s), 1}$.)

This topology is weaker than the topology on E defined by the norm $\|x\| = \sup \{|x(s)|: s \in S\}$. The identity mapping of E with the norm topology onto E with the topology of pointwise convergence provides an illustration promised in the text of a continuous image of a metrizable linear topological space which is not metrizable. (Cf. 9G.)

H   BOUNDED SETS AND FUNCTIONALS

(a) If E is a linear topological space containing a bounded set A which is a subset of no finite dimensional subspace of E, then there is a linear functional on E that is not bounded on A. Such a set exists in every infinite dimensional pseudo-metric space.

(b) Let E be a Hausdorff linear topological space. Then each bounded subset of E is a subset of a finite dimensional subspace of E if and only if every linear functional on E is bounded.

I   STRONGEST LOCALLY CONVEX TOPOLOGY I (see 12D, 14D, 20G)

If E is any linear space, the set of all convex circled subsets of E radial at 0 forms a local base for a Hausdorff locally convex topology, which is the strongest locally convex topology on E. Every linear mapping of E with this topology into any locally convex space is continuous; in particular every linear functional is continuous. With this strongest locally convex topology, an infinite dimensional linear space E is not metrizable and every bounded set is contained in a finite dimensional subspace (see the previous problem). The strongest locally convex topology coincides with the strongest vector topology (5E) if and only if the dimension of E is finite or $\aleph_0$. (If $\{e_\alpha: \alpha \in A\}$ is a non-countable basis of E, the set $\{x: x = \sum \lambda_\alpha e_\alpha, \sum |\lambda_\alpha|^{1/2} \leq 1\}$ is a neighborhood of zero in the strongest vector topology and cannot contain a convex circled set radial at 0.)

J   INNER PRODUCTS (see also 7H et seq.)

Let E be a complex linear space. A functional f from $E \times E$ to the complex numbers is called Hermitean bilinear, or sesquilinear iff for each y in E the mapping $x \to f(x,y)$ is linear and for each $x \in E$ the mapping $y \to f(x,y)$ is conjugate linear (that is, $f(x, ay + bz) = \bar{a} f(x,y) + \bar{b} f(x,z)$). It is called (Hermitean) symmetric iff $f(x,y) = \overline{f(y,x)}$ for all $x,y \in E$; then $f(x,x)$ is real for all $x \in E$. Finally the symmetric Hermitean bilinear functional f is called strictly positive, or positive definite, if $f(x,x) > 0$ for all $x \neq 0$. A strictly positive symmetric Hermitean bilinear functional f

on $E \times E$ is called an *inner product*, or a scalar product, for E. A complex linear space E with a fixed inner product f is called an inner product space and $f(x,y)$ is usually simplified to (x,y).

(a) (The CBS inequality.) For all x,y in E, $|(x,y)|^2 \leq (x,x)(y,y)$, with equality only if x and y are linearly dependent. (Note that the inner product of $(y,y)x - (x,y)y$ with itself is non-negative.)

(b) Put $\|x\| = \sqrt{(x,x)}$; this defines a norm for E called the inner product norm. The inner product can be recovered from the norm by means of the polarization identity

$$(x,y) = \tfrac{1}{4}(\|x + y\|^2 - \|x - y\|^2 + i\|x + iy\|^2 - i\|x - iy\|^2)$$

(c) (The parallelogram identity)

$$\|x + y\|^2 + \|x - y\|^2 = 2\|x\|^2 + 2\|y\|^2.$$

This identity characterizes inner product norms; any norm satisfying it corresponds to an inner product on E.

(d) If E has the inner product norm topology, the inner product is a continuous function on the topological product $E \times E$.

K   Spaces of integrable functions I (see 6N, 7M, 14M; also 9I, 16F, 22E)

Let $(X, \mathscr{S}, \mu)$ be a measure space. For each $p > 0$, $L^p(X, \mu)$, or more shortly $L^p(\mu)$, denotes the space of all real or complex valued measurable functions f on X such that $|f|^p$ is integrable. (For measure theory, we generally follow the definitions and notations used in the book by Halmos [4].) Then $L^p(\mu)$ is a linear space and if $p \geq 1$ the mapping $f \to \|f\|_p = (\int |f|^p d\mu)^{1/p}$ is a pseudo-norm on $L^p(\mu)$ such that $\|f\|_p = 0$ if and only if $f(x) = 0$ almost everywhere. It is usually called the L^p-norm and convergence relative to it is called *convergence in mean of order p*. In case $\mathscr{S}$ is the set of all subsets of X and $\mu(A)$ is the number of elements of A (with $\mu(A) = +\infty$ when A is an infinite set) the space $L^p(\mu)$ is usually denoted by $l^p(X)$. The function f belongs to $l^p(X)$ if and only if $f(x) = 0$ for all but a countable number of values of x and $\|f\|_p = (\sum \{|f(x)|^p : x \in X, f(x) \neq 0\})^{1/p} < \infty$. In particular, when X is the set of positive integers, we write simply l^p; it is the space of sequences $x = \{x_n\}$ with $\|x\|_p = (\sum \{|x_n|^p : n = 1, 2, \cdots\})^{1/p} < \infty$.

The proof that $\|.\|_p$ is a pseudo-norm for $p > 1$ requires the Hölder and Minkowski inequalities. One pattern of proof may be sketched as follows. Suppose $p > 1$ and $1/p + 1/q = 1$.

(a) If $a > 0$, $b > 0$, $\alpha > 0$, $\beta > 0$ and $\alpha + \beta = 1$, then $a^\alpha b^\beta \leq \alpha a + \beta b$. Hence, if $c > 0$ and $d > 0$, then $cd \leq c^p/p + d^q/q$.

(b) (Hölder) If $f \in L^p(\mu)$ and $g \in L^q(\mu)$, then

$$\left| \int fg d\mu \right| \leq \|f\|_p \cdot \|g\|_q.$$

(c) (Minkowski) If $f, g \in L^p(\mu)$, then

$$\|f + g\|_p \leq \|f\|_p + \|g\|_p.$$

(Note that $|f + g|^p \leq |f||f + g|^{p-1} + |g||f + g|^{p-1}$, and it is true that $|f + g|^{p-1} \in L^q(\mu)$.)

L   SPACES OF MEASURABLE FUNCTIONS I (see 7N)

Let $(X, \mathscr{S}, \mu)$ be a finite measure space, and let $S(X,\mu)$ be the space of all real valued measurable functions.   For each positive integer n, let U_n be the subset of $S(X,\mu)$ defined by $U_n = \{f: \mu(\{x: |f(x)| > 1/n\}) < 1/n\}$.

(a) The family of sets $\{U_n: n = 1, 2, \cdots\}$ is a local base for a vector topology $\mathscr{T}$, say, and $(S(X,\mu), \mathscr{T})$ is a pseudo-metrizable linear topological space.

(b) A sequence $\{f_n\}$ in $S(X,\mu)$ converges to 0 relative to the topology $\mathscr{T}$ iff $\mu(\{x: |f_n(x)| > r\}) \to 0$ for all $r > 0$.   The topology $\mathscr{T}$ is, therefore, referred to as *the topology of convergence in measure.*

(c) Let X be the closed interval $[0,1]$ and let μ be Lebesgue measure on X.   Then the convex extension of U_n is $S(X,\mu)$ for each n.   Consequently the only continuous linear functional on $S(X,\mu)$ is the functional which is identically zero.   Moreover, there is no non-negative linear functional which is not identically zero.   (For suppose that $f(x) = 1$.   Define by induction an increasing sequence $\{x_n(t)\}$ and a decreasing sequence $\{I_n\}$ of intervals, I_{n+1} being one half of I_n, so that $x_{n+1}(t) = x_n(t)$ except on I_n, $x_{n+1}(t) = 4x_n(t)$ on I_n, and $f(x_n\chi_n) \geq 2^{n-1}$, where χ_n is the characteristic function of I_n.   Then $f(\lim x_n) \geq 2^{n-1}$ for all n.)

M   LOCALLY BOUNDED SPACES

A linear topological space E is *locally bounded* if and only if for each x_0 in E and each open set G about x_0 there is a bounded open set B such that $x_0 \in B \subset G$.

(a) E is locally bounded if and only if it contains a non-void bounded open set; any locally bounded space is pseudo-metrizable.   With the strongest locally convex topology weaker than the given topology, E is pseudo-normable.

(b) If E is locally bounded there is a sequence $\{B_n\}$ of bounded sets such that $E = \bigcup \{B_n\}$.

(c) Parts (ii) and (iii) of theorem 6.2 remain true if "pseudo-normable" is replaced by "locally bounded".

(d) The space K^ω, consisting of all scalar sequences with the usual product topology, is not locally bounded.

N   SPACES OF INTEGRABLE FUNCTIONS II (see 6K, 7M, 14M)

When $0 < p < 1$, the mapping $f \to \|f\|_p = (\int |f|^p d\mu)^{1/p}$ (see 6K) is no longer a pseudo-norm, but for $n = 1, 2, \cdots$, the sets $\{f: \|f\|_p \leq 1/n\}$ form a local base for a pseudo-metrizable vector topology on $L^p(X,\mu)$.   In fact, an invariant pseudo-metric for the topology is $d(f,g) = \int |f - g|^p d\mu$.

(a) For each p with $0 < p < 1$, the space $L^p(X,\mu)$ is locally bounded (see the previous problem).

(b) If $0 < p < 1$, the space $L^p(X,\mu)$ is locally convex if and only if the set of values taken by μ is finite.   (Otherwise there are disjoint measurable sets A_n with $\mu(A_n) > 0$.   Given $U = \{f: \|f\|_p \leq e\}$ there are elements $f_n \in L^p$ with $\|f_n\|_p = e$ and $f_n(x) = 0$ for $x \notin A_n$.   If $g_n = n^{-1} \sum \{f_r: 1 \leq r \leq n\}$, $\|g_n\|_p = en^{(1-p)/p}$, and g_n belongs to the convex extension of U.)

(c) A measurable subset A of X is called an *atom* iff $0 < \mu(A) < \infty$ and, for every measurable subset B of A, either $\mu(B) = 0$ or $\mu(B) = \mu(A)$. If A is an atom, the mapping $f \to \int_A f d\mu$ is a continuous linear functional on L^p, for $0 < p < 1$. On the other hand, if X contains no atoms and $0 < p < 1$, the only continuous linear functional on L^p is the zero functional. (If ϕ is a non-zero linear functional on L^p, there is a bounded function f_0 vanishing outside a set A_0 of finite measure with $\phi(f_0) = 1$. There are disjoint subsets B and C of A_0 with $\mu(B) = \mu(C) = \frac{1}{2}\mu(A_0)$; let g and h be equal to $2f_0$ on B and C, respectively, and zero elsewhere. Then either $|\phi(g)| \geq 1$ or $|\phi(h)| \geq 1$. Suppose the former and put $A_1 = B, f_1 = g$. Continue to obtain a sequence $\{f_n\}$ with $\|f_n\|_p \to 0$ but $|\phi(f_n)| \geq 1$ for all n.)

(d) If $0 < p < 1$ and X is an infinite set, the strongest locally convex topology on $l^p(X)$ weaker than its non-locally convex topology is the relativization to $l^p(X)$ of the norm topology of $l^1(X)$, and for this locally convex topology $l^p(X)$ is dense in $l^1(X)$. (For each $x \in X$ let e_x be the function which takes the value 1 at the point x and is zero elsewhere. If $\|f\|_1 \leq 1$ and F is a finite subset of X, $\sum \{f(x)e_x : x \in F\}$ belongs to the convex circled extension of $\{f : \|f\|_p \leq 1\}$ and converges to f as F increases.)

7   COMPLETENESS

A product space is complete if and only if the factors are complete. For a subset, compactness is equivalent to completeness together with total boundedness. For a Hausdorff linear topological space, the space being finite dimensional is equivalent to the space being topologically isomorphic to Euclidean space, and to the existence of a totally bounded non-void open set. Finally, each space can be embedded densely in a complete space, called a completion, and the completion is essentially unique.

There is a natural extension of the notion of a Cauchy sequence. A net $\{x_\alpha, \alpha \in A, \geq\}$ in a linear topological space E is a **Cauchy net** if and only if $x_\alpha - x_\beta$ converges to zero; stated precisely, the net is a Cauchy net if and only if for each neighborhood U of 0 there is a γ in A such that if both α and β follow γ in the order $\geq$, then $x_\alpha - x_\beta \in U$. A **Cauchy sequence** is a sequence which is a Cauchy net. Just as for sequences, it follows immediately that each net which converges to a point of the space is a Cauchy net. A subset A of the space E which has the property that each Cauchy net in A converges to some point in A, is said to be **complete**. The image of a Cauchy net under a continuous linear function is a Cauchy net; hence, the image under a topological isomorphism of a complete space is complete. A closed subset of a complete set is complete, and if the space is a Hausdorff space, so that each net converges to at most one point, then a complete subset is necessarily closed.   This simple statement contains one of the most important and useful properties of completeness.

The relation between completeness of a linear topological space and completeness relative to a metric is simple, but a little delicate. Recall that, if d is a pseudo-metric for a set X, then X is complete relative to d if and only if each sequence $\{x_n\}$ which is a Cauchy sequence relative to d converges in the pseudo-metric topology to a member of X.   It is not hard to see that X is complete relative to d if and only if each net $\{x_\alpha,\ \alpha \in A\}$ which is a **Cauchy net relative to** d (that is, a net such that $d(x_\alpha, x_\beta)$ converges to 0) converges to a point of X.   However, it is possible to have a linear space E and a metric d for E such that E is not complete relative to d, but E with the topology of d is a complete linear topological space.   As an example, consider the set of real numbers, which (with the usual topology) is a complete, one dimensional space over the reals, but which is not complete relative to the metric $d(x,y) = |\arctan x - \arctan y|$, although that metric generates the usual topology.   If the pseudo-metric d is invariant (that is, $d(x,y) = d(x + z, y + z)$ for all x, y, and z), then this situation cannot occur, for if $d(x_\alpha, x_\beta)$ converges to 0, then $d(x_\alpha - x_\beta, 0)$ converges to 0, and completeness implies completeness relative to d.   We record this fact for future reference.[1]

7.1   THEOREM   *Let E be a linear topological space, and let d be an invariant pseudo-metric whose topology is that of E.   Then E is complete if and only if E is complete relative to d.*

Cauchy nets in a product $\mathsf{X}\,\{E_t\colon t \in A\}$ of linear topological spaces, and completeness of such a product, can be characterized in a simple manner.

7.2   COMPLETENESS IN PRODUCTS   *A net in a product of linear topological spaces is a Cauchy net if and only if its projection into each coordinate space is a Cauchy net.*

Consequently, the product of a non-void family of spaces is complete if and only if each of the factors is complete.

---

[1] A further note of explanation may be helpful.   Completeness is always defined, either explicitly or implicitly, relative to a uniform structure.   Given a metric, it is possible to define a metric uniform structure and a metric topology.   If E is a linear topological space, it is possible to define a uniform structure by using the topology and the algebraic structure.   This is the structure which is of primary concern in this study.   It is, in general, quite distinct from the metric uniform structure.

PROOF   If $\{x_\alpha, \alpha \in B, \geqq\}$ is a net in the product $\mathsf{X}\,\{E_t \colon t \in A\}$, then $x_\alpha - x_\beta$ converges to 0 relative to the product topology if and only if the projection into each coordinate space E_t converges to 0; that is, if and only if $\{x_\alpha(t), \alpha \in B, \geqq\}$ is a Cauchy net in E_t for each t. It follows that, if each E_t is complete, then each Cauchy net in the product converges to a point. The converse is easily established by using the injection map.|||

If E is a complete linear topological space, then each Cauchy sequence in E, being a Cauchy net, converges to some point of E. It is not true, however, that a space in which each Cauchy sequence converges to some point is necessarily complete. In fact, let E be the space of all real-valued functions on the real interval (0:1), topologized by pointwise convergence. Then E is simply the product space $\mathsf{X}\,\{K_t \colon t \in (0:1)\}$, where each K_t is the space of real numbers; with the usual product topology, E is a complete linear topological space. Let F be the subspace of E which consists of all functions which vanish except at a countable number of points. Then if a sequence $\{f_n\}$ in F converges to a function f', the function f' vanishes outside a countable set, and hence f' belongs to F. Thus each Cauchy sequence in F converges to a point in F. However, since every f in E can be approximated by an element in F on any finite set of points, it follows that F is dense in E relative to the product topology, and accordingly there are Cauchy nets in F which converge to members of E which are not in F; therefore, since E is a Hausdorff space, F is not complete.

A subset of a linear topological space is **sequentially complete** if and only if each Cauchy sequence in the set converges to a point in the set. In this terminology, the example above shows that a sequentially complete space may fail to be complete. For pseudo-metrizable spaces the concepts of completeness and sequential completeness actually coincide.

A locally convex space which is metrizable and complete is called a **Fréchet space**, and a linear space with a norm which is complete relative to the norm topology is called a **Banach space**. The quotient of a complete space may fail to be complete, but the quotient of a complete pseudo-metrizable space is complete (see problem 20D and 11.3).

As a first application of the notion of completeness it will be shown that a finite dimensional Hausdorff linear topological space must necessarily be topologically isomorphic to a product of a finite number of copies of the scalar field.

7.3 Uniqueness Theorem for Finite Dimensional Spaces *A finite dimensional subspace F of a linear topological Hausdorff space E is topologically isomorphic to a finite dimensional Euclidean space, and is therefore complete and closed in E.*

proof   Let x_i, $i = 1, \cdots, n$, be a (Hamel) base for F, and let T be the linear transformation from the product of n scalar fields $\times \{K: i = 1, \cdots, n\}$ onto F defined by $T(a_1, a_2, \cdots, a_n) = \sum \{a_i x_i : i = 1, \cdots, n\}$.   Then T is clearly one-to-one, and since E is a linear topological space, T is continuous.   It will be shown by induction on n that T^{-1} is continuous and that T is consequently a topological isomorphism.   If the dimension of F is one, and $\{x_1\}$ is a base for F (that is, $x_1 \neq 0$), then the function T, where $T(a) = ax_1$, is a continuous one-to-one map of the scalar field onto F.   Since T^{-1} is a linear functional with the closed null space $\{0\}$, it follows that T^{-1} is continuous.   If the dimension of F is $n + 1$, then, by the induction hypothesis, every maximal linear subspace H in F is topologically isomorphic to a Euclidean space, is therefore complete and closed in E, and hence every linear functional is continuous.   In particular the functions $x \to a_i(x)$ are continuous; thus T^{-1} is continuous.|||

There is a useful corollary to the foregoing theorem.

7.4 Corollary   *If F is a closed subspace of a linear topological space E and if G is a finite dimensional subspace, then F + G is closed.*

proof   Let Q be the quotient map of E onto the Hausdorff space E/F (see theorem 5.7).   Then $Q[G]$ is a finite dimensional subspace of E/F and is consequently closed by theorem 7.3.   Since Q is continuous, $G + F = Q^{-1}[Q[G]]$ is a closed subspace of E.|||

The notion of completeness can be reformulated in a fashion which is suggestive of compactness.   Compactness of a set A is equivalent to the following property: if $\mathscr{A}$ is a family of closed subsets of A with the **finite intersection property** (that is, each intersection of a finite number of members of $\mathscr{A}$ is not void), then the intersection of all of the members of $\mathscr{A}$ is not void.   A family $\mathscr{A}$ of subsets of a linear topological space E **contains small sets** if and only if for each neighborhood U of 0 there is a member B in $\mathscr{A}$ and a member x of E such that $B \subset x + U$; or, equivalently, if and only if, given U, there is B in $\mathscr{A}$ such that $B - B \subset U$, or such that $B^- - B^- \subset U$.

7.5 Small Set Characterization of Completeness   *A subset A of a linear topological space E is complete if and only if each family $\mathscr{A}$ of* *closed subsets of A such that $\mathscr{A}$ has the finite intersection property and contains small sets has a non-void intersection.*

PROOF   Let A be complete, and let $\mathscr{A}$ be a family of closed subsets of A such that $\mathscr{A}$ contains small sets and has the finite intersection property. Then the family $\mathscr{B}$, consisting of those sets which are the intersection of a finite number of members of $\mathscr{A}$, has the finite intersection property and is directed by $\subset$. Form a net $\{x_B, B \in \mathscr{B}, \subset\}$ by selecting an element x_B from each B in $\mathscr{B}$. Since $\mathscr{A}$ contains small sets, so does $\mathscr{B}$; and it follows that this net is a Cauchy net which converges to a point x' of A since A is complete by hypothesis. Since for each member C of $\mathscr{B}$ the net $\{x_B, B \in \mathscr{B}\}$ is eventually in C (in fact, if D follows C in the order, $x_D \in D \subset C$) and C is closed, x' belongs to every B in $\mathscr{B}$ and hence to their intersection. Consequently the intersection of all the sets in $\mathscr{A}$ is non-void. Thus, the condition is necessary.

To prove the converse, observe that, if $\{x_\alpha, \alpha \in A, \geqq\}$ is a Cauchy net, it is possible to construct a family containing small sets as follows. For each α in A let B_α be the set of all points x_β with $\beta \geqq \alpha$, and let $\mathscr{B}$ be the family of all such sets. Since the net is a Cauchy net, the family $\mathscr{B}$ contains small sets; and since A is directed, the intersection of finitely many members of $\mathscr{B}$ is non-void. The family $\mathscr{C}$ of closures of sets in $\mathscr{B}$ is a family of closed sets which has the properties proved above for $\mathscr{B}$. If it is assumed that such a family has a non-void intersection, then there is a point x' which belongs to the closure of every B_α. If U is a neighborhood of 0 and $B_\alpha{}^- - B_\alpha{}^- \subset U$, then $B_\alpha \subset x' + U$; hence, the net converges to x'. |||

The connection between completeness and compactness is simple. A compact subset of a linear topological space is automatically complete, for each Cauchy net has a cluster point, and a Cauchy net converges to each of its cluster points. Of course a set may be complete but not compact—for example, the space of real numbers; however, there is a condition which, combined with completeness, implies compactness. A subset B of a linear topological space E is **totally bounded** (sometimes called **precompact**) if and only if for each neighborhood U of 0 there exists a finite set N such that $B \subset N + U$. It is clear that each compact set is totally bounded, and each totally bounded set is bounded. It is pointed out in the problems at the end of this section that if A is totally bounded, then so are A^-, the smallest closed circled set containing A, and the product aA for each scalar a.

7.6 HAUSDORFF'S THEOREM ON TOTAL BOUNDEDNESS *A subset of a linear topological space is compact if and only if it is totally bounded and complete.*

PROOF It is easy to see that a compact set is totally bounded and complete. Conversely, if A is a totally bounded complete subset of a linear topological space, the proof that A is compact proceeds as follows. Let $\mathscr{A}$ be a family of closed subsets of A, and suppose that $\mathscr{A}$ has the finite intersection property. Then $\mathscr{A}$ is contained in a maximal family of this sort; for convenience, let $\mathscr{A}$ itself be such a maximal family. It must now be shown that the intersection of all the members of $\mathscr{A}$ is non-void. Since A is complete, the desired result will follow if it is proved that $\mathscr{A}$ contains small sets. Since A can be covered by a finite number of translates of an arbitrarily small open set, it is sufficient to show that if B_i, $i = 1, \cdots, n$, is a finite covering of A by closed subsets of A, then for some i, $B_i \in \mathscr{A}$. Since $\mathscr{A}$ is maximal, a closed subset B_i of A can fail to belong to $\mathscr{A}$ only because its adjunction to $\mathscr{A}$ destroys the finite intersection property; that is, there is a finite subfamily $\mathscr{B}_i$ of $\mathscr{A}$ whose intersection with B_i is void. If this is the case for each i, then there is a finite subfamily (namely $\bigcup \{\mathscr{B}_i : i = 1, \cdots, n\}$) of $\mathscr{A}$ whose intersection is disjoint from every B_i and is therefore void. This contradiction establishes the desired result.|||

There is a weakened form of compactness which still implies total boundedness. Recall that a set A is countably compact if and only if every covering of A by enumerably many open sets has a finite subcovering.

7.7 THEOREM *If A is a subset of a linear topological space E such that every sequence of points in A has a cluster point in E, then A is totally bounded.*

In particular, each countably compact subset of E is totally bounded, and hence each countably compact complete subset of E is compact.

PROOF Let U be a circled neighborhood of 0. Choose a subset B of A which is maximal with respect to the following property: if x and y are in B, then x is not in $y + U$. If B is finite, $A \subset B + U$; otherwise B is an infinite subset of A, and there is a sequence $\{x_n\}$ of distinct points of B. By hypothesis, there is a cluster point y of the sequence. Furthermore, if V is a neighborhood of 0 such that $V - V \subset U$, and if x_m and x_n belong to $V + y$, then $x_m - x_n \in V - V \subset U$; hence, $x_m \in x_n + U$. This contradiction establishes the theorem.|||

The notion of total boundedness yields an interesting characterization of finite dimensional spaces and hence, to a topological isomorphism, of Euclidean n-space.

7.8   CHARACTERIZATION OF FINITE DIMENSIONAL SPACES   *A Hausdorff linear topological space E is finite dimensional if and only if there is a totally bounded neighborhood of 0.*

PROOF   One half of the theorem has been established.   The converse will be proved with the aid of the following lemma: if Y is a proper closed subspace of E, if U is a bounded neighborhood of 0, and if V is a circled neighborhood of 0 such that $V + V \subset U$, then it is possible to find a point x in U such that $(x + V) \cap Y$ is empty. For, otherwise, $V + V \subset U \subset Y + V$.   By induction it follows that, for each positive integer m, $mV \subset Y + V$ or $V \subset Y + (1/m)V$. Since V is bounded and Y is closed, $V \subset \bigcap \{Y + (1/m)V : m = 1, 2, \cdots \} = Y$, which implies that Y is not a proper subspace.

Now suppose that U is a totally bounded neighborhood of 0, and that V is a circled neighborhood of 0 such that $V + V \subset U$.   Choose a non-zero member x_1 of U, and let Y_1 be the subspace generated by x_1.   Since E is Hausdorff, Y_1 is closed; hence, by the lemma, an element x_2 can be chosen from U such that $(x_2 + V) \cap Y_1$ is empty.   In general, choose x_{n+1} from U such that $(x_{n+1} + V) \cap Y_n$ is empty, where Y_n is the subspace generated by $x_1, \cdots, x_n$.   If some Y_n is E, the theorem is proved.   Otherwise the sequence $\{x_n\}$ in U has the property that $x_m \notin x_n + V$ for distinct m and n.   Finally, the set of all x_n is not totally bounded, but this set is a subset of U and hence U is not totally bounded.|||

In particular, each locally compact Hausdorff linear topological space is finite dimensional.

The section is concluded with a proof that each linear topological space can be completed, and in an essentially unique way.

If E is a linear topological space, if $E^\wedge$ is a complete linear topological space, and if T is a topological isomorphism of E into $E^\wedge$ such that the image of E under T is dense in $E^\wedge$, then $E^\wedge$ with the map T is a **completion** of E.

7.9   LEMMA   *If E is pseudo-metrizable, then there is a pseudo-metrizable completion $E^\wedge$ of E.*

PROOF   Let d be an invariant pseudo-metric, which gives the topology of E, and let $E^\wedge$ be the set of all Cauchy sequences in E.   For $\{x_n\}$

and $\{y_n\}$ in $E^\wedge$, let $d^\wedge(\{x_n\}, \{y_n\}) = \lim d(x_n, y_n)$. With the definitions $\{x_n\} + \{y_n\} = \{x_n + y_n\}$ and $a\{x_n\} = \{ax_n\}$, $E^\wedge$ is a linear topological space with invariant pseudo-metric $d^\wedge$. For an x in E let $T(x)$ be the sequence each of whose terms is x. Clearly T preserves distance, and the problem reduces to showing that $E^\wedge$ is complete relative to $d^\wedge$. Observe that $T[E]$ is dense in $E^\wedge$, because $T(x_n)$ is near $\{x_n\}$ if n is large. It follows that for each Cauchy sequence in $E^\wedge$ there is a Cauchy sequence in $T[E]$ so that the distance between the n-th term of the former and the n-th term of the latter converges to zero. Consequently $E^\wedge$ will be proved complete relative to $d^\wedge$ if it is shown that each Cauchy sequence in $T[E]$ converges to some member of $E^\wedge$. But if $\{T(x_n)\}$ is a Cauchy sequence, then $\{x_n\}$ is a Cauchy sequence and is therefore a member of $E^\wedge$, and $T(x_n)$ converges to $\{x_n\}$. Since $d^\wedge$ is an invariant pseudo-metric, $E^\wedge$ is complete by 7.1.|||

7.10   COMPLETION THEOREM   *Each linear topological space E can be mapped by a topological isomorphism onto a dense subspace of a complete linear topological space $E^\wedge$.*

PROOF   According to theorem 6.9 there is a topological isomorphism S which maps E into a product $\mathsf{X}\,\{E_t : t \in A\}$ of pseudo-metrizable linear spaces. Let $E_t^\wedge$ be a pseudo-metrizable completion of E_t, and let T_t embed E_t in $E_t^\wedge$. If T is the transformation such that for x in E, $T(x)(t) = T_t(S(x)(t))$, then T is a topological isomorphism of E into $\mathsf{X}\,\{E_t^\wedge : t \in A\}$. Since $\mathsf{X}\,\{E_t^\wedge : t \in A\}$ is complete by theorem 7.2, the closure $E^\wedge$ of $T[E]$ is complete and $T[E]$ is a dense subspace of $E^\wedge$.|||

It was shown earlier (theorem 5.11) that an arbitrary non-Hausdorff space $E^\wedge$ is topologically isomorphic to the product of two subspaces, F and G, where F is the closure of $\{0\}$ and G is an arbitrary subspace which is complementary to F. Moreover, G with the relative topology is Hausdorff. It is clear that if $E^\wedge$ is complete, then G is necessarily complete. Using these facts and the preceding theorem the following proposition becomes obvious.

7.11   COROLLARY   *Each linear topological Hausdorff space has a Hausdorff completion.*

Finally, the completion of a linear topological Hausdorff space is, in a sense which is made precise by the following theorem, unique.

7.12 Uniqueness of Completion　*If $E^\wedge$ and $E^\sim$ are linear topological Hausdorff spaces which are completions of a space E, and if T and S are the embedding mappings, then $T \circ S^{-1}$ has a unique continuous extension which is a topological isomorphism of $E^\sim$ onto $E^\wedge$.*

proof　By 5.5 both $T \circ S^{-1}$ and $S \circ T^{-1}$ have unique continuous extensions to all of $E^\sim$ and $E^\wedge$ respectively.　Since $(T \circ S^{-1}) \circ (S \circ T^{-1})$ is the identity, the composition of the extensions of these two maps is the identity map on a dense subspace of $E^\wedge$, and it is therefore the identity on $E^\wedge$.　It follows that the continuous extension of $T \circ S^{-1}$ is a topological isomorphism of $E^\sim$ onto $E^\wedge$.|||

In making use of the results of this section it is frequently convenient, when confusion is unlikely, to refer to E as a subspace of $E^\wedge$ without reference to the embedding map.　A more precise statement is made if the occasion demands it.

PROBLEMS

A　finite dimensional subspaces

If F is a finite dimensional subspace of a linear topological space E, then F is topologically isomorphic to $G \times H$ where G is a Euclidean space and H is a finite dimensional space with the trivial topology.

B　completion of a pseudo-metrizable, pseudo-normable, or locally convex space

(a) A completion of a pseudo-metrizable, pseudo-normable, or locally convex space is a space of the same sort.　In particular, the completion of a locally convex metrizable space is a Fréchet space, and the completion of a normed space is a Banach space.

(b) Let E be a pseudo-normed linear space with a pseudo-norm p, and let $E^\wedge$ be a completion of E relative to the p-topology.　Then p can be uniquely extended to a continuous pseudo-norm $p^\wedge$ on $E^\wedge$ and the $p^\wedge$-topology is the topology for the completion $E^\wedge$.

C　completeness for stronger topologies

Let E be a Hausdorff linear topological space with topology $\mathcal{T}$, and let $\mathcal{U}$ be a stronger vector topology having a local base of $\mathcal{T}$-closed $\mathcal{U}$-neighborhoods of 0.　Then a subset of E complete relative to $\mathcal{T}$ is also complete relative to $\mathcal{U}$.　(Cf. 18.3 and 18D.)

D　extension of a one-to-one mapping

Let E and F be a Hausdorff linear topological space and T a one-to-one continuous linear mapping of E into F.　Let $T^\wedge$ be the continuous extension of T, mapping $E^\wedge$ into $F^\wedge$.　A necessary and sufficient condition for $T^\wedge$ to be one-to-one is that whenever $\{x_\gamma \colon \gamma \in \Gamma\}$ is a Cauchy net in E and $T(x_\gamma) \to T(x_0)$, then $x_\gamma \to x_0$.

Let E be a Hausdorff linear topological space with topology $\mathscr{T}$, and let $\mathscr{U}$ be a stronger vector topology. Then $(E, \mathscr{U})^\wedge$ can be embedded in $(E, \mathscr{T})^\wedge$ if and only if every $\mathscr{U}$-Cauchy net which is $\mathscr{T}$-convergent is $\mathscr{U}$-convergent to the same limit.

This latter condition is satisfied if there is a local base of $\mathscr{T}$-closed $\mathscr{U}$-neighborhoods of 0.

E   COMPLEMENTARY SUBSPACES

(a) Let E be a Hausdorff linear topological space and let F be a closed subspace with finite co-dimension. Then any subspace complementary to F is isomorphic to E/F.

(b) Let E be a Hausdorff locally convex space and let F be a finite dimensional subspace. Then there exists a (closed) complementary subspace isomorphic to E/F.

(c) Let E be a Hausdorff linear topological space on which there are no continuous linear functionals not identically zero. Then no finite-dimensional subspace has a closed complement.

F   TOTALLY BOUNDED SETS

A subset B of a linear topological space is totally bounded iff
($*$) for each neighborhood U of 0, there exists a finite set N such that $B \subset N + U$.

(a) The set N may be supposed to be a subset of B.

(b) A subset of a totally bounded set is totally bounded.

(c) The closure of a totally bounded set is totally bounded.

(d) The image of a totally bounded set under a continuous linear map is totally bounded.

(e) A subset of a product is totally bounded if and only if each of its projections is totally bounded.

(f) A scalar multiple, and the closed circled extension, of a totally bounded set are totally bounded.

G   TOPOLOGIES ON A DIRECT SUM

Let $\{E_\alpha \colon \alpha \in A\}$ be an infinite family of non-zero Hausdorff linear topological spaces, let $F = \bigtimes \{E_\alpha \colon \alpha \in A\}$, and let E be the direct sum $\sum \{E_\alpha \colon \alpha \in A\}$. Let $\mathscr{T}_1$ be the usual product topology on F and $\mathscr{T}_2$ the box topology, defined in 5F. The relativization to E of $\mathscr{T}_2$ is a vector topology stronger than the relativization of $\mathscr{T}_1$; E is closed in F for $\mathscr{T}_2$ and dense in F for $\mathscr{T}_1$. If each E_α is complete, then so is E for $\mathscr{T}_2$. (The direct sum topology (see section 14) is stronger than the topology induced by $\mathscr{T}_2$, and E is also complete under this direct sum topology whenever each E_α is complete (14.6).)

H   HILBERT SPACES (see 14L)

An inner product space is called a *Hilbert space* if and only if it is complete with respect to the inner product norm. Let E be an inner product space and let H be its completion.   Then the inner product on E can be uniquely extended to $H \times H$ so that the extended function is an inner product for the completion H.   Therefore, any inner product space has a completion which is a Hilbert space.

I   HILBERT SPACES: PROJECTION

Let F be a closed convex set in a Hilbert space H.

(a) If x is any point in H, then there is a unique element $P(x)$ in F satisfying $\|x - P(x)\| = \inf\{\|x - z\| : z \in F\}$, the mapping P is called the *projection of H on F*.   (The parallelogram law can be used to show that an arbitrary minimizing sequence converges.)

(b) $\mathrm{Re}\,(x - P(x),\, y - P(x)) \leqq 0$ for each x in H and y in F, where $\mathrm{Re}\, z$ denotes the real part of the complex number z.   In particular, if F is a cone, then $\mathrm{Re}\,(x - P(x), y) \leqq 0$ and $\mathrm{Re}\,(x - P(x), P(x)) = 0$ for each x in H and each y in F; if F is a linear subspace, then $(x - P(x), y) = 0$ for each x in H and each y in F.

(c) $\|Px - Py\| \leqq \|x - y\|$ for all $x, y \in H$.

(d) If F is a linear subspace, then P is a continuous linear transformation with the properties

(i) $P \circ P = P$ (that is, P is *idempotent*).

(ii) $(P(x), y) = (x, P(y))$ for all x and y in H.

(iii) $F = \{x : P(x) = x\}$.

(e) If P is any continuous linear transformation from H into itself for which (i) and (ii) hold, then there is a closed linear subspace F of H such that P is the projection on F.

J   HILBERT SPACES: ORTHOGONAL COMPLEMENTS

Let A be a subset of a Hilbert space H.   A point x in H is said to be *orthogonal* to A, written $x \perp A$, if $(x, z) = 0$ for every z in A.   The *orthogonal complement of A in H*, denoted by $A^\perp$, is the set $\{x : x \perp A\}$.   In the following propositions A, B are subsets of H.

(a) $A^\perp$ is a closed linear subspace of H.

(b) $A \subset B$ implies $A^\perp \supset B^\perp$.

(c) $A^\perp$ is identical with the orthogonal complement of the closed linear extension of A.

(d) $A^{\perp\perp}$ is the closed linear extension of A.

(e) Let F be a closed linear subspace of H, and let P be the projection of H on $F^\perp$.   Then the projection of H on F is $I - P$, where I is the identity mapping of H onto itself.   Hence each element of H can be uniquely expressed as the sum of an element in F and an element in $F^\perp$; that is, F and $F^\perp$ are complementary in the sense of Section 1.

K   HILBERT SPACE: SUMMABILITY

Let $\{x_\alpha : \alpha \in A\}$ be a subset of a Hilbert space H indexed by A.   Let $\mathscr{A}$ be the family of all finite subsets of A; then $\mathscr{A}$ is directed by $\supset$.   For $F \in \mathscr{A}$ let $x_F = \sum\{x_\alpha : \alpha \in F\}$; then $\{x_F, F \in \mathscr{A}, \supset\}$ is a net in H.   If this net is convergent to a point x in H, the set $\{x_\alpha : \alpha \in A\}$ is *summable* and the limit x is denoted by $\sum\{x_\alpha : \alpha \in A\}$.

(a) A subset $\{x_\alpha: \alpha \in A\}$ of H is summable if and only if for each positive number e there is a finite subset B of A such that, whenever C is a finite subset of $A \sim B$, $\|\sum \{x_\alpha: \alpha \in C\}\| < e$.

(b) If a subset $\{x_\alpha: \alpha \in A\}$ is summable, then the set of indices α for which $x_\alpha \neq 0$ is countable. Furthermore, if those indices α for which $x_\alpha \neq 0$ are enumerated arbitrarily as $\alpha_1, \alpha_2, \cdots$, then $\sum \{x_\alpha: \alpha \in A\} = \lim_{n \to \infty} \sum \{x_{\alpha_r}: 1 \leq r \leq n\}$.

(c) If a subset $\{x_\alpha: \alpha \in A\}$ is orthogonal, that is, any two distinct elements of the set are orthogonal, it is summable if and only if the subset $\{\|x_\alpha\|^2: \alpha \in A\}$ of the real line is summable. If $x = \sum \{x_\alpha: \alpha \in A\}$, then $\|x\|^2 = \sum \{\|x_\alpha\|^2: \alpha \in A\}$.

L   HILBERT SPACES: ORTHONORMAL BASES

A subset A of a Hilbert space H is *orthonormal* iff A consists of elements of norm one and $(x,y) = 0$ whenever $x,y \in A$ and $x \neq y$.

(a) (Bessel's inequality) If $\{x_1, x_2, \cdots, x_n\}$ is a finite orthonormal subset of H then, for each x in H, $\sum \{|(x,x_r)|^2: 1 \leq r \leq n\} \leq \|x\|^2$. (The projection of x on the subspace generated by $x_1, \cdots, x_n$ is $\sum \{(x,x_r)x_r: 1 \leq r \leq n\}$.)

(b) Let A be an orthonormal subset of H. Then, for each x in H, $\{(x,y)y: y \in A\}$ is summable and, if P is the projection of H on the closed linear extension of A, then $P(x) = \sum \{(x,y)y: y \in A\}$.

(c) An *orthonormal basis* for H is an orthonormal subset A of H whose closed linear extension is H. An orthonormal subset A is a basis if and only if $A^\perp = \{0\}$. Given any orthonormal subset B of H, there exists an orthonormal basis for H containing B; in particular any Hilbert space admits an orthonormal basis.

(d) Let A be an orthonormal basis for H. Then for all x,y in H, $x = \sum \{(x,y)y: y \in A\}$, $(x,y) = \sum \{(x,z)(z,y): z \in A\}$ and $\|x\|^2 = \sum \{|(x,y)|^2: y \in A\}$.

(e) Any two orthonormal bases of a Hilbert space have the same cardinal number. Two Hilbert spaces are isomorphic, in the sense that there is an algebraic isomorphism between them which preserves the norm and hence also the inner product, if and only if the cardinal numbers of orthonormal bases for the two spaces are equal.

(f) Every separable Hilbert space H has a countable orthonormal basis $\{e_n: n = 1, 2, \cdots\}$ and the mapping $x \to \{(x,e_n): n = 1, 2, \cdots\}$ is an isomorphism of H onto l^2. In general, for any Hilbert space, if X is any set with the cardinal number of an orthonormal basis for H, then H is isomorphic to $l^2(X)$. For each cardinal number, there is a Hilbert space whose orthonormal basis has this cardinal number.

(g) Is the cardinal number of a Hamel base the same as that of an orthonormal basis?

M   SPACES OF INTEGRABLE FUNCTIONS III (see 6K, 6N, 14M)

For each $p > 0$ the space $L^p(X,\mu)$ is complete with respect to the topology of convergence in mean of order p. (Any Cauchy sequence contains a subsequence $\{f_n\}$ with $\sum \{\|f_{n+1} - f_n\|_p: n = 1, 2, \cdots\}$ convergent, and then $\{f_n(x)\}$ converges a.e., to $f(x)$, say. Fatou's lemma now shows that $f \in L^p$ and $\|f_n - f\|_p \to 0$.)

If $f, g \in L^2$, then $f\bar{g}$ is integrable and $(f,g) = \int f\bar{g}d\mu$ is an inner product defining the norm. Thus $L^2(X,\mu)$ is a Hilbert space (see the previous problems) and so every space $L^2(X,\mu)$ is isomorphic to a suitable space $l^2(Y)$.

N Spaces of measurable functions II (see 6L)

Let $(X,\mathscr{S},\mu)$ be a finite measure space; then the space $S(X,\mu)$ is complete relative to the topology of convergence in measure.

O The sum of closed subspaces

Let F and G be closed subspaces of a linear topological space E. The result 7.4 states that if G is finite dimensional, then $F + G$ is closed. The sum of two closed subspaces generally fails to be closed, even if E is a Hilbert space. In fact, if E is a Hilbert space and $F \cap G = \{0\}$, then $F + G$ is closed if and only if the angle between F and G is zero (that is, $\sup \{|(x,y)| : x \in F, y \in G, \|x\| = \|y\| = 1\} < 1$).

8 FUNCTION SPACES

This section is concerned with completeness and compactness properties of various families of functions on a set S to a linear topological space E, where the family is given the topology $\mathscr{T}_{\mathscr{A}}$ of uniform convergence on members of a collection $\mathscr{A}$ of subsets of S. In particular, we consider the family of functions which are bounded or totally bounded on members of $\mathscr{A}$, and we also study families of linear functions and families of continuous functions. A criterion for compactness of the latter (equicontinuity) and a surprising relationship to countable compactness are obtained.

Let S be any set, and let E be a linear topological space. The set $F(S,E)$ of all functions on S to E, with addition and scalar multiplication defined pointwise, is a linear space. For each subset A of S, and for each neighborhood U of 0 in E, let $N(A,U)$ be the family of all members f of $F(S,E)$ with the property that $f[A] \subset U$. Observe that $N(A,U)$ is circled if U is circled. A subset G of $F(S,E)$ is open relative to the **topology of uniform convergence** on A if and only if for each f in G there is a neighborhood U of 0 in E such that $f + N(A,U) \subset G$. It is easy to verify that this definition gives a topology for $F(S,E)$ such that the family of sets of the form $f + N(A,U)$ is a base for the neighborhood system of f. However, this topology need not be a vector topology for $F(S,E)$ because the neighborhoods of 0 need not be radial at 0. In fact, if U is circled, then $N(A,U)$ is radial at 0 if and only if for each f there is a scalar s such that $f \in sN(A,U)$, that is, $f[A] \subset sU$. It follows that if G is a subspace of $F(S,E)$ such that G with the relativized topology of uniform convergence on A is a linear topological space, then $f[A]$ is a bounded subset of E for each member f of G.   We shall say that a function f is **bounded on a set** A if and only if $f[A]$ is a bounded subset of E.

A net $\{f_\gamma, \gamma \in \Gamma\}$ in $F(S,E)$ converges to f **uniformly on a set** A if and only if the net converges to f relative to the topology of uniform convergence on A.   Clearly this is the case if and only if $f_\gamma - f$ is eventually in $N(A,U)$, for each neighborhood U of 0 in E.

If $\mathscr{A}$ is an arbitrary non-void family of subsets of S, then the topology $\mathscr{T}_{\mathscr{A}}$ of **uniform convergence on members of** $\mathscr{A}$ is defined to be the weakest topology which is stronger than that of uniform convergence on A, for every A in $\mathscr{A}$.   Convergence relative to $\mathscr{T}_{\mathscr{A}}$ can be described in a somewhat less esoteric fashion: a net converges to f relative to $\mathscr{T}_{\mathscr{A}}$ if and only if the net converges to f uniformly on each member of $\mathscr{A}$.   A base for the neighborhood system of 0 relative to $\mathscr{T}_{\mathscr{A}}$ is the family of all finite intersections of sets of the form $N(A,U)$, where A is a member of $\mathscr{A}$ and U is a neighborhood of 0 in E.   Since $N(A,U) \cap N(B,U) = N(A \cup B,U)$, the topology of uniform convergence on members of $\mathscr{A}$ is identical with the topology of uniform convergence on members of $\mathscr{B}$, where $\mathscr{B}$ is the family of all finite unions of members of $\mathscr{A}$.   If the union of two members of $\mathscr{A}$ is always contained in some member of $\mathscr{A}$ (that is, if $\mathscr{A}$ is directed by $\supset$), then the intersection of two sets of the form $N(A,U)$ contains a set of the same form, and consequently the family of such sets is a base for the neighborhood system of 0.

The pointwise topology, or the topology of pointwise convergence, is the topology $\mathscr{T}_{\mathscr{A}}$ where $\mathscr{A}$ is the family of all sets $\{t\}$ for all t in S.   The topology of uniform convergence on S is often called the **uniform topology**.   It is identical with the topology $\mathscr{T}_{\mathscr{A}}$ where $\mathscr{A} = \{S\}$.   Many of the theorems of this section concern the topology of uniform convergence on S.   Most of these theorems have corollaries which have to do with the topology $\mathscr{T}_{\mathscr{A}}$ of uniform convergence on each member of a family $\mathscr{A}$ of subsets of the domain set S.   The proofs of these corollaries are, for the most part, omitted; they are straightforward extensions of the theorems and of the foregoing remarks.   Throughout, E is a fixed linear topological space, and $\mathscr{A}$ is a fixed family of subsets of S.

8.1   FUNCTIONS BOUNDED ON S *The family $B(S,E)$ of all functions from S to E which are bounded on S is a subspace of $F(S,E)$ which is* *closed relative to the uniform topology, and the uniform topology is a vector topology for $B(S,E)$.*

PROOF   That $B(S,E)$ is a subspace of $F(S,E)$ is a consequence of the relations $(f + g)[S] \subset f[S] + g[S]$ and $(tf)[S] = tf[S]$ and of elementary computations with bounded sets.   Let $\{f_\gamma, \gamma \in \Gamma\}$ be a net in $B(S,E)$ which converges uniformly to f; it must be shown that $f \in B(S,E)$.   Let U be a neighborhood of 0 in E, and let V be a circled neighborhood of 0 such that $V + V \subset U$.   Choose γ such that $f(t) - f_\gamma(t) \in V$ for all t in S, and let a be such that $f_\gamma[S] \subset aV$. Then $f[S] \subset V + aV \subset \max(1,|a|)U$.   It follows that $f[S]$ is bounded.   The definition of $B(S,E)$ and the fact that E is a linear topological space show that $B(S,E)$ with the uniform topology is a linear topological space.|||

It follows from the foregoing theorem that if a net of functions, each of which is bounded on a subset A of S, converges uniformly on A to a function f, then f is also bounded on A.   The next proposition is then an easy corollary.

8.2   COROLLARY   *The class $B_{\mathscr{A}}(S,E)$ of all functions f on S to a linear topological space E such that f is bounded on each member of $\mathscr{A}$ is closed in $F(S,E)$ relative to the topology $\mathscr{T}_{\mathscr{A}}$ of uniform convergence on members of $\mathscr{A}$.*

Moreover, $\mathscr{T}_{\mathscr{A}}$ is a vector topology for $B_{\mathscr{A}}(S,E)$.

A function f on S to a linear topological space E is **totally bounded on a subset** A **of** S if and only if $f[A]$ is totally bounded; that is, a function f is totally bounded on S if and only if the range of f is a totally bounded subset of E.

8.3   FUNCTIONS WITH TOTALLY BOUNDED RANGES   *The class $T(S,E)$ of all functions from S to E, each of which has a totally bounded range, is a subspace of $B(S,E)$ which is closed in the uniform topology.*

PROOF   If C is totally bounded, then C is bounded; hence, $T(S,E) \subset B(S,E)$.   If C and D are totally bounded, then so are $C + D$ and aC for all scalars a.   It follows that $T(S,E)$ is a subspace of $B(S,E)$.

Suppose that the net $\{f_\gamma, \gamma \in \Gamma\}$ converges uniformly to f, and that each f_γ has a totally bounded range.   For a given neighborhood U of 0 in E, choose a neighborhood V of 0 such that $V + V \subset U$.   Then choose γ so that $f(t) - f_\gamma(t) \in V$ for all t in S, and choose a finite set C such that $f_\gamma[S] \subset C + V$.   Then $f[S] \subset C + U$, and $f[S]$ is totally bounded.|||

There is an obvious corollary.

8.4 COROLLARY *The class $T_{\mathscr{A}}(S,E)$ of all functions on S to E, each of which is totally bounded on each member of $\mathscr{A}$, is closed relative to $\mathscr{T}_{\mathscr{A}}$ in the space $B_{\mathscr{A}}(S,E)$ of all functions bounded on members of $\mathscr{A}$.*

The next theorem shows that certain completeness properties of E are inherited by $B(S,E)$ and $T(S,E)$. The critical fact needed for the demonstration is contained in the following lemma.

8.5 LEMMA *If $\{f_{\gamma}, \gamma \in \Gamma\}$ is a net of functions on S to E such that $\{f_{\gamma}(t), \gamma \in \Gamma\}$ converges to $f(t)$ for each t in a subset A of S, and if $\{f_{\gamma}, \gamma \in \Gamma\}$ is a Cauchy net relative to the topology of uniform convergence on A, then $\{f_{\gamma}, \gamma \in \Gamma\}$ converges to f uniformly on A.*

PROOF Let U be a closed neighborhood of 0 in E. If α is such that $f_{\alpha}(t) - f_{\beta}(t) \in U$ for all t in A and all $\beta \geq \alpha$, then $f_{\alpha}(t) - f(t) \in U$ for all t in A, because U is closed. Hence $\{f_{\gamma}, \gamma \in \Gamma\}$ converges to f uniformly on A.|||

If $\{f_{\gamma}, \gamma \in \Gamma\}$ is a net which is a Cauchy net relative to the uniform topology, then $\{f_{\gamma}(t), \gamma \in \Gamma\}$ is a Cauchy net in E for each t in S, and if E is complete, it is then possible to choose a limit $f(t)$ of $\{f_{\gamma}(t), \gamma \in \Gamma\}$ for each t. Then $\{f_{\gamma}, \gamma \in \Gamma\}$ converges to f uniformly, by the preceding lemma. Of course, the same argument applies to sequences, if E is sequentially complete. The following proposition is then clear.

8.6 COMPLETENESS RELATIVE TO THE UNIFORM TOPOLOGY *If E is complete or sequentially complete, then so is the space $B(S,E)$ of bounded functions on S to E, with the uniform topology.*

A particular consequence of this theorem is that $T(S,E)$ is complete or sequentially complete if E is complete or sequentially complete, for $T(S,E)$ is closed in $B(S,E)$.

The argument used in the proof of the foregoing theorem may be extended to the case of uniform convergence on each member of a family $\mathscr{A}$ of sets. If $\{f_{\gamma}(t), \gamma \in \Gamma\}$ converges to $f(t)$ for each t in $\bigcup \{A: A \in \mathscr{A}\}$, and if $\{f_{\gamma}, \gamma \in \Gamma\}$ is a Cauchy net relative to $\mathscr{T}_{\mathscr{A}}$, then this net converges to f uniformly on members of $\mathscr{A}$. In view of this fact it is easy to demonstrate the following proposition.

8.7 COMPLETENESS RELATIVE TO $\mathscr{T}_{\mathscr{A}}$ *If E is complete or sequentially complete, then $B_{\mathscr{A}}(S,E)$ has the same property relative to the topology $\mathscr{T}_{\mathscr{A}}$.*

There is another useful completeness property which is inherited by the function space $B_{\mathscr{A}}(S,E)$. The key to the proposition is the observation that, if t belongs to a member A of $\mathscr{A}$, then the map which carries f in $B_{\mathscr{A}}(S,E)$ into $f(t)$ is a continuous linear function on $B_{\mathscr{A}}(S,E)$, with the topology $\mathscr{T}_{\mathscr{A}}$, into E. (That is, uniform convergence on A implies convergence at each point of A.) Consequently, if G is a subset of $B_{\mathscr{A}}(S,E)$ which is bounded (or totally bounded) relative to $\mathscr{T}_{\mathscr{A}}$, then the set of all $f(t)$ for f in G is also bounded (totally bounded, respectively). It follows that if $\{f_{\gamma}, \gamma \in \Gamma\}$ is a Cauchy net in a bounded set G, and if E has the property that closed bounded sets are complete, then $\{f_{\gamma}(t), \gamma \in \Gamma\}$ converges to a limit $f(t)$ for each t in $\bigcup \{A : A \in \mathscr{A}\}$. The argument given before then demonstrates the following theorem.

8.8   Bounded Completeness   *If E has the property that each closed bounded (or totally bounded) set is complete, then $B_{\mathscr{A}}(S,E)$, with the topology $\mathscr{T}_{\mathscr{A}}$, has the same property.*

Of course $T_{\mathscr{A}}(S,E)$, being a closed subspace of $B_{\mathscr{A}}(S,E)$, shares the property stated in the foregoing theorem.

If S has a topology, it is possible to consider continuous functions from S to E. Although the topology of uniform convergence is independent of any topology for S, the class of continuous functions on S has special properties relative to uniform convergence.

8.9   Continuous Functions   *The class $C(S,E)$ of continuous functions from a topological space S to a linear topological space E is a closed linear subspace of $F(S,E)$ relative to the uniform topology.*

PROOF   It is clear that $C(S,E)$ is a linear subspace of $F(S,E)$. To show that $C(S,E)$ is closed, assume that $\{f_{\gamma}, \gamma \in \Gamma\}$ converges uniformly to f, and that all f_{γ} are continuous. The proof will be completed by showing that f is continuous. In order to show that f is continuous at t_0, let U be a neighborhood of 0 in E and V a neighborhood of 0 such that $-V + V + V \subset U$. Choose γ such that $f_{\gamma}(t) - f(t) \in V$ for all t in S. Let W be a neighborhood of t_0 such that $f_{\gamma}(t) - f_{\gamma}(t_0) \in V$ for all t in W; then $f(t) - f(t_0) = f(t) - f_{\gamma}(t) + f_{\gamma}(t) - f(t_0) + f_{\gamma}(t_0) - f(t_0) \in (-V + V + V) \subset U$ for all t in W. Thus f is continuous.|||

8.10   Corollary   *The class of bounded continuous functions on S to E is a closed linear subspace of $B(S,E)$ with respect to the uniform topology.*

As in the case of the earlier theorems, there is a corollary which involves the topology $\mathcal{T}_{\mathcal{A}}$. A function f is **continuous on a set** A if and only if the function f, restricted to A, is continuous with respect to the relativized topology. (The reader is reminded that a function may be continuous on A but fail to be continuous at any point of A.) The foregoing theorem implies that if a net of functions continuous on A converges uniformly on A to a function f, then f is continuous on A. The following proposition is now manifest.

8.11 Functions Continuous on Members of $\mathcal{A}$ *The family $C_{\mathcal{A}}(S,E)$ of all functions which are continuous on each member of a family $\mathcal{A}$ of subsets of a topological space S is closed relative to $\mathcal{T}_{\mathcal{A}}$ in the space $F(S,E)$ of all functions on S to E.*

A consequence of the preceding theorem is the fact that the family of all functions which are bounded and continuous on each member of $\mathcal{A}$ is closed relative to $\mathcal{T}_{\mathcal{A}}$ in $B_{\mathcal{A}}(S,E)$, and is complete if E is complete.

The notion of equicontinuity of a family of functions on a topological space S to a linear topological space E is a natural extension of the usual definition for real valued functions, and the fundamental theorems about equicontinuity are established by means of familiar arguments. A family M of functions on S to E is **equicontinuous at a point** s of S if and only if for each neighborhood U of 0 in E there exists a neighborhood V of s in S such that $f(t) - f(s) \in U$ for all t in V and for all f in M; that is, $f[V] \subset f(s) + U$ for all f in M. The family M of functions is **equicontinuous** if it is equicontinuous at each point of S. If the functions in a net in $F(S,E)$ form an equicontinuous family, the net is called an **equicontinuous net**.

8.12 Pointwise Closure of an Equicontinuous Family *The closure relative to the pointwise topology of a family M which is equicontinuous at a point t is equicontinuous at t.*

proof Let t_0 in S and a member U of the neighborhood system of 0 in E be given. Choose a circled neighborhood V of 0 in E such that $V + V + V \subset U$ and a neighborhood W of t_0 such that $f(t) - f(t_0) \in V$ for all t in W and all f in M. It will be shown that, for all f in the pointwise closure of M, $f(t) - f(t_0) \in U$ whenever $t \in W$.

Let $\{f_\gamma, \gamma \in \Gamma\}$ be a net in M which converges pointwise to f_0. For t in W choose γ such that $f_0(t) - f_\gamma(t) \in V$ and $f_\gamma(t_0) - f_0(t_0) \in V$. Then $f_0(t) - f_0(t_0) = f_0(t) - f_\gamma(t) + f_\gamma(t) - f_\gamma(t_0) + f_\gamma(t_0) - f_0(t_0)$, which is an element in $V + V + V \in U$.|||

The foregoing theorem implies that the pointwise closure of an equicontinuous family of functions consists of continuous functions. This is an important fact.

8.13 TOPOLOGIES FOR AN EQUICONTINUOUS FAMILY *If M is an equicontinuous family of functions on S to a linear topological space E, then the topology for M of pointwise convergence on a subset A of S is identical with the topology of pointwise convergence on the closure A^- of A.*

PROOF If U is a circled neighborhood of 0 in the range space E and s is a point of A^- then, by equicontinuity, there is a neighborhood V of s such that $g(t) \in g(s) + U$ for all members g of M and all t in V. If t is a member of $A \cap V$, and if f and g are members of M such that $g(t) - f(t) \in U$, then a simple computation shows that $g(s) - f(s) \in U + U + U$. Hence $\{g : g(t) \in f(t) + U\} \subset \{g : g(s) \in f(s) + U + U + U\}$. It follows that each of the neighborhoods of f relative to the topology of pointwise convergence on A^- contains a neighborhood of f relative to the topology of pointwise convergence on A.|||

8.14 JOINT CONTINUITY *Let M be an equicontinuous family of functions on S to E, and let M have the topology of pointwise convergence on S. Then f(s) is jointly continuous in f and in s, in the sense that the map of M × S into E defined by $(f,s) \to f(s)$ is continuous relative to the product topology.*

PROOF Suppose that $(f,s) \in M \times S$ and that U is a neighborhood of 0 in E. If V is a neighborhood of s such that $g(t) \in g(s) + U$ for all g in M and all t in V, and if h is a member of M such that $h(s) \in f(s) + U$, then $h(t) \in h(s) + U \subset f(s) + U + U$. That is, if $(h,t) \in \{g : g(s) \in f(s) + U\} \times V$, then $h(t) \in f(s) + U + U$. Joint continuity follows.|||

The results which are most useful in the study of linear topological spaces concern families of *linear* functions which are continuous or equicontinuous. When E and F are linear spaces, the linear space of all linear functions on E into F is denoted by $L(E,F)$.

8.15 THEOREM *Let E and F be linear topological spaces, let F be complete, and let $\mathscr{A}$ be a family of bounded subsets of E. Then, with the topology $\mathscr{T}_{\mathscr{A}}$, the space $L(E,F) \cap B_{\mathscr{A}}(E,F)$ (the family of linear functions on E to F which are bounded on each member of $\mathscr{A}$) is a complete linear topological space.*

PROOF   Consider first the particular case where F is Hausdorff and E is the linear extension of the union of the members of $\mathscr{A}$. Then $\mathscr{T}_{\mathscr{A}}$-convergence of a net of functions implies pointwise convergence, and since the pointwise limit of linear functions is evidently linear, the space $L(E,F) \cap B_{\mathscr{A}}(E,F)$ is a closed subspace of $B_{\mathscr{A}}(E,F)$ and hence complete.

The proof in the general case requires a little more manipulation. One first shows that $L(E,F)$ is complete relative to the topology of pointwise convergence by decomposing F into the product of a Hausdorff space G and a trivial space H by means of theorem 5.11 on non-Hausdorff spaces, and using the fact that G is complete. A $\mathscr{T}_{\mathscr{A}}$-Cauchy net of linear functions will be necessarily a Cauchy net relative to the topology of pointwise convergence on $A_0 = \bigcup \{A : A \in \mathscr{A}\}$ and will hence converge pointwise on the linear extension E_0 of A_0, to some linear function f on E_0 into F. This convergence will be uniform on members of $\mathscr{A}$ in view of 8.5, and the net will then converge relative to $\mathscr{T}_{\mathscr{A}}$ to any linear function on E to F which is an extension of f. The details of this proof are left to the reader.|||

The notion of an equicontinuous family M is particularly simple in case the functions are linear. In this case the definition can be phrased: M is equicontinuous at a point s of E if and only if for each neighborhood U of 0 in F there is a neighborhood V of 0 in E such that, if $t \in s + V$, then $f(t) \in f(s) + U$ for all f in M. Rephrased, for U there is V such that $f(r) \in U$ whenever $r \in V$ and $f \in M$. The following proposition is then straightforward.

8.16   EQUICONTINUITY AND UNIFORM EQUICONTINUITY THEOREM *Let M be a family of linear functions on a linear topological space E to a linear topological space F. Then the following conditions are equivalent:*

(i) *M is an equicontinuous family;*

(ii) *M is equicontinuous at some point;*

(iii) *for each neighborhood U of 0 in F there is a neighborhood V of 0 in E such that $f[V] \subset U$ for all f in M; and*

(iv) *if U is a neighborhood of 0 in F, then $\bigcap \{f^{-1}[U] : f \in M\}$ is a neighborhood of 0 in E.*

The discussion of equicontinuity is concluded with a proposition concerning equicontinuity and uniform convergence on totally bounded sets. The result relies on the fact, noted in the preceding theorem, that an equicontinuous family of linear functions is "uniformly" equicontinuous.

8.17   THEOREM   *If M is an equicontinuous family of linear functions from a linear topological space E to a linear topological space F, then the pointwise topology and the topology of uniform convergence on totally bounded sets coincide on M.*

PROOF   It is clear that the pointwise topology is weaker than the topology of uniform convergence on totally bounded sets.   In order to prove the reverse, it is necessary to prove that, if a net $\{T_\gamma, \gamma \in \Gamma\}$ in M converges pointwise to T_0 in M, then $\{T_\gamma, \gamma \in \Gamma\}$ converges uniformly to T_0 on totally bounded sets.   Let U be a neighborhood of 0 in F; let V be a circled neighborhood of 0 in F such that $V + V + V \subset U$; and let W be a neighborhood of 0 in E such that T in M and x in W imply that $T(x) \in V$.   If B is a totally bounded subset of E, then choose $x_1, x_2, \cdots, x_n$ such that $B \subset \bigcup \{x_i + W: i = 1, 2, \cdots, n\}$.   Next, choose an α such that $T_\gamma(x_i) - T_0(x_i) \in V$ for all $i = 1, 2, \cdots, n$, and for all $\gamma \geq \alpha$.   Then any x in B belongs to some $x_i + W$, and $T_\gamma(x) - T_0(x) = T_\gamma(x) - T_\gamma(x_i) + T_\gamma(x_i) - T_0(x_i) + T_0(x_i) - T_0(x) \in V + V + V \subset U$.   It follows that T converges uniformly to T_0 on B, and the proof is complete.|||

The final propositions of the section concern compactness, relative to the topology of pointwise convergence, of a family of continuous functions on a compact topological space.   The most surprising result is that countable compactness is a sufficient condition for compactness for certain families of functions—a theorem which is of considerable importance in the later study of the weak topology for a linear topological space.

8.18   CONTINUITY OF POINTWISE LIMITS   *Let S be a compact topological space, and let G be the space of all continuous functions on S to a compact metric space (Z,d) with the topology of pointwise convergence.   Then for each subset F of G, the following conditions are equivalent.*

(i) *Each sequence in F has a cluster point in G.*

(ii) *For all sequences $\{s_m\}$ in S and $\{f_n\}$ in F, it is true that*
$$\lim_m \lim_n f_n(s_m) = \lim_n \lim_m f_n(s_m) \text{ whenever each of the limits}$$
exists.

(iii) *The closure of F in G is compact (equivalently, each limit, relative to the topology of pointwise convergence, of members of F is continuous).*

PROOF   Assume (i), and let $\{s_m\}$ and $\{f_n\}$ be sequences in S and F, respectively, for which $\lim_{m} \lim_{n} f_n(s_m)$ and $\lim_{n} \lim_{m} f_n(s_m)$ exist.   Let s in S and f in F be cluster points of $\{s_m\}$ and $\{f_n\}$; then $\lim_{m} \lim_{n} f_n(s_m) = \lim_{m} f(s_m) = f(s) = \lim_{n} f_n(s) = \lim_{n} \lim_{m} f_n(s_m)$.

Next assume (ii), and let H be the space of all functions on S into Z with the topology of pointwise convergence.   By the theorem of Tychonoff (see 4.1 or 4.2), the space H is compact.   Therefore, in order to prove (iii), it suffices to show that the closure of F in H consists of continuous functions.   Let f be an element of H belonging to the closure of F, and assume that f is not continuous at a point s in S.   Then there is a neighborhood U of $f(s)$ in Z such that each neighborhood of s contains a point t of S with $f(t)$ not belonging to U.   Take any f_1 in F; then there is an s_1 in S such that $d(f_1(s), f_1(s_1)) < 1$ and $f(s_1) \notin U$.   Take f_2 in F so that $d(f_2(s_1), f(s_1)) < 1$ and $d(f_2(s), f(s)) < 1$.   Now choose s_2 in S such that $d(f_i(s), f_i(s_2)) < 1/2$ $(i = 1, 2)$ and $f(s_2) \notin U$.   Then take f_3 in F so that $d(f(s_j), f_3(s_j)) < 1/2$ and $d(f(s), f_3(s)) < 1/2$.   Proceeding in this way, one obtains sequences $\{f_n\}$ and $\{s_m\}$ in F and S such that, for each n, $d(f_i(s), f_i(s_n)) < 1/n$ $(i = 1, 2, \cdots, n)$, $d(f(s_j), f_{n+1}(s_j)) < 1/n$ $(j = 1, 2, \cdots, n)$, $d(f(s), f_{n+1}(s)) < 1/n$, and $f(s_n) \notin U$.   Then $\lim_{n} \lim_{m} f_n(s_m) = \lim_{n} f_n(s) = f(s)$, and $\lim_{n} f_n(s_m) = f(s_m) \notin U$.   Since it is possible to take a subsequence of $\{s_m\}$ so that the corresponding subsequence of $\{f(s_m)\}$ converges to a point outside of U, the assumption that f is not continuous contradicts the iterated limit condition of (ii).

That (iii) implies (i) is evident from the net characterization of compact spaces (see Section 4).|||

The next two propositions are essentially lemmas needed for the proof of compactness of certain countably compact families of functions.   However, the results are not without interest in themselves.

8.19   LEMMA   *Let F be a family of continuous functions on a compact topological space S to a metric space Z, and let f be a continuous function on S to Z which belongs to the closure of F relative to the topology of pointwise convergence.   Then f is a cluster point of some sequence in F.*

PROOF   For each positive integer n let S^n be the space of n-tuples of members of S, with the product topology.   For each member $(s_1, \cdots, s_n)$ of S^n there is a member g of F such that the distance $d(g(s_i), f(s_i)) < 1/n$ for $i = 1, \cdots, n$, because f is in the closure of F.

Because both g and f are continuous, it is true that $\max_i d(g(t_i),$
$f(t_i)) < 1/n$ for all points $(t_1, \cdots, t_n)$ in a sufficiently small neighborhood of $(s_1, \cdots, s_n)$. But S^n is compact, and it follows that there is a finite subfamily G_n of F such that for every $(s_1, \cdots, s_n)$ in S^n, for some g in G_n, it is true that $\max_i d(g(s_i), f(s_i)) < 1/n$. If $\{g_n\}$ is a sequence containing the union of a sequence of such finite sets G_n, then clearly f is a cluster point of $\{g_n\}$.|||

As a preliminary to the next theorem we recall that if F is a countable family of functions on a set S to a metric space Z, then the weakest topology $\mathscr{T}$ for S which makes each member of F continuous is pseudo-metrizable. (Convergence of a net relative to $\mathscr{T}$ is equivalent to convergence of its natural image in the countable product $\times \{Z : f \in F\}$ of metric spaces.) If each member of F is continuous relative to some compact topology $\mathscr{U}$ for S then $\mathscr{T}$ is weaker than $\mathscr{U}$ and hence S, with $\mathscr{T}$, is compact.

8.20　Theorem　*Let G be the family of all continuous functions on a compact space S to a compact metric space Z, let G have the topology of pointwise convergence, and let F be a subfamily of G such that each sequence in F has a cluster point in G. Then each function f which belongs to the closure of F is a cluster point of a sequence in F, and each cluster point of a sequence in F is the limit of a subsequence.*

Proof　The closure F^- of F in the space of all functions on S into Z consists of continuous functions, by virtue of theorem 8.18, and it is necessary to show that each member f of this closure is a limit of a sequence in F. Because of theorem 8.19 it may be assumed that F is countable, and, by 8.18 again, the set F satisfies the iterated limit condition: $\lim_m \lim_n f_n(s_m) = \lim_n \lim_m f_n(s_m)$ whenever $\{s_m\}$ and $\{f_n\}$ are sequences in S and F, respectively, such that each of these limits exist. Let $\mathscr{T}$ be the weakest topology for S such that each member of F is continuous. In view of the remarks preceding the theorem, the set S with the topology $\mathscr{T}$ is compact and pseudo-metric. In brief, it may be supposed that S is compact and pseudo-metric, and that the closure F^- of F consists of continuous functions (this last because of the iterated limit condition). There is a countable dense subset S_0 of S, and we will show that the topology for F^- of pointwise convergence on S_0 is identical with the topology of pointwise convergence on S. This fact will establish the theorem, for the space F^- with the topology of pointwise convergence on S_0 is pseudo-metric (it is the weakest topology which makes evaluation at each of the countable number of members of S_0 continuous), and there is consequently a sequence in F which converges to f.   Finally, if a net in F^- converges to g relative to the topology of pointwise convergence on S_0, and h is a member of F^- which is a cluster point of the net relative to the topology of pointwise convergence on S, then $g = h$ on S_0 and therefore $g = h$ on S, because both g and h are continuous. It follows without difficulty that pointwise convergence on S_0 implies pointwise convergence on S.|||

The final result of the section is an immediate consequence of the two preceding theorems.

8.21   THEOREM ON COUNTABLE COMPACTNESS   *Let G be the family of all continuous functions on a compact space S to a compact metric space Z, let G have the topology of pointwise convergence, and let F be a subfamily of G such that each sequence in F has a cluster point in G.   Then the closure of F in G is compact, and each member of this closure is the limit of a sequence in F.*

In particular, each countably compact family of continuous functions on S to Z is compact and sequentially compact.

PROBLEMS

A   CONVERSE OF 8.1

Let G be a subspace of the space $F(S,E)$ of all functions on S to E.   If the uniform topology for G is a vector topology, then each function of G is bounded.

B   MAPPINGS IN PSEUDO-NORMED SPACES II (see 6B)

Let E and F be pseudo-normed linear spaces and let $\mathscr{A} = \{S\}$, where S is the unit sphere in E.   Then $L(E,F) \cap B_{\mathscr{A}}(E,F)$ is the space, G say, of all continuous linear mappings of E into F and $\mathscr{T}_{\mathscr{A}}$ is the pseudo-norm topology for G(6B).   Hence if F is complete, G is complete relative to the pseudo-norm topology, and if F is a Banach space, so is G.   In particular, the space of continuous linear functionals on a pseudo-normed space is a Banach space.

Suppose that E and F are Banach spaces.   A linear mapping of E into F is *completely continuous* iff it maps the unit sphere of E into a relatively compact subset of F.   The set $L(E,F) \cap T_{\mathscr{A}}(E,F) = H$, say, is the set of all completely continuous mappings of E into F, and is a closed subspace of G.   If $E = F$, then H is an ideal in G (that is, if $T \in G$ and $S \in H$, then $T \circ S \in H$ and $S \circ T \in H$).   (Cf. 21A–D.)

C   POINTWISE CAUCHY NETS

Let E and F be linear topological spaces and $\{f_\alpha : \alpha \in A\}$ an equicontinuous net of linear mappings of E into F. If $\{f_\alpha(x): \alpha \in A\}$ is a Cauchy net for each x in a dense subset of E, then it is a Cauchy net for every x in E.

D   PRODUCT OF $\mathcal{T}_\mathscr{A}$ AND $\mathcal{T}_\mathscr{B}$

Let E and F be linear spaces, let G be a linear topological space, and let $\mathscr{A}$ and $\mathscr{B}$ be families of non-void subsets of E and F, respectively. Then the natural algebraic isomorphism on $L(E \times F, G)$ onto $L(E,G) \times L(F,G)$ is a homeomorphism relative to $\mathcal{T}_\mathscr{C}$ and $\mathcal{T}_\mathscr{A} \times \mathcal{T}_\mathscr{B}$ where $\mathscr{C} = \{(A \cup \{0\}) \times (B \cup \{0\}): A \in \mathscr{A}, B \in \mathscr{B}\}$.

If $\mathscr{A}$ is a family of non-void subsets of E and if F and G are linear topological spaces, there is a similar result for $L(E, F \times G)$.

E   FUNCTIONAL COMPLETION

Let E be a linear topological space which is composed of real valued functions on a set S, addition and scalar multiplication being defined as they are usually defined in function spaces. Assume that E has the property

(P) If $\{f_\alpha\}$ is a net in E converging to f,
      then $f_\alpha(t) \to f(t)$ for each t in S.

(In other words, the topology of E is stronger than that of pointwise convergence.) Is it always possible to embed E in a complete linear topological space which is composed of functions on S and which has the property (P)?

F   ADDITIVE SET FUNCTIONS

Let $\mathscr{A}$ be a ring of sets in the sense of Halmos [4]; a function ϕ on $\mathscr{A}$ to the complex numbers is *additive* if $\phi(A \cup B) = \phi(A) + \phi(B)$ whenever A and B in $\mathscr{A}$ are disjoint.

(a) Let $BA(\mathscr{A})$ be the set of all bounded additive functions; then $BA(\mathscr{A})$ is a closed subspace of the space of all bounded complex-valued functions on $\mathscr{A}$ with the uniform topology.

(b) When $\mathscr{A}$ is a σ-ring, let $CA(\mathscr{A})$ denote the set of all countably additive functions on $\mathscr{A}$ to the complex numbers; ϕ is *countably additive* iff $\sum \{\phi(A_k)\} = \phi(\bigcup \{A_k\})$ whenever $\{A_k\}$ is a sequence of pairwise disjoint elements of $\mathscr{A}$. Then $CA(\mathscr{A})$ is a closed subspace of $BA(\mathscr{A})$.

G   BOUNDEDNESS IN $B_\mathscr{A}(S,E)$

Let S be a set, E be a linear topological space, and $\mathscr{A}$ be a family of subsets of S. Then a subset G of $B_\mathscr{A}(S,E)$ is $\mathcal{T}_\mathscr{A}$ bounded if and only if it is uniformly bounded on each A in $\mathscr{A}$; this is the case if and only if $\{g(s): g \in G, s \in A\}$ is bounded for each A in $\mathscr{A}$.

H   COMPACTNESS OF SETS OF FUNCTIONS

(a) Let S be a set and let E be a Hausdorff linear topological space. Then a subset G of the space $F(S,E)$ of all functions on S to E is compact relative to the topology of pointwise convergence if and only if G is closed and for each t in S, $\{g(t): g \in G\}^-$ is compact.

(b) Let S be a compact space and let E be a Hausdorff linear topological space. Then a subset G of the space $C(S,E)$ of all continuous functions on S to E is compact relative to the uniform topology if and only if G is closed relative to the uniform topology, G is equicontinuous and $\{g(t): g \in G\}^-$ is compact for each t in S. (On an equicontinuous family of functions the uniform topology agrees with the topology of pointwise convergence.)

(c) A topological space S is called a *k-space* if and only if a subset A of S is closed whenever $A \cap B$ is closed in B for all compact subsets B of S. A Hausdorff space S which is either locally compact or satisfies the first countability axiom (that is, at each point x of S there is a countable base for the neighborhood system of x) is a k-space. A function f on a k-space S into a topological space is continuous if and only if $f|B$ is continuous for each compact subset B of S.

(d) A family G of continuous functions on a topological space S to a linear topological space E is called equicontinuous on a set A if and only if the family of restrictions of members of G to A is equicontinuous. The following theorem is a generalization of the classical Ascoli theorem.

Theorem   Let S be a k-space, let E be a Hausdorff linear topological space, and let $\mathscr{A}$ be the family of all compact subsets of S. Then a subset G of $C(S,E)$ is compact relative to the topology $\mathscr{T}_{\mathscr{A}}$, the *topology of uniform convergence on compact sets*, if and only if G is $\mathscr{T}_{\mathscr{A}}$-closed, G is equicontinuous on each compact subset of S, and $\{g(t): g \in G\}^-$ is compact for each t in S.

I   SPACES OF CONTINUOUS FUNCTIONS I (see 14J)

If X is a Hausdorff space, the space of all real or complex valued continuous functions on X is denoted by $C(X)$. Those functions of $C(X)$ which are bounded on X form a subspace denoted by $B(X)$. A scalar valued function on X is said to *vanish at infinity* iff to each $e > 0$ corresponds a compact subset A of X such that $|f(x)| < e$ whenever $x \in X \sim A$. The continuous functions which vanish at infinity form a subspace of $B(X)$ denoted by $C_0(X)$. The *support* or carrier of a scalar valued function f on X is the smallest closed subset of X outside which f vanishes. The continuous functions of compact support on X form a subspace of $C_0(X)$, denoted by $K(X)$.

(a) With the topology of uniform convergence on X, $B(X)$ is a Banach space, $C_0(X)$ is a closed subspace, and $K(X)$ is a dense subspace of $C_0(X)$. The norm in $B(X)$, defined by $\|f\| = \sup\{|f(x)|: x \in X\}$, is often referred to as the *supremum norm*.

(b) With the topology $\mathscr{T}$ of uniform convergence on compact subsets of X, $C(X)$ is a Hausdorff locally convex linear topological space. If X has a sequence $\{B_n\}$ of compact subsets, such that every compact subset of X is contained in some B_n, the topology $\mathscr{T}$ is metrizable. The converse assertion holds when X is completely regular and Hausdorff. If X is a k-space (see 8H(c)), $C(X)$ is complete relative to the topology $\mathscr{T}$. If X is locally compact, $K(X)$ is dense in $C(X)$.

(c) For each compact subset B of X, let $K_B(X)$ be the subspace of $K(X)$ consisting of those functions whose supports are contained in B, and consider on $K_B(X)$ the topology of uniform convergence on B. The convex circled subsets of $K(X)$ which intersect each $K_B(X)$ in a neighborhood of the origin form a local base for a Hausdorff locally convex topology on $K(X)$ (called the inductive limit topology—see 16C(j)). It is the strongest topology coinciding on each $K_B(X)$ with the topology of uniform convergence.

J DISTRIBUTION SPACES I (see 20I; also 16C(k), 16D(k))

Let $x = (x_1, x_2, \cdots, x_n)$ be a point of R^n and let $p = \{p_1, p_2, \cdots p_n\}$ be a set of n non-negative integers; put $|p| = p_1 + p_2 + \cdots + p_n$. For each p, define the differential operator $D^p = \partial^{|p|}/\partial x_1^{p_1} \partial x_2^{p_2} \cdots \partial x_n^{p_n}$.

(a) Let $\mathscr{E}^{(m)}$ be the space of real or complex valued functions on R^n which are at least m times continuously differentiable. The topology of *compact convergence for all derivatives* is defined by the family of pseudo-norms $\{q_K^{(m)}: K \text{ compact}, K \subset R^n\}$, where

$$q_K^{(m)}(f) = \sup \{|D^p f(x)|: x \in K, 0 \le |p| \le m\}.$$

The space $\mathscr{E}^{(m)}$ is locally convex and metrizable; if $K(r) = \{x: |x_i| \le r, 1 \le i \le n\}$ for each positive integer r, the corresponding pseudo-norms determine a countable local base. This topology is the weakest that makes continuous each mapping $D^p(0 \le |p| \le m)$ of $\mathscr{E}^{(m)}$ into $C(R^n)$, the space of continuous functions on R^n with the topology of compact convergence (8I(b)).

(b) Let $\mathscr{E}$ be the space of repeatedly differentiable real or complex valued functions on R^n (i.e., functions which are infinitely many times differentiable). The topology of compact convergence for all derivatives is defined by the family of all pseudo-norms of the form $q_K^{(m)}$ above, for each compact subset K of R^n and for each $m = 0, 1, 2, \cdots$. The space $\mathscr{E}$ is locally convex and metrizable. The topology is the weakest that makes each $D^p(|p| = 0, 1, \cdots)$ a continuous mapping of $\mathscr{E}$ into $C(R^n)$. For each p, the differential operator D^p is a continuous mapping of $\mathscr{E}$ into itself.

(c) The spaces $\mathscr{E}^{(m)}$ and $\mathscr{E}$ are complete. (If $\{f_\alpha: \alpha \in A\}$ is a Cauchy net in $\mathscr{E}^{(m)}$, then each $\{D^p f_\alpha: \alpha \in A\}$ is Cauchy and so $D^p f_\alpha \to g_p$ for each p. Show that $g_0 \in \mathscr{E}^{(m)}$ and $D^p g_0 = g_p$.)

(d) Let $\mathscr{D}_K$ be the space of repeatedly differentiable real or complex valued functions on R^n whose supports lie in a compact subset K of R^n. The relativization to $\mathscr{D}_K$ of the topology of $\mathscr{E}$ has a countable base $\{q_K^{(m)}: m = 0, 1, \cdots\}$ of pseudo-norms. With this topology, $\mathscr{D}_K$ is a Fréchet space (a complete metrizable locally convex space).

(e) Let $\mathscr{D}$ be the space of repeatedly differentiable real or complex valued functions on R^n having compact support. It is a dense subspace of $\mathscr{E}$. A more useful stronger topology for $\mathscr{D}$ is defined by taking for a local base the family of all convex circled subsets U of $\mathscr{D}$ such that $U \cap \mathscr{D}_K$ is a neighborhood of 0 in $\mathscr{D}_K$ for each compact subset K of R^n. This is the strongest locally convex topology with the property that its relativization to each $\mathscr{D}_K$ coincides with the original Fréchet space topology for $\mathscr{D}_K$. For each p, D^p is a continuous mapping of $\mathscr{D}$ into itself.

Chapter 3

THE CATEGORY THEOREMS

This short chapter is concerned with the concept of category and with its application to the theory of linear topological spaces. The results include some of the most profound and most useful theorems of the subject of linear topological spaces, and are among the most important of the applications of category.

In the first section, which treats category in topological spaces, three primary tools are established. The first two are well known —Banach's condensation theorem and a classical result of Baire. The third is a result of Osgood that in essence is a double limit theorem.

The major part of the chapter is devoted to four theorems on category in linear topological spaces. The first two of the theorems concern subsets of a linear topological space, the third deals with continuity (and dually, openness) of a linear function, and the fourth treats equicontinuity and uniform boundedness of a family of linear functions. It may be remarked that, although certain of the results of the chapter hold in a more general situation than that prescribed here, the proofs of the generalizations depend on the theorems of this chapter.

A remark on classification may be of interest. Two of the four theorems (the difference theorem and the closed graph theorem) are essentially propositions about topological groups. The other two results depend on scalar multiplication (via convexity) in a very simple fashion. Intensive use of scalar multiplication is not made until the succeeding chapter, where convexity and a closely related topic, continuous linear functionals, are investigated.

9 CATEGORY IN TOPOLOGICAL SPACES

This section contains two theorems on category which are essential to the discussion of linear topological spaces. The first of these is the condensation theorem; from the fact that a set A is of the second category it is inferred that there is a non-void open set of points at which A is of the second category. The second theorem is the classic result of Baire on the category of a complete metric space (or a locally compact Hausdorff space). The section concludes with Osgood's theorem on the points of equicontinuity of a convergent sequence of functions.

A subset A of a topological space X is **nowhere dense** in X (**rare**) if and only if the interior of the closure of A is void. Equivalently, A is nowhere dense in X if and only if the interior of the complement of A is dense in X. The set A is of the **first category in X** (**meager, exhaustible**) if and only if A is the union of a countable family of sets each of which is nowhere dense in X, and A is of the **second category (inexhaustible) in** X if and only if it is not of the first category. A set A is said to be of the **first** or **second category** if and only if A is, respectively, of the first or second category in itself (that is, in A).

The following theorem implies that if A is of the second category in X, then there is a non-void open set W, such that every non-void open subset of W intersects A in a set which is of the second category in X.

9.1 Theorem *Let A be a subset of a topological space X, and let W be the union of all open sets which intersect A in a set which is of the first category in X. Then the intersection of A with the closure of W, $A \cap W^-$, is of the first category in X.*

proof Choose a disjoint family $\mathscr{U}$ of open sets such that $\mathscr{U}$ is maximal with respect to the following property: for V in $\mathscr{U}$, $V \cap A$ is of the first category in X. Then—and this is the critical point of the proof—the intersection of A with the union U of the members of $\mathscr{U}$ is of the first category in X. For if, for each V in $\mathscr{U}$, $V \cap A$ is the union of a sequence $\{B_{V,n}\}$ of nowhere dense sets, then the union $B_n = \bigcup \{B_{V,n} : V \in \mathscr{U}\}$ is, for each positive integer n, nowhere dense. It follows that $A \cap U$ is of the first category in X, and since $U^- \sim U$ is nowhere dense, $A \cap U^-$ is of the first category in X. Finally, because $\mathscr{U}$ is maximal, U^- contains every open set which intersects A in a set of the first category in X; therefore $U^- = W^-$.|||

The theorem above is sometimes phrased in a slightly different fashion. A subset A of a topological space is of the **first category at a point** x if and only if there is a neighborhood W of x such that $W \cap A$ is of the first category in X; if A is not of the first category at x, then it is, by definition, of the **second category** at x. In this case, for every neighborhood U of x, $U \cap A$ is of the second category in X. It is clear from the definition that the set of points at which A is of the first category is open; it is precisely the set W of theorem 9.1, and in view of this theorem, W is identical with the interior of W^-. Consequently A is of the second category at the points of the complement of the interior of W^-, which is the closure of $X \sim W^-$. The following proposition is then clear.

9.2 CONDENSATION THEOREM *The set of points at which a subset A of a topological space X is of the second category is the closure V^- of an open set V. Moreover, the intersection of A with the complement of V is of the first category in X.*

The existence of sets which are of the second category is a problem of the first importance. In spite of a great deal of work on the subject, there is essentially only one method known for showing that a set is of the second category. Some refinement of the following results is possible, but the next two propositions are the basic existence theorems.

9.3 THEOREM *If X is a topological space which is either a Hausdorff or a regular space, and if A is a compact subset with non-void interior, then A is of the second category in X.*

The proof of theorem 9.3 is omitted; it is an obvious variant of the proof of the Baire theorem, 9.4.

A particular consequence of theorems 9.3 is that a locally compact Hausdorff space is of the second category in itself, and each non-void open subset of a compact, regular (or Hausdorff) space is of the second category in the space.

The other result required depends on a non-topological concept. Let X be a set and d a pseudo-metric on X. A sequence $\{x_n\}$ is a **Cauchy sequence relative to** d if and only if $d(x_m,x_n)$ converges to zero; that is, if and only if for every $e > 0$ there is an integer p such that $d(x_m,x_n) < e$ whenever $m \geq p$ and $n \geq p$. A subset A of X is **complete relative to** d if and only if each sequence in A which is a Cauchy sequence relative to d converges to some point of A.

9.4   BAIRE THEOREM   *If X is a space with a pseudo-metric d, and if A is a complete subset of X with non-void interior, then A is of the second category in X.*

PROOF   Suppose A is the union of a countable number of nowhere dense sets: $A = \bigcup \{F_n : n = 1, 2, \cdots\}$.   Since F_1 is nowhere dense, the open set $A^i \sim F_1^-$ is non-void, and it follows that there is a closed sphere V_1 of radius at most one such that $V_1 \subset A^i \sim F_1^-$. Recursively, select a closed sphere V_n of radius at most 2^{-n} such that $V_n \subset (V_{n-1})^i \sim F_n^-$.   This selection is possible because each F_n is nowhere dense.   The intersection of the sets V_n is non-void, because A is complete, and a point of this intersection is clearly a member of A which belongs to no F_n.   This is a contradiction.|||

The statement of the Baire theorem has a rather curious form, since a topological conclusion (that a set A is of second category in X) is deduced from a non-topological premise (that the set A is complete relative to a pseudo-metric d).   It is clear that a formally stronger result of the following form can be stated: if X is a topological space for which there exists a pseudo-metric d such that X has the pseudo-metric topology, then a subset A of X which has a non-void interior and is complete relative to d is of the second category in X.   This question is examined further in a problem on metric topological completeness at the end of the next section.

The final theorem of this section concerns the limit, relative to the topology of pointwise convergence, of a sequence of continuous functions.   This proposition has to do with the set of all points of equicontinuity of a sequence, although the original statement of the theorem was in terms of "almost uniform convergence."   We recall that a class F of functions on a topological space X to a pseudo-metric space (Y,d) is equicontinuous at a point x if and only if for each $e > 0$ there is a neighborhood U of x such that $d(f(x),f(y)) < e$ for all y in U and all f in F.   The closure, relative to the topology of pointwise convergence, of a class which is equicontinuous at x is equicontinuous at x, and, in particular, each member of the closure is continuous at x.   A subset Z of a topological space X is **residual in** X if and only if $X \sim Z$ is of the first category in X.

9.5   OSGOOD THEOREM   *If $\{f_n\}$ is a sequence of continuous functions on a topological space X into a pseudo-metric space (Y,d), and if the sequence $\{f_n\}$ converges pointwise to a function f on X to Y, then $\{f_n\}$ is equicontinuous at the points of a residual subset of X.*

In particular, f is continuous at the points of a residual subset of X.

PROOF   Because the sequence $\{f_n(x)\}$ is convergent, the requirement of equicontinuity of $\{f_n\}$ at x can be rephrased: for $e > 0$ there are a neighborhood U of x and a positive integer k such that $d(f_p(y),$ $f_k(y)) < e$ for y in U and for $p \geq k$. Let $K_{n,k}$ be the closed set defined by $K_{n,k} = \{x: d(f_p(x), f_k(x)) \leq 1/n \text{ for } p \geq k\}$, and let $K_n = \bigcup \{K_{n,k}{}^i: k = 1, 2, \cdots\}$, the union of the interiors. Then the set of points of equicontinuity is the intersection $\bigcap \{K_n: n = 1, 2, \cdots\}$ of the sets K_n. On the other hand, $\bigcup \{K_{n,k}: k = 1, 2, \cdots\} = X$, since the sequence $\{f_n\}$ is convergent. Hence $X \sim K_n \subset \bigcup \{K_{n,k} \sim K_{n,k}{}^i: k = 1, 2, \cdots\}$, which is obviously of the first category in X, and hence $X \sim \bigcap \{K_n: n = 1, 2, \cdots\}$ is of the first category.|||

PROBLEMS

A   EXERCISE ON CATEGORY

The set of rational numbers is not the intersection of a countable number of open subsets of the space of all real numbers.

B   PRESERVATION OF CATEGORY

Let X and Y be topological spaces, and let f be a continuous mapping of X into Y such that the image of each open subset of X is a somewhere dense subset of Y. Let A be a subset of X and B a subset of Y.

(a) If B is nowhere dense in Y, $f^{-1}[B]$ is nowhere dense in X.

(b) If B is of the first category in Y, then $f^{-1}[B]$ is of the first category in X.

(c) If A is of the second category in X, $f[A]$ is of the second category in Y.

C   LOWER SEMI-CONTINUOUS FUNCTIONS

Let E be a topological space and let Φ be a family of lower semi-continuous real functions on E, that is, the set $\{x: x \in E, \phi(x) \leq a\}$ is closed in E for each real number a and each ϕ in Φ. Suppose that the set of x in E for which sup $\{\phi(x): \phi \in \Phi\}$ is finite is of the second category in E. Then there is a non-void open subset G of E and a positive integer k such that sup $\{\phi(x): \phi \in \Phi, x \in G\} \leq k$.

D   GENERALIZED BAIRE THEOREM

A subset A of a topological space X is of the second category in X if and only if A is somewhere dense in X and there exists a sequence $\{\mathscr{E}_n: n = 1, 2, \cdots\}$ of families of open subsets of X with these properties: (i) for any closed nowhere dense set N and any non-void open set G in X such that $G \subset A^-$ and $G \cap N$ is void, there exists for each positive integer n an element E_n of $\mathscr{E}_n$ such that $N \cap E_n$ is void and $G \cap E_n$ is not void;

(ii) if $\{E_n : n = 1, 2, \cdots\}$ is a sequence with E_n in $\mathscr{E}_n$ and $A \cap P_n$ not void for each n, where $P_n = \bigcap \{E_i : i = 1, 2, \cdots, n\}$, then $A \cap P$ is not void, where $P = \bigcap \{E_i : i = 1, 2, \cdots\}$.

E EMBEDDING OF A FINITE DIMENSIONAL COMPACT METRIC SPACE INTO AN EUCLIDEAN SPACE

A most elegant application of the Baire category theorem is the proof by W. Hurewicz of the "embedding theorem" for finite dimensional separable metric spaces. This theorem does not have a direct connection with linear space theory, but the beauty of the proof more than justifies the inclusion of the theorem here. Let X be a set, then a *finite covering of X* is a finite family of subsets of X whose union is X. If X is a topological space and each member of a finite covering is open, then the covering is called *open*. The *order* of a finite covering of X is the smallest non-negative integer n such that any $n + 2$ distinct members of the covering have an empty intersection. Equivalently, the order of a finite covering is the largest integer n such that some $n + 1$ distinct members of the covering have a non-empty intersection.

In order to illustrate the method, we restrict attention to compact metric spaces. A finite covering of a metric space is called an *e-covering* if and only if the diameter of each member of the covering is less than e. A compact metric space X is said to be of *dimension r*, where r is an integer, if and only if (i) for each positive number e, there is a finite open e-covering of X of order less than or equal to r, and (ii) there is a positive number e such that each finite open e-covering of X is of order greater than $r - 1$. Let E_n denote the Euclidean n-space, i.e., the n-dimensional real linear topological space in which a topology is given by the norm defined by $\|x\| = (\sum \{x_i{}^2 : i = 1, 2, \cdots, n\})^{1/2}$ where x_i is the i-th component of the vector x in E_n. A map on a metric space X into a space Y is said to be an *e-map* if and only if the diameter of $f^{-1}[\{y\}]$ is less than e for each y in Y.

Lemma Let X be a compact metric space of dimension less than or equal to n, and let f be a continuous map of X into E_{2n+1}. Then, for any two positive numbers e and d, there is a continuous e-map g on X into E_{2n+1} such that $\sup \{\|f(x) - g(x)\| : x \in X\} < d$.

(Since f is uniformly continuous there is a positive number e_1, such that $e_1 < e$ and $\|f(x) - f(y)\| < d/2$ whenever $d(x,y) \leqq e_1$ (d is the metric for X). Let $\{U_1, U_2, \cdots, U_k\}$ be a finite open e_1-covering of X of order less than or equal to n. Choose vectors p_i, $i = 1, 2, \cdots, k$, in E_{2n+1} so that, for each i, dist $(p_i, f[U_i]) < d/2$ and $\{p_1, p_2, \cdots, p_k\}$ are in general position in E_{2n+1}, that is, if t_i's are real numbers such that $t_1 p_{i1} + \cdots + t_j p_{ij} = 0$, $j \leqq 2n + 2$, and $t_1 + \cdots + t_j = 0$, then $t_1 = \cdots = t_j = 0$. [This is equivalent to: for $m \leqq 2n + 2$, no m points of $p_1, \cdots, p_k$ are contained in a translate of a linear subspace of dimension $m - 2$.] For each i, let $\psi_i = d(x, \sim U_i)$, and let $\phi_i = (\sum \{\psi_i : i = 1, \cdots, k\})^{-1}\psi_i$; then $\sum \{\phi_i : i = 1, 2, \cdots, k\} = 1$, $\phi_i \geqq 0$, and $\phi_i(x) > 0$ if and only if $x \in U_i$. Now let $g(x) = \sum \{\phi_i p_i : i = 1, 2, \cdots, k\}$; then g is a continuous e-map such that

$$\sup \{\|f(x) - g(x)\|\ x \in X\} < d.)$$

Theorem   Let X be a compact metric space of dimension less than or equal to n.   Then X can be homeomorphically mapped into E_{2n+1}.

(Let E be the space of all continuous maps on X into E_{2n+1}, with the topology of uniform convergence.   Let B_e be the set of all continuous e-maps on X into E_{2n+1}.   Then B_e is an open dense subset of E.)

F   LINEAR SPACE OF DIMENSION $\aleph_0$

An infinite dimensional Hausdorff linear topological space that is of the second category cannot be the union of a countable number of finite-dimensional subspaces.   In particular, an infinite dimensional complete metrizable space cannot be of linear dimension $\aleph_0$.

G   IMAGE OF A PSEUDO-METRIZABLE LINEAR SPACE

The example in 6G shows that a continuous image of a pseudo-metrizable linear topological space need not be pseudo-metrizable.

(a) Let T be an additive mapping of a linear topological space E into a linear topological space F such that the image of each neighborhood of 0 in E is a somewhere dense set in F.   Then the closure of the image of each neighborhood of 0 in E is a neighborhood of 0 in F.   If, in addition, T is continuous and E is pseudo-metrizable, then F is pseudo-metrizable.

(b) If E is a pseudo-metrizable linear topological space, then the image of E under a continuous linear mapping is either pseudo-metrizable or of the first category.

(c) If E is a pseudo-metrizable subspace of a linear topological space F, then the closure of E is pseudo-metrizable.   If F is not pseudo-metrizable, then E is nowhere dense in F.

H   ADDITIVE SET FUNCTIONS

Let $\mathscr{S}$ be a σ-ring of subsets of a set X and $\{\phi_n\}$ a sequence of finitely additive scalar-valued functions on $\mathscr{S}$; suppose that $\phi_n(A) \to \phi(A)$ for each $A \in \mathscr{S}$.

(a) If each ϕ_n is absolutely continuous with respect to a measure μ on $\mathscr{S}$, then $\{\phi_n\}$ is equiabsolutely continuous [called uniformly absolutely continuous by Halmos] and ϕ is absolutely continuous with respect to μ.

(b) If each ϕ_n is countably additive, so is ϕ.

(c) Let $\mathscr{S}_0$ be a subset of $\mathscr{S}$ such that (i) if $A, B \in \mathscr{S}_0$, then $A \cup B \in \mathscr{S}_0$ and (ii) if $A \in \mathscr{S}_0$ and $B \subset A$, then $B \in \mathscr{S}_0$, and let $\mathscr{S}_0$ be directed by $\subset$. (This will be applied to the set of subsets of a measure space which have finite measure.)   If each net $\{\phi_n(A): A \in \mathscr{S}_0\}$ converges, then the convergence is uniform in n and ϕ also converges.

(Let $\mathscr{S}_\mu$ be the set of subsets of $\mathscr{S}$ of finite measure; put $d(A,B) = \mu(A \sim B) + \mu(B \sim A)$ for $A, B \in \mathscr{S}_\mu$.   Then $\mathscr{S}_\mu$ is a complete metric space. A finitely additive function on $\mathscr{S}$ is absolutely continuous if and only if it is a continuous function on $\mathscr{S}_\mu$; for this, it is sufficient that it be continuous at one point of $\mathscr{S}_\mu$.   A set Φ of such functions is equiabsolutely continuous if and only if it is an equicontinuous set of mappings on $\mathscr{S}_\mu$.   Osgood's theorem 9.6 gives (a).

To prove (b), it is sufficient to construct a finite measure ν so that each

ϕ_n is absolutely continuous with respect to ν. For then applying (a) with this measure ν, it follows that ϕ is countably additive, ν being finite. Now if μ_n is the total variation of ϕ_n (Halmos [4] p. 122; also 14I), then $c_n = \sup\{\mu_n(A): A \in \mathscr{S}\} < \infty$ and $\nu = \sum\{2^{-n}(1 + c_n)^{-1}\mu_n: n = 1, 2, \cdots\}$ will suffice.

For (c), apply (a) again with the measure ν to deduce that $\phi_n(B)$ is small uniformly in n if $\mu_n(B)$ is small enough for a finite number of n's, say $1 \leq n \leq m$; use the convergence along $\mathscr{S}_0$ of ϕ_n for $1 \leq n \leq m$ to find $A \in \mathscr{S}_0$ so that $B \cap A = \phi$ and $B \in \mathscr{S}_0$ imply that $\mu_n(B)$ is small for $1 \leq n \leq m$.)

I SEQUENTIAL CONVERGENCE IN $L^1(X,\mu)$ (see 16F)

Let $\{f_n\}$ be a sequence of elements of $L^1(X,\mu)$ such that $\{\int_A f_n d\mu\}$ converges for each measurable subset A of X.

(a) There exists $f \in L^1(X,\mu)$ with $\lim \int_A f_n d\mu = \int_A f d\mu$ for each measurable $A \subset X$. (Use (a) and (b) of the previous problem with $\phi_n(A) = \int_A f_n d\mu$ and the Radon-Nikodym theorem.)

(b) Suppose also that $\{f_n\}$ is convergent in measure on every subset of finite measure. Then $\{f_n\}$ is convergent in norm.

(We may suppose, in view of (a), that $\int_A f_n d\mu \to 0$ for each A. Apply (c) of the previous problem with $\mathscr{S}_0 = \mathscr{S}_\mu$, the set of subsets of finite measure, to show the existence of a set $A \in \mathscr{S}_\mu$ with $\int_{X \sim A} |f_n| d\mu$ small uniformly in n. It follows from (a) of that problem that $\int_B |f_n| d\mu$ is small uniformly in n if $\mu(B)$ is small.)

10 THE ABSORPTION THEOREM AND THE DIFFERENCE THEOREM

The results of this section have to do with the deduction of boundedness and continuity properties from premises about category. The two principal theorems are the following. If A is a closed convex symmetric set in a linear topological space E such that A absorbs each point of a convex set B, and if B is bounded and of the second category in itself, then A absorbs B (the absorption theorem). If A is a subset of E that is of the second category in E and that satisfies the condition of Baire, then $A - A$ is a neighborhood of 0 (the difference theorem). There are corollaries concerning the absorption of compact convex sets and the continuity of linear functions.

A subset A of a linear space **absorbs** a set B, and B is **absorbed by** A, if and only if there is a positive number r such that $B \subset sA$ for all numbers s such that $s \geq r$. This definition is clearly an extension of the notion whereby boundedness was defined; a set is bounded if and only if it is absorbed by each neighborhood of 0.

10.1 ABSORPTION THEOREM *Let A and B be convex sets in a linear topological space E such that A is closed and absorbs each point of*

$B \cup (-B)$.   *If B is bounded and of the second category in itself, then A absorbs B.*

PROOF   By the hypotheses, there is some positive integer N such that $\bigcap \{nA \cap B : n \geq N\}$ is somewhere dense in B.   Consequently there is a point y of B and a neighborhood U of 0 in E such that $nA = (nA)^- \supset (U + y) \cap B$ for $n \geq N$.   Since B is bounded, so is $B - B$; thus there is a real number b, $0 < b < 1$, such that $U \supset b(B - B)$ and hence $nA \supset (y + b(B - B)) \cap B$ for $n \geq N$.   The set $y(1 - b) + bB$ is contained in $y + b(B - B)$ and, since B is convex, it is also contained in B; therefore, $nA \supset y(1 - b) + bB$ for $n \geq N$.   The set A absorbs each point of $-B$, and consequently there is an integer m, which may be supposed to be larger than N, such that $-y(1 - b) \in rA$ for $r \geq m$.   Then rA contains $(1/2)[y(1 - b) + bB] + (1/2)[-y(1 - b)] = (1/2)bB$, and $(2r/b)A \supset B$, for all $r \geq m$.|||

There is a very useful corollary to the absorption theorem; it is obtained by recalling from theorem 9.3 that each compact subset of a regular topological space is of the second category in itself.

10.2   COROLLARY   *A closed convex set A absorbs a compact convex set B if A absorbs each point of B $\cup$ (− B).*

The following is also a useful consequence of the absorption theorem.   A set A is called **sequentially closed** if A contains the limit points of each sequence in A.

10.3   COROLLARY   *Let A be a closed convex set and B be a sequentially closed, convex, circled and bounded set which is sequentially complete. Then A absorbs B if A absorbs each point of B.*

PROOF   Without loss of generality, one can assume that B is radial at 0.   In view of the absorption theorem it is sufficient to exhibit a linear topology which is stronger than the original one $\mathcal{T}$, and with respect to which the set B is closed and of the second category in itself.   Let p be the Minkowski functional of B; then p is a pseudo-norm, and the pseudo-norm topology is stronger than $\mathcal{T}$ since B is bounded.   Clearly B is closed relative to the new topology.   If B is shown to be complete with respect to the pseudo-metric associated with p, then B is of the second category in itself relative to the pseudo-norm topology (Baire theorem 9.4).   Let $\{x_n\}$ be a Cauchy sequence relative to p in B; then it is a $\mathcal{T}$-Cauchy sequence.   Since B is sequentially complete, the sequence $\{x_n\}$ converges to a point y in B

relative to $\mathcal{T}$. Given a positive number e, there is an integer k such that $p(x_m - x_n) \leqq e$ whenever $m,n \geqq k$, or equivalently $x_m - x_n \in eB$ whenever $m,n \geqq k$. Because B is sequentially closed, $x_m - y \in eB$, that is $p(x_m - y) \leqq e$ for all $m \geqq k$. Therefore, the sequence $\{x_n\}$ converges to y relative to the pseudo-norm topology.|||

A subset A of a topological space X satisfies the **condition of Baire** if and only if there is an open set U such that both $U \sim A$ and $A \sim U$ are of the first category in X. Roughly speaking, a set satisfies the condition of Baire if it is open—give and take a set of the first category. Each closed set satisfies the condition of Baire because it is equal to the union of its interior and its boundary, and the latter is nowhere dense. More generally, the complement of any set satisfying the condition of Baire also satisfies the condition. It is evident that the union of a countable number of sets satisfying the condition of Baire is a set of the same sort, and it follows that the family of all sets satisfying the condition is closed under complements, countable unions, and countable intersections. In particular, every Borel set satisfies the condition of Baire.

The proof of the difference theorem will require one elementary fact about category in linear topological spaces. If a subset of such a space E is of the second category in E, then E is of the second category. Each set A which is radial at 0 is then of the second category in E because: $E \subset \bigcup \{nA: n = 1, 2, \cdots\}$; hence nA is of the second category in E for some n, and therefore A itself is of the second category in E. It follows by a translation argument that each set which is radial at some point, and in particular every non-void open set, is of the second category in E whenever E is of the second category.

10.4 DIFFERENCE THEOREM *If a subset A of a linear topological space E is of the second category in E and satisfies the condition of Baire, then $A - A$ is a neighborhood of 0.*

PROOF For convenience, for this proof only, let us agree that a set A *fills* an open set U if $U \sim A$ is of the first category. Each set which is of the second category in E and satisfies the condition of Baire fills some non-void open set. If A fills an open set U and B fills an open set V, then $A \cap B$ fills $U \cap V$; in particular, if $U \cap V$ is non-void, then $A \cap B$ is non-void, because each open set is of the second category. Consequently, if A fills U and $(x + U) \cap U$ is non-void, then $(x + A) \cap A$ is non-void. But the set of all points x such that $(x + U) \cap U$ is non-void is precisely the set $U - U$, by an elementary calculation ($x = u - v$ if and only if $x + v = u$). Hence,

$A - A$ contains the open neighborhood $U - U$ of 0 whenever A fills the non-void open set U, and the theorem follows.|||

There are a number of important results which are straightforward corollaries to the difference theorem.

10.5 Banach's Subgroup Theorem   *If G is a subset of a linear topological space E such that $G - G \subset G$ (that is, G is a subgroup of E), and if G contains a set A that is of the second category in E and satisfies the condition of Baire, then $G = E$. Hence if G is any proper subgroup of E, then either G is of the first category in E or else G fails to satisfy the condition of Baire.*

PROOF   Let $N = (A - A)^i$; by the difference theorem N is a neighborhood of 0. Since $G \supset G - G \supset N$ and $0 \in N$, we have $G \supset G - G \supset G - N \supset G$, and hence $G = G - N$. Since N is open, so is $-N$; it follows that $G - N$ is open and that $G^- \subset G - N$. Hence $G = G^i = G^-$. From the connectedness of E it follows that $G = E$.|||

The first assertion in the following corollary has been proved earlier; the second is a direct consequence of the difference theorem.

10.6 Corollary   *If E is a linear topological space of the second category and A is a subset of E that is radial at some point, then A is of the second category in E. If A also satisfies the condition of Baire, then $A - A$ is a neighborhood of 0.*

10.7 Continuity and Openness   *Let E and F be linear topological spaces and T an additive map of E into F. If for each neighborhood V of 0 in F the set $T^{-1}[V]$ contains a set that is of the second category in E and that satisfies the condition of Baire, then T is continuous.*

Dually, if for each neighborhood U of 0 in E the set $T[U]$ contains some set that is of the second category in F and that satisfies the condition of Baire, then T is an open mapping onto F.

PROOF   Let V be any neighborhood of 0 in F, and choose a neighborhood W of 0 in F such that $W - W \subset V$. Then $T^{-1}[V] \supset T^{-1}[W] - T^{-1}[W]$, where $T^{-1}[W]$ contains some set A that is of the second category in E and satisfies the condition of Baire. Then $A - A$ is a neighborhood of 0 that is contained in $T^{-1}[V]$, and hence T is continuous. The remaining half of the theorem is proved dually.|||

Proofs of the following two corollaries are quite easy to construct and will be omitted. Both use the simple observation that any additive operation T is partially homogeneous, in the sense that $T\left(\dfrac{x}{n}\right) = \dfrac{T(x)}{n}$ for every non-zero integer n and every x in the domain of T.

10.8 COROLLARY *Let E and F be linear topological spaces and T an additive map of E into F. If E is of the second category, then T is continuous if and only if $T^{-1}[V]$ satisfies the condition of Baire for every open neighborhood V of 0 in F.*

Dually, if F is of the second category, then T is an open mapping onto F if and only if $T[U]$ is radial at 0 and satisfies the condition of Baire for every open neighborhood U of 0 in E.

10.9 COROLLARY *Let E and F be linear topological spaces and T an additive map of E into F. If $T[A]$ is bounded in F for some set A of the second category in E that satisfies the condition of Baire, then T is continuous.*

Dually, if $T[A]$ is of the second category in F and satisfies the condition of Baire for some bounded set A in E, then T is an open mapping.

The assumption on T in the first part of the last corollary can be varied in case F is the scalar field K. Recall that a real valued function r is **upper semi-continuous** if and only if $\{x : r(x) \geqq a\}$ is closed for each real number a. A function r is **lower semi-continuous** if $-r$ is upper semi-continuous.

10.10 CONTINUITY OF LINEAR FUNCTIONALS *Let f be a linear functional on a linear topological space E, let r be the real part of f, and let A be a set which is of the second category in E and satisfies the condition of Baire. Then f is continuous if*

(i) *r is bounded from above (or below) on A, or*
(ii) *r is upper (or lower) semi-continuous on A.*

PROOF Suppose that r is bounded from above on A. Because A is of the second category and satisfies the condition of Baire, there is a non-void open set U such that $(E \sim A) \cap U$ is of the first category. Making an obvious translation argument, it may be supposed that U is a symmetric neighborhood of 0. Then $[-(E \sim A)] \cap (-U) = [E \sim (-A)] \cap U$ is of the first category, and it follows that $V = A \cap (-A)$ is of the second category and satisfies the condition of Baire. But V is symmetric, and r is therefore bounded from both above and below on V. Then r is also bounded on $V - V$, which, by the difference theorem, is a neighborhood of 0. Hence r is continuous and therefore f is continuous. This establishes the first statement of the theorem.

To establish the second statement, suppose that r is upper semi-continuous on A. For each positive integer n let $A_n = \{x : x \in A$ and $r(x) < n\}$. Then A_n is open in A, and therefore, being the intersection of A and an open set, A_n satisfies the condition of Baire. Finally, since A is of the second category, it is true that A_n is of the second category for some n, and the result of the preceding paragraph implies that f is continuous.$\|\|$

PROBLEMS

A   CONTINUITY OF ADDITIVE MAPPINGS

(a) Let T be an additive function on one linear topological space E to another, F. Suppose that there is a set A of the second category in E and satisfying the condition of Baire such that T considered as a function only on A is continuous. Then T is continuous as a function on E. (By a translation argument and the condensation theorem, A can be assumed to be of the second category at 0.)

(b) Any pseudo-norm defined on a linear topological space that is bounded on some non-void open set, or on some set of the second category that satisfies the condition of Baire, is continuous.

(c) Let T be an additive function on a linear topological space E to a pseudo-normed space F with pseudo-norm p. Then T is continuous if either of the following two conditions holds:

   (i) there is a real function ϕ defined on some set A of the second category in E such that $\phi(x) \geq p(T(x))$ for each x in A, and the set $A_n = \{x : x \in A, \phi(x) \leq n\}$ satisfies the condition of Baire in E for each positive integer n;

   (ii) E is pseudo-metrizable and of the second category, and

$$\limsup p(T(x_n)) \geq p(T(x))$$

whenever $\lim x_n = x$ in E.

B   SUBSPACES OF THE SECOND CATEGORY

Let E be any infinite dimensional linear topological space of the second category. Then E contains a maximal proper linear subspace F that is of the second category in E. Moreover:

(a) F does not satisfy the condition of Baire.

(b) There exists a discontinuous linear functional on E. Thus if E is an infinite dimensional complete metric (or normed) space it must contain a subspace that is infinite dimensional, of the second category and metric (or normed) but not complete.

(There is a sequence $\{e_n\}$ of linearly independent elements of E. Let $B \cup \{e_n\}$ be a Hamel base of E and let H_n be the (maximal proper) linear subspace spanned by $B \cup \{e_m : m \neq n\}$. Then $E = \bigcup \{H_n : n = 1, 2, \cdots\}$.)

C   LINEAR SPACES WITH PSEUDO-METRIZABLE TOPOLOGY

Let E be a linear space and $\mathscr{T}$ a pseudo-metrizable topology for E such that $x + y$ and ax are continuous in each variable and E is of the second category relative to $\mathscr{T}$. Then $\mathscr{T}$ is a vector topology for E. (By means of Osgood's theorem, show that $x + y$ is continuous in x and y simultaneously, and similarly for ax.)

D   MIDPOINT CONVEX NEIGHBORHOODS

(a) In a linear topological space E any subset of the second category in E which satisfies the condition of Baire must contain a translate of a symmetric set which also satisfies the conditions.

(b) Any set A which is midpoint convex ($A + A = 2A$) in a linear topological space E and which contains a symmetric set that is of the second category in E and satisfies the condition of Baire is necessarily a neighborhood of 0.

(c) Any midpoint convex set that contains a set which is of the second category in E and which satisfies the condition of Baire has a non-void interior.

E   SETS OF SEQUENTIAL CONVERGENCE

(a) If $\{T_n\}$ is a sequence of continuous linear maps of a linear topological space E into a pseudo-metrizable linear topological space, then the set of points x for which $\{T_n(x)\}$ is Cauchy is either of the first category in E or identical with E.

(b) Let $\{T_{m,n}: m,n = 1, 2, \cdots\}$ be a double sequence of continuous linear maps of a linear topological space E into a pseudo-metrizable linear topological space, and suppose that for each m there is an x_m in E such that the sequence $\{T_{m,n}(x_m): n = 1, 2, \cdots\}$ is not Cauchy. Then the set of all points x of E, for which $\{T_{m,n}(x): n = 1, \cdots\}$ is not Cauchy for all m, is residual.

F   PROBLEMS IN TOPOLOGICAL COMPLETENESS AND METRIC COMPLETION

A metrizable space $(X,\mathscr{T})$ is called *topologically complete* if and only if it is homeomorphic to a complete metric space. A topological space $(X,\mathscr{T})$ is an *absolute* G_δ if and only if it is metrizable, and is a G_δ (a countable intersection of open sets) in every metric space in which it is topologically embedded. In the following, (a), (b), and (c) are lemmas for (d) and (e).

(a) Let (X,d) be a complete metric space, let U be an open subset of X, and let $\mathscr{T}$ be the induced metric topology on U. Then $(U,\mathscr{T})$ is topologically complete. (For x in U, let $f(x) = (\text{dist } (x, X \sim U))^{-1}$, and let $d^*(x,y) = d(x,y) + |f(x) - f(y)|$. Then (U,d^*) is a complete metric space.)

(b) A G_δ subset of a complete metric space with the relative topology is topologically complete. (If $U = \bigcap \{U_n: n = 1, 2, \cdots\}$, consider the map of U into the product of the complete metric spaces (U_n, d_n^*), where d_n^* is constructed from d and U_n as in (a).)

(c) If a topological subspace Y of a Hausdorff space X is homeomorphic to a complete metric space (Z,d), and if Y is dense in X, then Y is a G_δ in

X. (One can assume that $Y = Z$. Let U_n be the set of all points x in X such that, for some neighborhood V of x, the diameter of $V \cap Y$ is less than $1/n$. It is easy to show that $Y = \bigcap \{U_n : n = 1, 2, \cdots\}$.)

(d) A metrizable space (X, d) is topologically complete if and only if it is an absolute G_δ.

(e) If a metrizable linear topological space is topologically complete, then it is complete.

11   THE CLOSED GRAPH THEOREM

The closed graph theorem and the open mapping theorem, the two principal theorems of this section, are among the most celebrated results of linear space theory. In these theorems continuity and openness of mappings are deduced from formally weaker conditions by means of arguments involving category. There is an important corollary on the continuity of the inverse of a continuous map.

As a preliminary to the proof of the closed graph theorem we recall that if A is a subset of a linear topological space E, and if E is of the second category in itself, and if A is radial at 0 (or at any other point), then A is of the second category in E. In particular, if T is an arbitrary linear map of a space E which is of the second category in itself into a linear topological space F, and if V is a neighborhood of 0 in F, then $T^{-1}[V]$ is of the second category in E. Then the interior of $T^{-1}[V]^-$ is non-empty and $T^{-1}[V]^- - T^{-1}[V]^-$ is a neighborhood of 0. But $(T^{-1}[V - V])^- = (T^{-1}[V] - T^{-1}[V])^- \supset T^{-1}[V]^- - T^{-1}[V]^-$, and it follows that $T^{-1}[V - V]$ is dense in some neighborhood of 0. Consequently such a map has the property that the inverse of each neighborhood of 0 in F is dense in some neighborhood of 0 in E.

11.1   CLOSED GRAPH THEOREM   *A linear transformation T of a linear topological space E into a complete pseudo-metrizable linear topological space F is continuous provided*

(i) *the graph of T is closed in $E \times F$, and*

(ii) *for each neighborhood V of 0 in F the closure of $T^{-1}[V]$ is a neighborhood of 0 in E.*

Condition (ii) is automatically satisfied if E is of the second category.

PROOF   By the metrization theorem 6.7, there is an invariant pseudo-metric d for F, whose topology is that of F. For a positive number e, let $S(e) = \{y : d(0, y) \leqq e\}$. We prove the theorem by showing that $T^{-1}[S(e)]^- \subset T^{-1}[S(2e)]$. Let $x \in T^{-1}[S(e)]^-$; then we construct inductively a sequence $\{y_n: n = 0, 1, \cdots\}$ in F such that (a) $y_0 = 0$, $d(y_n, y_{n+1}) \leq e/2^n$ $(n = 1, 2, \cdots)$, and (b) $x \in T^{-1}[y_n + S(e/2^n)]^-$ $(n = 0, 1, 2, \cdots)$. Clearly $y_0(=0)$ satisfies (a) and (b). Next assume that $y_0, \cdots, y_n$ have been chosen. Then by (b) and (ii), $x \in T^{-1}[y_n + S(e/2^n)] + T^{-1}[S(e/2^{n+1})]^-$. Hence there is an element x' of E such that $T(x') \in y_n + S(e/2^n)$ and $x \in x' + T^{-1}[S(e/2^{n+1})]^- \subset T^{-1}[T(x') + S(e/2^{n+1})]^-$. Let $y_{n+1} = T(x')$; then (a) and (b) are clearly satisfied. For $n = 0, 1, 2, \cdots$, let $W_n = y_n + S(e/2^{n-1})$. Then by (a), $W_n \supset W_{n+1}$. Since F is complete, there is an element y in $\bigcap \{W_n: n = 0, 1, \cdots\}$ by theorem 7.5. Obviously $y \in W_0 = S(2e)$.

In view of (i), the proof is complete if it is shown that the point (x,y) is in the closure of the graph of T. Let U and V be arbitrary neighborhoods of x and y respectively. Then there is an n such that $W_n \subset V$. It follows from (b) that $x \in T^{-1}[W_n]^- \subset T^{-1}[V]^-$. Hence $U \cap T^{-1}[V] \neq \phi$ or equivalently $U \times V$ intersects the graph of T.|||

There are some notable consequences of the closed graph theorem —the following proposition is one of the most useful.

11.2   CONTINUITY OF INVERSE MAPS   *A one-to-one continuous linear map of a complete pseudo-metrizable linear topological space onto a Hausdorff linear topological space of the second category is necessarily a topological isomorphism.*

PROOF   The inverse of such a transformation has a closed graph, the closed graph theorem applies, and the inverse is therefore continuous.|||

A particular case of the foregoing result concerns two Hausdorff topologies, $\mathscr{T}_1$ and $\mathscr{T}_2$, for a single linear space E. If $\mathscr{T}_1 \subset \mathscr{T}_2$, then the identity map of E with $\mathscr{T}_2$ into E with $\mathscr{T}_1$ is continuous, and hence its inverse (also the identity) has a closed graph. Consequently, if E with $\mathscr{T}_1$ is a linear topological space of the second category, and if E with $\mathscr{T}_2$ is a complete metrizable linear topological space, then the identity map is a topological isomorphism, and $\mathscr{T}_1 = \mathscr{T}_2$.

The argument used in the preceding paragraph may be extended to discuss the relation between two distinct topologies $\mathscr{T}_1$ and $\mathscr{T}_2$ on a linear space E by considering the topology $\mathscr{T}_1 \cap \mathscr{T}_2$ and the topology whose subbase is $\mathscr{T}_1 \cup \mathscr{T}_2$. This line of investigation is pursued in a problem at the end of this section, the most striking result being: if E with each of $\mathcal{T}_1$ and $\mathcal{T}_2$ is a complete metrizable linear topological space, and if $\mathcal{T}_1 \neq \mathcal{T}_2$, then there is a non-zero vector x which belongs to $U + V$ whenever U is a $\mathcal{T}_1$-neighborhood of 0 and V is a $\mathcal{T}_2$-neighborhood of 0.

The open mapping theorem is dual to the closed graph theorem; actually, either theorem may be derived fairly easily from the other. As a preliminary to the open mapping theorem we prove a lemma which is of some interest in itself.

11.3 LEMMA   *The image under an open, continuous linear map T of a complete pseudo-metrizable linear topological space E is complete and pseudo-metrizable.*

PROOF   It is easy to see that the family of images of the members of a local base for E is a local base for the range space F, and consequently the range space is pseudo-metrizable. Suppose $\{y_k\}$ is a Cauchy sequence in F which has no limit point, and suppose that $\{U_n\}$ is a local base for E such that $U_1 = E$, and $U_{n+1} + U_{n+1} \subset U_n$ for each n.   Since the family of images of the sets U_n is a local base for F, it may be supposed (by choosing a subsequence of $\{y_k\}$ if necessary) that $y_k - y_{k-1} \in T[U_k]$ for all k.   Choose z_1 such that $T(z_1) = y_1$, and for $k > 1$ choose a member z_k of U_k such that $T(z_k) = y_k - y_{k-1}$, and let $x_k = \sum \{z_j : j = 1, \cdots, k\}$.   It is clear that $T(x_k) = y_k$ for all k, and $\{x_k\}$ is a Cauchy sequence because $x_k - x_{k-1} \in U_k$ and hence $x_{k+p} - x_k \in U_{k+1} + U_{k+2} + \cdots + U_{k+p} \subset U_k$.   But then $\{x_k\}$ converges to a point x of E, and therefore $T(x_k) = y_k$ converges to a point of F.   This is a contradiction.$|||$

11.4 OPEN MAPPING THEOREM   *Let T be a continuous linear map of a complete pseudo-metrizable linear topological space E into a Hausdorff linear topological space F.   If the range of T is of the second category in F, then T maps E onto F, F is complete and metrizable, and T is an open mapping.*

PROOF   The null space N of T is closed in E because $\{0\}$ is closed in F.   Consequently the quotient space E/N is Hausdorff, and since the quotient map Q of E onto E/N is open and continuous, the preceding lemma implies that E/N is complete and metrizable.   Finally, there is a continuous induced map S of E/N into F such that $S \circ Q = T$ (see theorem 5.8).   Since the range $T[E]$ of T is of the second category in F, it is dense in F and of the second category in itself.

Hence, by theorem 11.2, it follows that S is a topological isomorphism of E/N onto $T[E]$. In particular, $T[E]$ is complete, and hence $T[E] = F$. It follows that T is open, and the space F is complete and metrizable.|||

PROBLEMS

A   COMPARISON OF TOPOLOGIES

Let $\mathscr{T}_1$ and $\mathscr{T}_2$ be topologies for a linear space E such that $(E,\mathscr{T}_1)$ and $(E,\mathscr{T}_2)$ are complete metrizable linear topological spaces. Then the following are equivalent:
  (i) $\mathscr{T}_1 \neq \mathscr{T}_2$.
  (ii) The vector topology, whose subbase is $\mathscr{T}_1 \cup \mathscr{T}_2$ is not complete.
  (iii) The topology $\mathscr{T}_1 \cap \mathscr{T}_2$ is not Hausdorff.
  (iv) There is a non-zero vector x such that $x \in U + V$ for each $\mathscr{T}_1$-neighborhood U of 0 and for each $\mathscr{T}_2$-neighborhood V of 0.
  (v) There is a sequence $\{x_n\}$ in E such that $x_n \to 0$ relative to $\mathscr{T}_1$ and $x_n \to y \neq 0$ relative to $\mathscr{T}_2$.
(Prove in the order (i) $\Rightarrow$ (ii) $\Rightarrow$ (iii) $\Rightarrow$ (iv) $\Rightarrow$ (v) $\Rightarrow$ (i).)

B   SUBSPACE OF $L^p \cap L^q$

Let E be a closed linear subspace of each of $L^p(\mu)$ and $L^q(\mu)(0 < p, q \leq \infty)$. Then the two topologies induced on E coincide. (Show that (v) of the previous problem cannot hold by extracting a subsequence convergent almost everywhere.)

C   SYMMETRIC OPERATORS

A linear operator T (defined everywhere) on a Hilbert space H is *symmetric* if and only if $(Tx,y) = (x,Ty)$ for all x and y in H.

A symmetric linear operator is continuous.

Part (e) of 7I can now be stated as follows. An idempotent symmetric linear operator (defined everywhere) on a Hilbert space H is the projection of H on some closed linear subspace.

D   AN OPEN MAPPING THEOREM

Let E_n be a complete metrizable linear topological space for $n = 1, 2, \cdots$ and let u_n be a continuous linear mapping of E_n into a linear topological Hausdorff space E. If $E = \bigcup \{u_n[E_n]: n = 1, 2, \cdots\}$, then every continuous linear mapping of E onto a complete metrizable space is open. (Let f be a continuous linear mapping of E onto a complete metrizable space F; then, for some n, the range of $f \circ u_n$ is of the second category in F.) Cf. 19B.

E   CLOSED RELATION THEOREM

A *relation* between two sets X and Y is a subset of $X \times Y$. Let R be a relation between X and Y and let A be a subset of X; then $R[A]$ denotes the set $\{y: y \in Y$ and, for some x in A, $(x,y) \in R\}$. If E and F are linear spaces, then a *linear relation* between E and F is a linear subspace of $E \times F$.

Let E and F be complete metrizable linear topological spaces and let R be a linear relation between E and F such that R is closed in $E \times F$ and $R[E]$ is of the second category in F. Then $R[U]$ is open in F whenever U is open in E, and $R[E] = F$. (Notice that R is a complete metrizable space, and consider the projection of R into F.)

F   CONTINUOUSLY DIFFERENTIABLE FUNCTIONS

Let C be the space of all continuous real valued functions on $[0,1]$, with the supremum norm, and let E be a closed subspace of C consisting of functions which have continuous derivatives. Then E is finite dimensional. (Show that differentiation is continuous on E, and hence that the unit sphere of E is equicontinuous.)

G   MAPPINGS INTO THE SPACE L^1

Let E be a metrizable linear space and let T be a linear mapping of E into the space $L^1(X,\mu)$. If, for each measurable subset A of X the linear functional $x \to \int_A T(x)d\mu$ on E is continuous, then T is continuous. (First extend T to the completion of E, using 9I.)

H   CONDITION FOR A CLOSED GRAPH

Let T be a linear mapping of a linear topological space E into a linear topological space F. Then the graph of T in $E \times F$ is closed if and only if $\bigcap \{\overline{T[U]} + V: U \in \mathcal{U}, V \in \mathcal{V}\} = \{0\}$, where $\mathcal{U}$ and $\mathcal{V}$ are local bases in E and F.

I   CLOSED GRAPH THEOREM FOR METRIZABLE SPACES?

It is an immediate consequence of the closed graph theorem 11.1 and the Baire category theorem that a linear mapping of one complete metrizable linear topological space into another is continuous if (and only if) its graph is closed. This is no longer true if the hypothesis of completeness on either one of the two spaces involved is dropped out (but cf. 12E). One way of seeing this is to notice first that, on any infinite dimensional normed space, there is a strictly stronger and a strictly weaker norm topology. (Let $\{x_n: n = 1, 2, \cdots\}$ be a linearly independent set, with $\|x_n\| = 1$ for each n, and let $\{y_\alpha: \alpha \in A\}$ be a set, with $\|y_\alpha\| = 1$ for each α, which, together with $\{x_n: n = 1, 2, \cdots\}$, makes up a Hamel base. Take for the strictly smaller unit sphere the convex circled extension of the set $\{n^{-1}x_n: n = 1, 2, \cdots\} \cup \{y_\alpha: \alpha \in A\}$. A strictly larger unit sphere may be reached by carrying out this first construction in the adjoint space.)

J   CONTINUITY OF POSITIVE LINEAR FUNCTIONALS

With certain restrictions on a linear topological space with a vector ordering, every positive linear functional is continuous. Let E be a metrizable linear topological space of the second category and C a complete cone with $C - C = E$. If $f(x) \geqq 0$ on C, then f is continuous. (Let $\{U_n : n = 1, 2, \cdots\}$ be a local base with $U_{n+1} + U_{n+1} \subset U_n$. Use the open mapping theorem 11.4 to show that $\{U_n \cap C - U_n \cap C : n = 1, 2, \cdots\}$ is also a local base. If f is not bounded by 1 on $U_n \cap C$, there is a sequence of points $x_n \in U_n \cap C$ so that $y_n = \sum_{1 \leqq r \leqq n} x_r \to y$ and $f(y) \geqq f(y_n) \geqq n$ for all n.)

12   EQUICONTINUITY AND BOUNDEDNESS

Suppose that F is a family of continuous linear functions on a linear topological space E to a linear topological space H. It is not hard to see that, if F is equicontinuous, then F is bounded relative to the topology of uniform convergence on bounded subsets of E, and that boundedness relative to this topology implies boundedness relative to the topology of pointwise convergence. The surprising fact (and this is the import of this section) is that the converses of these propositions also hold provided suitable conditions are imposed.

Throughout, F will be a family of linear functions, each on a linear topological space E to a linear topological space H. Recall that such a family F is equicontinuous if and only if for each neighborhood V of 0 in H there is a neighborhood U of 0 in E such that $f[U] \subset V$ for all f in F. Restated, the family F is equicontinuous if and only if, for each neighborhood V of 0 in H, it is true that the set $\bigcap \{f^{-1}[V] : f \in F\}$ is a neighborhood of 0 in E.

The family F is **uniformly bounded on a subset** A of E if and only if the set of all points of the form $f(x)$, for f in F and x in A, is a bounded subset of H. The set of all such points $f(x)$ will be denoted by $F[A]$; in this terminology, F is uniformly bounded on A if and only if $F[A]$ is a bounded subset of H. Restating again, F is uniformly bounded on A if and only if each neighborhood V of 0 in H absorbs $F[A]$. The family of sets of the form $\{f : f \text{ linear on } E \text{ to } H, f[A] \subset V\}$, where V is a neighborhood of 0 in H, is a local base for the topology of uniform convergence on A, and clearly V absorbs $F[A]$ if and only if $\{f : f[A] \subset V\}$ absorbs F. It follows that F is uniformly bounded on A if and only if F is bounded relative to the topology of uniform convergence on A. Finally, a neighborhood V of 0 in H absorbs $F[A]$ if and only if the set $\bigcap \{f^{-1}[V] : f \in F\}$ absorbs A. It follows that if F is equicontinuous and A is bounded, then F is necessarily uniformly bounded on A.

The family F is said to be **pointwise bounded** on a set A if and only if for each point x of A it is true that the set $F[x]$ of all points $f(x)$, for f in F, is a bounded subset of H. Equivalently, F is pointwise bounded if and only if F is uniformly bounded on $\{x\}$ whenever $x \in A$. It is clear that if F is uniformly bounded on A, then F is pointwise bounded on A. Finally, the family of all sets of the form $\{f: f(x) \in V\}$, where $x \in A$ and V is a neighborhood of 0 in H, is a subbase for the neighborhood system of 0 relative to the topology of pointwise convergence on A, and it follows without difficulty that F is pointwise bounded if and only if it is bounded relative to the topology of pointwise convergence on A. The foregoing facts may be summarized as follows.

12.1 ELEMENTARY RELATIONS BETWEEN EQUICONTINUITY AND BOUNDEDNESS

Let F be a family of linear functions, each on a linear topological space E to a linear topological space H. Then the equivalences listed in (ii), (iii), and (iv) hold. Moreover, the condition (i) implies each of the conditions listed in (ii), and each of the conditions in (iii) implies each of the conditions of (iv).

(i) *The family F is equicontinuous.*

(ii) *The family F is uniformly bounded on each bounded set.*
    ⇔ *F is bounded relative to the topology of uniform convergence on bounded sets.*

(iii) *The family F is uniformly bounded on a set A.*
    ⇔ *The set $F[A]$ is bounded.*
    ⇔ *F is bounded relative to the topology of uniform convergence on A.*
    ⇔ *For each neighborhood V of 0 in H the set $\bigcap \{f^{-1}[V]: f \in F\}$ absorbs A.*

(iv) *The family F is pointwise bounded on A.*
    ⇔ *$F[x]$ is bounded for each x in A.*
    ⇔ *F is bounded relative to the topology of pointwise convergence on A.*
    ⇔ *For each neighborhood V of 0 in H the set $\bigcap \{f^{-1}[V]: f \in F\}$ absorbs each point of A.*

The major theorem of this section states that uniform boundedness on each bounded set, and in fact equicontinuity, may be deduced from pointwise boundedness under suitable circumstances.

12.2 BANACH-STEINHAUS THEOREM *If F is a family of continuous linear functions, each on a linear topological space E to a linear topological space G, and if F is pointwise bounded on some set which is of the second category in E, then F is equicontinuous.*

PROOF Let U be a neighborhood of 0 in G, let V be a closed neighborhood such that $V - V \subset U$, and let A be a subset of E which is of the second category in E and such that F is pointwise bounded on A. The set $W = \bigcap \{f^{-1}[V] : f \in F\}$ absorbs each point of A in view of 12.1 (iv); hence some integer multiple of W is of the second category in E; and hence W is of the second category in E. But the set W is closed, and consequently the interior of W is non-void and $W - W$ is a neighborhood of 0. Finally, $\bigcap \{f^{-1}[U] : f \in F\} \supset \bigcap \{f^{-1}[V] - f^{-1}[V] : f \in F\} \supset W - W$, and the family F is therefore equicontinuous.|||

A very useful formulation of the basic principle is geometric in nature. A **barrel** (or **disk**) in a linear topological space E is a closed, circled, convex set D which is radial at 0. If E is of the second category, then each barrel D is of the second category (because D is radial at 0), and hence the interior of D is non-void (because D is closed). Consequently $D - D$ is a neighborhood of 0, and it is true that $D - D = D + D = 2D$ because D is convex and circled. It follows that each barrel in a space E which is of the second category in itself is necessarily a neighborhood of 0. This simple fact is noteworthy.

A locally convex space with the property that each barrel is a neighborhood of 0 is said to be **barrelled.** There are linear topological spaces which are of the first category which, nevertheless, have this barrel property (see problem 12D). The following theorem applies to these.

12.3 BARREL THEOREM *Let E be a linear topological space such that each barrel is a neighborhood of 0, and let F be a pointwise bounded family of continuous linear functions on E to a locally convex space H. Then the family F is equicontinuous.*

Consequently, in this case, F is uniformly bounded on each bounded subset of E.

PROOF If V is a closed convex circled neighborhood of 0 in H, then $f^{-1}[V]$ is a barrel, for each f in F. The intersection $\bigcap \{f^{-1}[V] : f \in F\}$ is evidently closed, convex and circled, and because F is pointwise bounded, this intersection absorbs each point of E (part (iv) of proposition 12.1). In brief, $\bigcap \{f^{-1}[V] : f \in F\}$ is a barrel, hence a neighborhood of 0 in E, and F is proved to be equicontinuous.|||

The following theorem, which yields uniform boundedness in certain cases where equicontinuity does not hold, is essentially a corollary to the absorption theorem.

12.4 UNIFORM BOUNDEDNESS THEOREM *Let F be a family of continuous linear functions, each on a linear topological space E to a locally convex space H. If A is a subset of E which is convex, circled, bounded, sequentially closed and sequentially complete, and if F is pointwise bounded on A, then F is uniformly bounded on A.*

PROOF Let V be a closed, convex and circled neighborhood of 0 in H. Then the set $W = \bigcap \{f^{-1}[V] : f \in F\}$ is a closed, convex and circled set which absorbs each point of A (theorem 12.1(iv)). In view of corollary 10.3, the set W absorbs A; hence F is uniformly bounded on A (theorem 12.1(iii)).|||

PROBLEMS

A BOUNDEDNESS OF NORMS OF TRANSFORMATIONS

Let $\{T_\alpha : \alpha \in A\}$ be a family of continuous linear mappings of a pseudo-normable space E of the second category into a pseudo-normed space (F,p). If $\sup \{p(T_\alpha(x)) : \alpha \in A\} < \infty$ for each x in E, then $\sup \{\|T_\alpha\| : \alpha \in A\} < \infty$.

B THE PRINCIPLE OF CONDENSATION OF SINGULARITIES

Let $\{T_{m,n} : m,n = 1, 2, \cdots\}$ be a double sequence of continuous linear maps of a pseudo-normable space E of the second category into a pseudo-normed space (F,p) such that, for each m, $\sup \{\|T_{m,n}\| : n = 1, 2, \cdots\} = \infty$. Then there is a residual subset A of E such that, for each x in A, $\sup \{p(T_{m,n}(x)) : n = 1, 2, \cdots\} = \infty$ for all m. (See 10E.)

C BANACH-STEINHAUS THEOREM

Let E and F be linear topological spaces and suppose that $\{T_n : n = 1, 2, \cdots\}$ is a sequence of continuous linear mappings of E into F such that $T(x) = \lim T_n(x)$ exists for all x in a subset of E of the second category. If F is complete, then $T(x) = \lim T_n(x)$ exists everywhere, and T is continuous and the convergence is uniform on every totally bounded set. (Use 12.2, 8.13, and 8.17.)

If E and F are locally convex spaces, the same conclusion holds if $\lim T_n(x)$ exists for all x in E and E is barrelled (by 12.3).

D STRONGEST LOCALLY CONVEX TOPOLOGY II (see 6I, 14D, 20G)

A linear space E is barrelled when it has its strongest locally convex topology. It is then of the second category if and only if it is finite dimensional. (Use 10B and 7.3.)

E CLOSED GRAPH THEOREM I (see 13G, 18J, 19B; also 11I, 20J)

In the statement of the closed graph theorem 11.1, the condition (ii) is also automatically satisfied when the spaces are locally convex and E is barrelled. Thus any linear mapping, with a closed graph, of a barrelled space into a complete locally convex pseudo-metrizable space is continuous. It is natural to try to generalize these conditions on the two spaces E and F. In the case of F, this may be done (see 13G, 18J). On the other hand, the condition that E be barrelled is, in a certain sense, necessary: suppose that E is a locally convex space with the property that every linear mapping, with a closed graph, of E into any Banach space is continuous. Then E is barrelled. (Let B be a barrel and $N = \bigcap \{(1/n)B : n = 1, 2, \cdots\}$. Let F be the completion of E/N, normed by taking $Q[B]$ as its unit sphere, Q being the quotient mapping. It is sufficient to prove that the graph of Q is closed; one way of doing this is to use the condition in 11H, and the fact that $Q[B]$ is a closed subset of E/N.)

The conditions on the two spaces E and F cannot be reversed: even if E is a Banach space and F a complete barrelled space, the theorem fails. This situation may be realized by anticipating a result of 14D, stating that any space is complete for its strongest locally convex topology (see the previous problem).

F CONTINUOUS FUNCTIONS NON-DIFFERENTIABLE ON SETS OF POSITIVE MEASURE

Let E be the space of continuous real-valued functions on $[0,1]$, with the norm given by $\|x\| = \sup \{|x(t)| : 0 \leq t \leq 1\}$, and let F be the space of Lebesgue measurable functions on $[0,1]$ with the topology of convergence in measure (6L). Suppose the functions x extended beyond $t = 1$, for example by periodicity, and for each $n = 1, 2, \cdots$ let T_n be the linear mapping of E into F defined by $T_n(x)(t) = n(x(t + 1/n) - x(t))$.

(a) Each T_n is continuous.

(b) There are functions of E which are non-differentiable on sets of positive Lebesgue measure. (If each x were differentiable almost everywhere, then $\lim T_n(x)$ would exist for each x: use 12C.)

G BILINEAR MAPPINGS

Let E, F, and G be three linear topological spaces. A mapping f of $E \times F$ into G is called *bilinear* iff, for each $y \in F$, the mapping $x \to f(x, y)$ is a linear mapping of E into G and, for each $x \in E$, the mapping $y \to f(x, y)$ is a linear mapping of F into G. When G is the scalar field, f is called a *bilinear functional*.

(a) A bilinear mapping f is continuous if and only if it is continuous at the origin of $E \times F$. The continuous bilinear mappings of $E \times F$ into G form a linear space B. If E, F, and G are pseudo-normed spaces, with pseudo-norms p, q, and r, f is continuous iff there is a constant k with $r(f(x, y)) \leq kp(x)q(y)$. The smallest such constant is $\sup \{r(f(x, y)) : p(x) \leq 1, q(y) \leq 1\}$; denoting it by $s(f)$, s is a pseudo-norm for B. If r is a norm, so is s.

(b) If E and F are barrelled metrizable spaces and G is locally convex, every pointwise bounded family of continuous bilinear mappings of $E \times F$

into G is equicontinuous. (Let $\{f_\alpha : \alpha \in A\}$ be the family. For each closed convex circled neighborhood W of 0 in G, and each $x \in E$, $\{y : f_\alpha(x,y) \in W \text{ for all } \alpha\}$ is a barrel in F. Hence if $y_n \to 0$, $\{f_\alpha(x,y_n) : \alpha \in A, n = 1, 2, \cdots\}$ is bounded in G. Thus $\{x : f_\alpha(x,y_n) \in W \text{ for all } \alpha \text{ and all } n\}$ is a barrel in E, so that, if $x_n \to 0$, $\{f_\alpha(x_n,y_n) : \alpha \in A, n = 1, 2, \cdots\}$ is bounded in G.)

(c) If E, F, and G are locally convex, E and F are metrizable and E is barrelled, every separately continuous bilinear mapping f of $E \times F$ into G is continuous (f is called separately continuous iff all of the linear mappings $y \to f(x,y)$ and $x \to f(x,y)$ are continuous). (The proof is similar to that of (b).)

(d) Let E, F, and G be locally convex spaces and f a separately continuous bilinear mapping of $E \times F$ into G. Also let A be a bounded subset of E and B a complete, sequentially closed, convex, circled bounded subset of F. Then $f[A \times B]$ is bounded in G. (If W is a closed convex circled neighborhood of 0 in G, $\{y : f(x,y) \in W \text{ for all } x \in A\}$ is a barrel in F. Now use 10.3.)

Chapter 4

CONVEXITY IN LINEAR TOPOLOGICAL SPACES

This chapter, which begins our intensive use of scalar multiplication in the theory of linear topological spaces, marks the definite separation of this theory from that of topological groups. The results obtained here do not have generalizations or even analogues in the theory of groups.

The first section is devoted to technical matters concerning convex subsets of linear topological spaces and of locally convex linear topological spaces. We find, for example, that the convex extension of a bounded or totally bounded subset of a locally convex space is of the same sort.

The second section is devoted primarily to the existence of continuous linear functionals; the principal theorems of the section are simple consequences of the separation and extension theorems of section 3. However, these propositions signal the beginning of the study of the duality which is the central feature of the theory of linear topological spaces. The power of linear space methods lies in the fact that many natural and important problems can be attacked by directing primary attention to the family of all continuous linear functionals on the space. The class of all continuous linear functionals is itself a linear space, and in many important cases this class, with a suitable topology, determines the original space to a topological isomorphism. In such cases a powerful duality exists, and each problem concerning the linear topological space has an equivalent formulation which is a statement about the space of continuous linear functionals. It may happen that an apparently difficult problem may have a formulation which is dual in this sense and which is much easier to investigate. (As an example of the usefulness of this duality, recall that one of the simplest ways of showing that the square of a measurable function is integrable is to show that its product with each square-integrable function is integrable.) But in order to construct a duality between a linear topological space and the space of its continuous linear functionals it is necessary that continuous linear functionals exist, and exist in some abundance.   Because of this fact the propositions of Section 14 play a critical role in the development.

It is also the duality theory which lends importance to the concept of local convexity.   There are linear topological spaces such that the only continuous linear functional is the functional which is identically zero.   The space of Borel functions on the unit interval, with the topology of convergence in Lebesgue measure, is an example of such a space.   The theorems of section 14 ensure that this cannot happen for locally convex spaces (unless the topology is trivial); local convexity is sufficient for the existence of continuous linear functionals. It is also true that local convexity is, in a weakened sense, necessary for the existence of such functionals.   If $\mathscr{T}$ is an arbitrary vector topology for a linear space E, then the family of all convex circled $\mathscr{T}$-neighborhoods of 0 is a local base for a locally convex topology $\mathscr{U}$, which might be called the locally convex topology **derived** from $\mathscr{T}$. Alternatively, the topology $\mathscr{U}$ can be described as the strongest locally convex topology which is weaker than the topology $\mathscr{T}$.   Each linear functional which is $\mathscr{U}$-continuous is also $\mathscr{T}$-continuous, because $\mathscr{U} \subset \mathscr{T}$.   But conversely, if f is a linear functional which is $\mathscr{T}$-continuous, then f is bounded on a convex circled $\mathscr{T}$-neighborhood of 0 (namely, $f^{-1}[\{t : |t| < 1\}]$), and consequently f is $\mathscr{U}$-continuous. Thus the topology $\mathscr{T}$ and the topology $\mathscr{U}$ yield exactly the same class of continuous linear functionals.

There are several consequences of the remarks of the preceding paragraph.   A class C of linear functionals on a linear space E is said to **distinguish points** of E if and only if for each pair x and y of distinct points of E it is true that there is a member f of the class C such that $f(x) \neq f(y)$. If $\mathscr{T}$ is a vector topology for E and the class of $\mathscr{T}$-continuous linear functionals distinguishes points of E, then, in view of the foregoing paragraph, the class of linear functionals which are continuous relative to the derived topology $\mathscr{U}$ has the same property.   In this case it is clear that $\mathscr{U}$ is a Hausdorff topology. The theorem of this chapter will show that if $\mathscr{U}$ is a Hausdorff topology, then there are enough $\mathscr{T}$-continuous functionals to distinguish points.   Consequently the class of $\mathscr{T}$-continuous linear functionals distinguishes points if and only if the class of $\mathscr{U}$-continuous linear functionals distinguishes points, and this is the case if and only if the derived locally convex topology $\mathscr{U}$ is Hausdorff. There are many cases in which these facts can be used to extend a proposition which is established for locally convex linear topological spaces to a result which holds for any linear topological space for which there are enough continuous linear functionals to distinguish points. A case in point is the extreme point theorem of Section 15 (see the problems which follow that section).

Section 15, the last section of this chapter, exhibits the first consequences of the theorems on the existence of continuous linear functionals. The principal theorem of the section asserts that, under suitable conditions, a convex set has extreme points (a point is an extreme point of a convex set A if it is not an interior point of any line segment contained in A). This theorem has been used to demonstrate the existence of irreducible representations of locally compact groups and, more generally, the existence of such representations for certain Banach algebras. The theorem also occurs as an essential element in various calculations of a purely Banach space nature; in brief, it has taken its place with the Hahn-Banach theorem and the Tychonoff theorem as an indispensable theorem of algebraic analysis.

13   CONVEX SUBSETS OF LINEAR TOPOLOGICAL SPACES

The section is devoted to a number of technical propositions concerning convex sets. In particular, we consider the closures and the interiors of convex sets, and the convex extensions of sets which are compact, or bounded, or totally bounded. Continuity of the restriction of a linear functional to a convex set is characterized in several equivalent ways.

The first theorem of the section concerns the construction of convex sets from arbitrary sets by means of combined topological and algebraic operations. A **convex body** is a convex set with non-void interior.

13.1   CONVEX EXTENSIONS AND COMBINATIONS OF CONVEX SETS. *Let E be a linear topological space, let A and B be subsets of E, and let a and b be scalars. Then:*

(i) *If A is convex, so are A^- and A^i; if A is a convex body, then $tA^- + (1 - t)A^i \subset A^i$ for $0 \leq t < 1$, A^i is the radial kernel of A, $A^{i-} = A^-$, and $A^{-i} = A^i$.*

(ii) *The smallest closed convex set containing B is the closure of the convex extension $\langle B \rangle$ of B.*

(iii) *If $\langle A \rangle^-$ and $\langle B \rangle^-$ are compact, then $\langle A \cup B \rangle^-$ and $\langle aA + bB \rangle^-$ are compact.*

(iv) *If $\langle A \rangle$ is compact and $\langle B \rangle$ is closed, then $\langle aA + bB \rangle$ is closed; if, further, E is Hausdorff and $\langle B \rangle$ is bounded, then $\langle A \cup B \rangle$ is closed.*

PROOF (i) If A is convex, then by theorem 5.2(xiii) A^- is convex. If A^i is void, clearly it is convex. Now assume that A^i is non-void. If it is shown that $tA^- + (1 - t)A^i \subset A^i$ for $0 \leqq t < 1$, the convexity of A^i also follows. Because $tA^- + (1 - t)A^i$ is open, it is sufficient to prove that $tA^- + (1 - t)A^i \subset A$. Let $x \in A^i$; then $(1 - t)(A^i - x)$ is an open neighborhood of 0. Therefore, $tA^- \subset (tA)^- \subset tA + (1 - t)(A^i - x) = tA + (1 - t)A^i - (1 - t)x \subset A - (1 - t)x$, and $tA^- + (1 - t)A^i \subset A$ follows.

Let A be a convex body. Since an open set is radial at each of its points, A^i is contained in the radial kernel of A. Conversely, if z is in the radial kernel and x is any point in A^i, there is y in A such that $z \in (y:x]$. But $(y:x] \subset A^i$ by what has just been shown; hence, $z \in A^i$.

The inclusions $A^{i-} \subset A^-$ and $A^{-i} \supset A^i$ are obvious. Let A be a convex body. To see that $A^{i-} \supset A^-$, it is sufficient to show that any element x of $A^- \sim A^i$ is in A^{i-}. If y is in A^i, clearly the set $(x:y]$ is non-void and contained in A^i by what was proved earlier, and the closure of $(x:y]$ contains the end point x. Hence x is in A^{i-}. To establish $A^{-i} \subset A^i$, again choose y in A^i. Any point x in A^{-i} is in the radial kernel of A^-, so that there is an element z in A^- such that $x \in [y:z)$. Since $y \in A^i$ and $z \in A^-$, the earlier result implies $[y:z) \subset A^i$; therefore, $x \in A^i$.

(ii) This is an obvious corollary to part (i).

(iii) If $\langle A \rangle^-$ and $\langle B \rangle^-$ are compact, then their join (the union of all line segments with one end point in $\langle A \rangle^-$ and the other in $\langle B \rangle^-$) is also compact because it is the image of $[0:1] \times \langle A \rangle^- \times \langle B \rangle^-$ under the continuous map: $(t,x,y) \to tx + (1 - t)y$. But it is easy to see that this join is closed and is precisely $\langle A \cup B \rangle^-$. The proposition concerning $aA + bB$ is proved in similar fashion.

(iv) It is true that $\langle aA + bB \rangle = \langle aA \rangle + \langle bB \rangle$, and if $\langle A \rangle$ is compact and $\langle B \rangle$ is closed, then $\langle aA \rangle + \langle bB \rangle$ is the sum of a compact set and a closed set and is closed, in view of proposition 5.2(vii). Assume that $\langle B \rangle$ is bounded and that E is a Hausdorff space. The set $\langle A \cup B \rangle$ is the image of $[0{:}1] \times \langle A \rangle \times \langle B \rangle$ under the continuous mapping $(a,x,y) \to ax + (1 - a)y$. If $\{z_\alpha, \alpha \in C\}$ is a net in $\langle A \cup B \rangle$ which converges to the point z in E, and if $z_\alpha = a_\alpha x_\alpha + (1 - a_\alpha)y_\alpha$, then by the compactness of $[0{:}1]$ and $\langle A \rangle$ there are subnets $\{a_\beta, \beta \in D\}$ and $\{x_\beta, \beta \in D\}$ which converge to a in $[0{:}1]$ and x in $\langle A \rangle$, respectively. If $a = 1$, then $(1 - a_\beta)y_\beta \to 0$ since $\langle B \rangle$ is bounded; hence, since E is a Hausdorff space, $z = x \in \langle A \rangle \subset \langle A \cup B \rangle$. If $a < 1$, then $y_\beta \to (z - ax)/(1 - a)$, which is a member y of $\langle B \rangle$ since $\langle B \rangle$ is closed. Then $z = ax + (1 - a)y \in \langle A \cup B \rangle$ is closed.|||

The Minkowski functional of a convex body having 0 in its interior has special properties, which are listed in the following proposition. It should be mentioned that this proposition, the proof of which is omitted, can be used to derive some of the results on convexity which have been deduced here by other methods.

13.2   MINKOWSKI FUNCTIONAL OF A CONVEX BODY   *Let A be a convex body such that $0 \in A^i$, and let p be the Minkowski functional of A. Then:*

(i) *p is uniformly continuous;*

(ii) *the interior of A is $\{x: p(x) < 1\}$ and the closure of A is $\{x: p(x) \leq 1\}$.*

If A and B are bounded (totally bounded) subsets of a linear topological space E and a is a scalar, then, as has already been shown, each of aA, $A + B$, and the closed circled extension of A are bounded (totally bounded). It is not in general true that the convex extension of a bounded or of a totally bounded set is of the same sort (see problem C of this section). However, this proposition does hold for locally convex spaces.

13.3   CONVEX EXTENSIONS OF BOUNDED AND TOTALLY BOUNDED SETS   *If the linear topological space E is locally convex, then the closed convex circled extension of a bounded (totally bounded) set is again bounded (totally bounded).*

PROOF   Let B be the closed convex circled extension of A, and let U be a neighborhood of 0. Then there exists a closed convex circled neighborhood V which is contained in U. Hence, if $tA \subset V$, then $tB \subset V \subset U$. It follows that B is bounded if A is bounded. Assume now that A is totally bounded, and let C be the circled extension of A; then C is totally bounded also. Corresponding to the given neighborhood U of 0 there exists a convex neighborhood V of 0 such that $V + V \subset U$. If R is a finite set in E such that $C \subset V + R$, then $\langle C \rangle \subset \langle V + R \rangle = \langle V \rangle + \langle R \rangle = V + \langle R \rangle$. Since R is finite, $\langle R \rangle$ is compact, and hence there exists a finite set S in E such that $\langle R \rangle \subset V + S$. Then $\langle C \rangle \subset V + V + S \subset U + S$; that is, $\langle C \rangle$ is totally bounded. Hence, $\langle C \rangle^-$ is totally bounded, and the closed convex circled extension of A is totally bounded.|||

The following theorem is an immediate consequence of the preceding result.

13.4   Convex Extension of a Compact Set   *In a complete locally convex space the closed circled convex extension of any totally bounded set is compact.*

PROOF   Recall that a set A is compact if and only if A is totally bounded and complete, and apply the preceding theorem.|||

A subfamily $\mathscr{B}$ of a family $\mathscr{A}$ is called a **co-base** (or **dual base**) **for** $\mathscr{A}$ if and only if each member of $\mathscr{A}$ is contained in some member of $\mathscr{B}$. In this terminology, the family of closed bounded (totally bounded) convex circled subsets of a locally convex linear topological space is a co-base for the class of bounded sets (totally bounded sets).

The remaining theorem of this section concerns linear functions which are continuous on a convex set. The motivation for the theorem is the fact that the space of all linear functions which are bounded and continuous on a convex set A is complete relative to uniform convergence on A, whereas the space of linear functions which are continuous on E and bounded on A may fail to be complete relative to the same topology. Recall that $f|A$ is the function f restricted to A, so that f is continuous on A if and only if $f|A$ is continuous.

13.5   Functions Continuous on a Convex Set   *Let A be a convex subset of a linear topological space E and let f be a linear functional. Then:*

(i) *if E is a real linear space, then $f|A$ is continuous if and only if $f^{-1}[a] \cap A$ is closed in A for each scalar a;*

(ii) *if E is a real linear space and $f[A]$ is symmetric, then $f|A$ is continuous if and only if $f^{-1}[0] \cap A$ is closed in A;*

(iii) *if E is a complex linear space and A is circled as well as convex, then $f|A$ is continuous if and only if $f^{-1}[0] \cap A$ is closed in A;*

(iv) *if g is an arbitrary linear function and A is circled and convex, then $g|A$ is uniformly continuous if and only if $g|A$ is continuous at 0.*

PROOF   (i) The necessity of the condition is obvious. For the sufficiency, it is enough to prove that a set of the form $\{x: f(x) \leq a\} \cap A$ or $\{x: f(x) \geq a\} \cap A$ is closed in A. Put $B = \{x: f(x) \leq a\} \cap A$ and let y be a point in $A \sim B$. Then $f(y) > a$ and there is a circled neighborhood U of 0 such that $(y + U) \cap f^{-1}[a] \cap A$ is empty. Then $(y + U) \cap B$ is also empty since, if $x \in y + U$, then $[x:y] \subset y + U$. Hence B is closed in A. Similarly a set of the form $\{x: f(x) \geq a\} \cap A$ is closed in A.

(ii) Let a be a real number for which $f^{-1}[a] \cap A$ is not empty. Then there is y in A such that $f(y) = -a$. Consider the map h on A into A defined by $h(x) = (x + y)/2$. The map h is continuous and the set $f^{-1}[a] \cap A$ is the inverse image of $f^{-1}[0] \cap A$ under h. Assertion (ii) now follows in view of (i).

(iv) This is a consequence of the equation $g(x) - g(y) = 2g((x - y)/2)$ and the fact that $(x - y)/2$ belongs to A whenever x and y belong to A.

(iii) Assume that $f^{-1}[0] \cap A$ is closed in A. Let r be the real part of f, and let B be the convex symmetric subset $\{x: f(x) \text{ is real}\}$ of E. Then $f^{-1}[0] \cap A \cap B = r^{-1}[0] \cap A \cap B$. Hence by (ii) r is continuous on $A \cap B$. It follows that, for a given positive number e, there is a circled neighborhood U of 0 such that $|r(x)| < e$ for all x in $A \cap B \cap U$; hence, $|f(x)| < e$ whenever $x \in A \cap U$. Therefore, f is continuous at 0 on A, and by (iv) f is continuous on A. The converse is obvious.|||

PROBLEMS

A   MIDPOINT CONVEXITY

Let E be a linear topological space. A subset A of E is called midpoint convex iff $\frac{1}{2}(x + y) \in A$ wherever $x \in A$ and $y \in A$.

(a) If A is a subset of E such that $A + A \subset 2A^-$, then A^- is convex; if $A^i + A^i \subset 2A$, then A^i is convex. (First show that A^- and A^i are midpoint convex. For x and y in A^- and $0 \leq t \leq 1$, let $\phi(t) = tx + (1 - t)y$, and use 2A.)

(b) The closure A^- and interior A^i of a midpoint convex set A are convex. If A is also circled, then A^- and A^i are circled.

B   CONDENSATION COROLLARY (cf. 9.2)

If A is a convex set of the second category in a linear topological space E, then A is of the second category at each point of A^-. (Let x be a point of A at which A is of the second category and let y be a point of A^-. If V is an open neighborhood of y, choose t so that $0 < t \leq 1$ and $tx + (1 - t)y \in V$ and let $\phi(z) = tz + (1 - t)y$. Then ϕ is a homeomorphism of E

into itself, and there is a neighborhood U of x such that $\phi[U] \subset V$. Then $\phi[A \cap U] \subset A \cap V$, $\phi[A \cap U]$ is of the second category in E and therefore $A \cap V$ is of the second category in E.)

C   CONVEX EXTENSION OF BOUNDED AND TOTALLY BOUNDED SETS

In a linear topological space which is not locally convex, the convex extension of a bounded or totally bounded set may fail to have the same property.

In the space $l^{1/2}$ (6N), the set $\{x: \|x\|_{1/2} < 1\}$ is bounded but its convex extension is not bounded. Let $x_{11} = (1,0,0,\cdots)$, $x_{12} = (0,\frac{1}{2},0,\cdots)$, $x_{22} = (0,0,\frac{1}{2},0,\cdots)$, $x_{13} = (0,0,0,\frac{1}{3},0,\cdots)$ and in general let $x_{mn}(1 \leq m \leq n)$ be the sequence with all terms zero except the $(\frac{1}{2}n(n-1) + m)$-th, which is $1/n$. Then $A = \{x_{mn}: 1 \leq m \leq n, n = 1, 2, \cdots\}$ is totally bounded, but $\langle A \rangle$, which contains all the points $y_n = (1/n) \sum \{x_{mn}: 1 \leq m \leq n\}$, is not totally bounded.

D   TRANSLATES OF CONVEX SETS

Let B be a non-empty subset of a Hausdorff linear topological space E, x_1 and x_2 be two points in E, and t_1 and t_2 be two non-negative real numbers.

(a) If B is convex, $t_1 \leq t_2$, and $x_1 - x_2 \in (t_2 - t_1)B$, then $x_1 + t_1 B \subset x_2 + t_2 B$.

(b) If B is bounded and sequentially closed, $t_1 \leq t_2$, and $x_1 + t_1 B \subset x_2 + t_2 B$, then $x_1 - x_2 \in (t_2 - t_1)B$.

(c) If B is bounded and contains more than one point, and if $x_1 + t_1 B \subset x_2 + t_2 B$, then $t_1 \leq t_2$.

(d) If B is bounded and convex and contains more than one point, then some translate of $t_1 B$ is contained in $t_2 B$ if and only if $t_1 \leq t_2$.

E   EXTENSION OF OPEN CONVEX SETS

Let E be a locally convex linear topological space and let F be a subspace of E.

(a) Let A be a relatively open convex non-void subset of F. Then for each convex open set B in E such that $B \supset A$ there is a convex open set C in E such that $B \supset C$ and $C \cap F = A$. If A is circled, C may be chosen to be circled. (By a translation argument it can be supposed that $0 \in A$. Choose a convex circled open neighborhood U of 0 in B, and take C to be the convex extension of $A \cup U$.)

(b) Any convex subset A of F that is open relative to F is the intersection of F with some open convex set C in E; if A is circled, C may be taken to be circled.

(c) If F is closed and A is any convex set in F that is open relative to F, then for any x_0 in E not in A there is an open convex set C in E such that $C \cap F = A$ and x_0 is not in C. If A is circled, C may be taken to be circled.

(d) Let p be a pseudo-norm defined in F and continuous; then there exists a pseudo-norm p^- defined and continuous on E that is an extension of p. If F is closed and x_0 is any point of E not in F, p^- can be chosen so that $p^-(x_0) \geq 1$.

F   HYPERCOMPLETE SPACES

The set of non-void subsets of a linear topological space E can be made into a uniform space in the following way.   Let $\mathscr{U}$ be a local base and for each $U \in \mathscr{U}$ put

$$\mathbf{V}_U = \{(A,B) : A \subset B + U) \quad and \quad B \subset A + U\}.$$

Then $\{\mathbf{V}_U : U \in \mathscr{U}\}$ is the base of a uniformity, called the *Hausdorff uniformity*, because it is the uniformity of the Hausdorff metric (see 4D) when E is a metrizable space.

With this uniformity, the space of subsets of E does not satisfy Hausdorff's separation axiom, because the uniformity fails to distinguish a set from its closure.   To avoid this superfluity, we restrict attention to the set $\mathscr{E}$ of non-void closed subsets of E, with the Hausdorff uniformity.

A net $\{A_\gamma : \gamma \in \Gamma\}$ in $\mathscr{E}$ is convergent to $A \in \mathscr{E}$ if for each $U \in \mathscr{U}$ there is some $\gamma(U)$ with $A_\gamma \subset A + U$ and $A \subset A_\gamma + U$ for all $\gamma \geqq \gamma(U)$; it is Cauchy if for each $U \in \mathscr{U}$ there is some $\gamma(U)$ with $A_\alpha \subset A_\beta + U$ for all $\alpha,\beta \geqq \gamma(U)$.

When E is complete and metrizable, then (by 4D) $\mathscr{E}$ is complete.   However, without the hypothesis of metrizability, $\mathscr{E}$ may fail to be complete, even when E is a locally convex Hausdorff space; numerous examples will soon transpire.

A locally convex Hausdorff space E is called *hypercomplete* if the set $\mathscr{C}$ of convex circled non-void closed subsets of E is complete under the Hausdorff uniformity.   A hypercomplete space is complete.   Since $\mathscr{C}$ is always a closed subset of $\mathscr{E}$, a complete metrizable locally convex space is hypercomplete.

To ensure that $\mathscr{C}$ is complete, it is sufficient that every decreasing Cauchy net should converge.   (For if $\{A_\gamma : \gamma \in \Gamma\}$ is any Cauchy net, let C_γ be the closed convex extension of $\bigcup \{A_\alpha : \alpha \geqq \gamma\}$.   Then $\{C_\gamma : \gamma \in \Gamma\}$ is a decreasing Cauchy net; it therefore converges, to $C = \bigcap \{C_\gamma : \gamma \in \Gamma\}$. Now for each convex circled neighborhood U of 0, there is some $\gamma(U)$ with $C_\gamma \subset C + U$, $C \subset C_\gamma + U$ and $A_\alpha \subset A_\beta + U$ for all α, β, $\gamma \geqq \gamma(U)$. Hence $A_\gamma \subset C_\gamma \subset C + U$; also $\bigcup \{A_\alpha : \alpha \geqq \gamma\} \subset A_\gamma + U$ and so $C_\gamma \subset (A_\gamma + U)^- \subset A_\gamma + 2U$ Thus $C \subset A_\gamma + 3U$ and so $\{A_\gamma : \gamma \in \Gamma\}$ converges to C.   The device used here for dealing with the closure is often useful: if A is any set and $\mathscr{U}$ a local base, then $A^- = \bigcap \{A + U : U \in \mathscr{U}\}$.)

The image by an open continuous linear mapping of a hypercomplete space is hypercomplete (cf. 11.3), and therefore so is the quotient by a closed subspace.   Hypercompleteness is also inherited by closed subspaces.   An example of a non-metrizable hypercomplete space is given in 18H.

G   CLOSED GRAPH THEOREM II (see 12E, 18J, 19B)

A linear mapping T of a locally convex Hausdorff space E into a hypercomplete space F (see the previous problem) is continuous, provided that
(i) the graph of T in $E \times F$ is closed
(ii) for each neighborhood V of 0 in F, the closure of $T^{-1}[V]$ is a neighborhood of 0 in E.   (Cf. 11.1)

(Scheme for proof: let $\mathscr{U}$ and $\mathscr{V}$ be local bases of convex circled neighborhoods of 0 in E and F.

(a) The set $\{(T[U])^- : U \in \mathscr{U}\}$ is Cauchy.   For if $V \in \mathscr{V}$ there is some $U \in \mathscr{U}$ with $2U \subset (T^{-1}[V])^-$.   It follows that if $U_1 \subset U$ and $U_2 \subset U$, $(T[U_1])^- \subset (T[U_2])^- + V$.

(b) Since F is hypercomplete, the set converges to $B \subset \bigcap \{T[U] + V : U \in \mathscr{U}, V \in \mathscr{V}\}$.   For if $V \in \mathscr{V}$, there is some $U_V \in \mathscr{U}$ with $B \subset (T[U])^- + \frac{1}{2}V$ for all $U \subset U_V$.   Hence $B \subset \bigcap \{(T[U])^- + \frac{1}{2}V : U \in \mathscr{U}\} \subset \bigcap \{T[U] + V : U \in \mathscr{U}\}$, and this holds for all $V \in \mathscr{V}$.

(c) Since the graph is closed, $B = \{0\}$ by 11H.   Hence $T[U] \subset V$. Condition (ii) is satisfied whenever E is barrelled.)   (Cf. 12E.)

14   CONTINUOUS LINEAR FUNCTIONALS

Here the fundamental theorems on existence and extension of continuous linear functionals are presented as simple consequences of the separation and extension theorems of Section 3.   The continuous linear functionals on subspaces, quotient spaces, products, and direct sums are described.   The problems at the end of the section contain other representation theorems for continuous linear functionals on various linear topological spaces.

This section begins with theorems on the existence and extension of continuous linear functionals.   All of these theorems are simple consequences of the separation and extension theorems of Section 3 and of proposition 5.4 on the continuity of linear functionals.   The theorems are collected partly for convenient reference, but mainly to emphasize the fact that for locally convex spaces there always exists a rich supply of continuous linear functionals.   The section ends with an explicit description of the continuous linear functionals on a product or direct sum.   A number of other results concerning the form of continuous linear functionals for various spaces are outlined in the problem set.

Recall that if p is a pseudo-norm for a linear space E and f is a linear functional, then f is continuous if and only if $p^*(f) = \sup \{|f(x)| : p(x) \leq 1\}$ is finite.   The number $p^*(f)$ is the **norm of** f **on** E, and p^* is the **conjugate norm** to p.

14.1   EXTENSION THEOREM   *Let E be a linear space, F a subspace, and f a linear functional on F.   Then*

   (i) *if E is a pseudo-normed space and f is continuous on F, there is a linear functional f^- on E, an extension of f, such that the norm of f^- on E is equal to the norm of f on F;*

   (ii) *if K is a convex circled set which is radial at 0, and if f is a linear functional on F such that $|f(x)| \leq 1$ for x in $K \cap F$, then there*

*is a linear functional f^- on E, an extension of f, such that $|f^-(x)| \leq$
1 for x in K; and*

(iii) *If $\mathcal{T}$ is a locally convex topology for E, and f is $\mathcal{T}$-continuous on
F, then there is a $\mathcal{T}$-continuous linear functional f^- on E which
is an extension of f.*

PROOF   Part (i) is a direct consequence of the Hahn-Banach theorem
3.4, part (ii) is the geometric formulation 3.6 of the same result, and
(iii) follows from (ii) in view of the fact that there is a convex circled
neighborhood K of 0 in E such that $|f|$ is bounded on $K \cap F$,
because f is continuous on F.|||

Recall that a linear functional f on a linear space E separates the
sets A and B if f is non-identically-zero and $r(x) \leq r(y)$ for x in A
and y in B, where r is the real part of f.   Moreover, f is continuous if
and only if r is continuous.

14.2   Separation Theorem   *Suppose that A and B are non-void
convex subsets of a linear topological space E and that the interior of A
is non-void.   Then there is a continuous linear functional f on E separating A and B if and only if B is disjoint from the interior of A.*

PROOF   The interior of A is identical with the radial kernel by 13.1,
and the extension theorem 3.8 applies.   Any functional separating A
and B is necessarily continuous because its real part is bounded from
above on the interior of A, which is a non-void open set.|||

There is a fact concerning the foregoing theorem which is frequently
useful.   A non-zero real linear functional f maps the radial kernel of a
convex set A into the radial kernel of $f[A]$.   It follows that if x belongs
to the interior of A and f separates A and B, then $f(x) < \inf \{f(y):
y \in B\}$.

Recall that a linear functional f on a linear space E strongly separates A and B if and only if $\sup \{r(x): x \in A\} < \inf \{r(y): y \in B\}$,
where r is the real part of f.

14.3   Theorem on Strong Separation   *Let A and B be non-void
disjoint convex subsets of a locally convex linear topological space E.
Then:*

(i) *there is a continuous linear functional f strongly separating A and
B if and only if 0 is not a member of the closure of $B - A$; and*

(ii) *if B is circled, then there is a continuous linear functional f such
that $\sup \{|f(x)|: x \in B\} < \inf \{|f(y)|: y \in A\}$ if and only if 0 is
not a member of the closure of $B - A$.*

PROOF   This is clearly a variant of theorem 3.9 on strong separation. If f strongly separates A and B, then f is bounded away from 0 on the set $B - A$ because sup $\{r(x): x \in A\} <$ inf $\{r(y): y \in B\}$ where r is the real part of f. Hence, if f is continuous, then 0 does not belong to $(B - A)^-$. To show the converse, suppose that $0 \notin (B - A)^-$ and that E is locally convex. Then there is a convex circled open neighborhood U of 0 such that $0 \notin B - A + U$ and the preceding theorem shows that there is a continuous linear functional f separating $B - A + U$ and $\{0\}$. If r is the real part of f, then sup $\{r(x): x \in A\} \leqq$ inf $\{r(y): y \in B\} +$ inf $\{r(z): z \in U\}$. It follows that f is a continuous functional which strongly separates A and B. A simple argument serves to establish part (ii).|||

It has already been shown that if A is compact and B closed, then $B - A$ is closed (theorem 5.2, part (vii)). The foregoing theorem then yields the following as a corollary.

14.4   COROLLARY   *Let A and B be non-void disjoint convex subsets of a locally convex linear topological space, and suppose that A is compact and B is closed. Then there is a continuous linear functional strongly separating A and B, and if B is circled, there is a continuous linear functional f such that* sup $\{|f(x)|: x \in B\} <$ inf $\{|f(y)|: y \in A\}$.

In particular, a point which does not belong to a closed convex subset of a locally convex space can always be strongly separated from the set.

The preceding theorems show the existence of a rich supply of continuous linear functionals for a locally convex space. In applications one frequently needs to know not only the existence, but some sort of explicit description, of the continuous linear functionals. A number of such descriptions (representation theorems) are given in the problems at the end of the section: for the present we shall describe the continuous linear functionals on subspaces, quotient spaces, product spaces, and on direct sums.

The **adjoint (conjugate, dual)** of a linear topological space E is the space E^* of all continuous linear functionals on E. If several topologies for E are being considered, then $(E,\mathscr{T})^*$ will denote the $\mathscr{T}$-continuous linear functionals. If F is a subspace of E, then the set of all members f of E^* which vanish identically on F is the **anni-hilator** of F, or the space **orthogonal** to F, and is denoted by $F^\perp$. Each member f of $F^\perp$ induces a continuous linear functional on the quotient space E/F, in view of the induced map theorem 5.8, and it is evident that each continuous linear functional g on E/F is induced by a member of $F^\perp$ (namely, by the composition of g with the quotient map). Consequently $(E/F')^*$ is isomorphic to $F^\perp$; this isomorphism will be called, with remarkable lack of originality, a **canonical** isomorphism. The space $F^\perp$ may also be described as the null space of the map which restricts each member of E^* to F. Consequently there is an induced isomorphism of $E^*/F^\perp$ into F^*. In general this is not an isomorphism of $E^*/F^\perp$ onto F^*—for example, F may be a one-dimensional subspace of a linear topological space which has no non-trivial linear functionals. However, if E is locally convex, then, by the extension theorem 14.1(iii), each continuous linear functional on F is the restriction of some continuous linear functional on E. Thus:

14.5 ADJOINT OF A SUBSPACE AND A QUOTIENT SPACE   *Let E be a linear topological space, let F be a subspace, and let $F^\perp$ be the annihilator of F in E^*. Then $(E/F)^*$ is canonically isomorphic to $F^\perp$, and, if E is locally convex, $E^*/F^\perp$ is canonically isomorphic to F^*.*

14.6 ADJOINT OF A PRODUCT   *A linear functional ϕ on a product $\times \{E_t : t \in A\}$ of linear topological spaces is continuous relative to the product topology if and only if ϕ can be represented in the form $\phi(x) = \sum \{g_t(x_t) : t \in A\}$ where g is a member of the direct sum $\sum \{E_t^* : t \in A\}$ of the adjoints.*

PROOF   Recall that a local base for the product topology is the family of sets U of the following form: U is the set of all x in the product such that, for each t in a fixed finite set B of indices, x_t belongs to a given neighborhood U_t of zero in E_t. If f is a continuous linear functional which is bounded on such a neighborhood U, and if x is a point of the product such that x_t is 0 for t in B, then every scalar multiple of x belongs to U, and it follows that $f(x) = 0$. Consequently, for each continuous linear functional f there is a finite set B of indices such that, if $x_t = 0$ for t in B, then $f(x) = 0$. Finally, if I_t is the injection of E_t into the product $\times \{E_t : t \in A\}$, and if $y = x - \sum \{I_t(x_t) : t \in B\}$, then y_t is 0 for t in B; hence, $f(y) = 0$, and therefore $f(x) = \sum \{f(I_t(x_t)) : t \in B\}$. But, for each t, $f \circ I_t$ is a continuous linear functional on E_t: that is, a member of E_t^*. It follows that each continuous linear functional f on the product $\times \{E_t : t \in A\}$ can be represented in the form: $f(x) = \sum \{g_t(x_t) : t \in A\}$ where g is a member of the direct sum of the spaces E_t^*. It is evident that, conversely, any functional of this form is an element of the adjoint of the product.|||

There is a natural topologization of the direct sum $\sum \{E_t: t \in A\}$ of locally convex spaces E_t. Recall that for each t in A there is an algebraic isomorphism, called an injection, of E_t into the direct sum. The injection I_t is defined by letting $I_t(z)$, where $z \in E_t$, be the member of the direct sum which is 0 for $s \neq t$ and z for $s = t$. The direct sum may be given· the topology induced by the mappings I_t (see Section 4); explicitly, a set is open relative to this topology if and only if its inverse under each I_t is open.   Unfortunately, this topology is usually not a vector topology for the direct sum (see problem 5B), so a variant of the induced topology is desirable. · Let $\mathcal{U}$ be the family of all convex circled subsets U of the direct sum such that $I_t^{-1}[U]$ is a neighborhood of 0 for each t. By the convexity, each member of $\mathcal{U}$ is radial at 0.   Therefore, in view of theorem 6.5, the family $\mathcal{U}$ is a local base of a unique locally convex topology for the direct sum; this topology is called the **direct sum topology**. If, for the moment, we overlook the distinction between E_t and its isomorphic image $I_t[E_t]$, then a convex circled subset U of the direct sum is a neighborhood of 0 if and only if its intersection with each E_t is a neighborhood of 0.   For each t, let U_t be a neighborhood of 0 in E_t; then the convex extension $\langle \bigcup \{U_t: t \in A\} \rangle$ is a neighborhood of 0 in the direct sum, and each neighborhood of 0 relative to the direct sum topology contains a neighborhood of this form.

Before discussing the elementary properties of the direct sum topology we digress to observe that the construction of this topology is a special case of a more general method.   If F is a family of linear functions, each member f of F being on a linear topological space E_f to a fixed linear· space H, then the **inductive topology** for H (the F-**inductive topology**) has as a local base the family of all convex circled subsets U of H, which are radial at 0, such that $f^{-1}[U]$ is a neighborhood of 0 in E_f for each f in F.   It is evident that the F-inductive topology is the strongest *locally convex* topology which makes each member f of F continuous, and that a linear function g on H to a locally convex space is continuous relative to the inductive topology if and only if $g \circ f$ is continuous for each f in F.   An important special case is that in which each E_f is a subspace of H and each f is the identity map.   Then the inductive topology can be described as the strongest locally convex topology whose relativization to each E_f is weaker than the topology of E_f.

The elementary properties of the direct sum topology are easy to establish; in particular, it is almost self-evident that, dually to 14.6, the adjoint of a direct sum is the product of the adjoints.   Part (iv) can also be proved using the completeness criterion of 17.7 (see problem 17F).

14.7   ELEMENTARY PROPERTIES OF THE DIRECT SUM TOPOLOGY

(i) *The direct sum topology is the strongest locally convex topology with the property that each injection is continuous.*

(ii) *Each injection is a topological isomorphism of a factor onto a subspace of the direct sum.*

(iii) *A linear function on a direct sum to a locally convex space is continuous if and only if its composition with each injection is continuous.*

(iv) *The direct sum of complete locally convex spaces is complete relative to the direct sum topology.*

(v) *A linear functional ϕ on a direct sum $\sum \{E_t : t \in A\}$ of locally convex spaces is continuous relative to the direct sum topology if and only if there is a member f of $\times \{E_t^* : t \in A\}$ such that $\phi(x) = \sum \{f_t(x_t) : t \in A\}$ for all x in the direct sum.*

PROOF   We prove only part (iv) of the proposition.   In this proof we write $x(t)$ for x_t.   If $\{x_\alpha : \alpha \in D\}$ is a Cauchy net in the direct sum $E = \sum \{E_t : t \in A\}$, then, because projection into each coordinate space E_t is continuous, $\{x_\alpha(t) : \alpha \in D\}$ is a Cauchy net in E_t and hence converges to one or more points of E_t.   Select $x(t)$ to be one of the points to which $\{x_\alpha(t)\}$ converges, taking care to select 0 if possible. For each t such that $x(t) \neq 0$, let U_t be a closed convex neighborhood of 0 in E_t such that $x(t) \notin U_t$, and, for t for which $x(t) = 0$, let $U_t = E_t$.   Since $\{x_\alpha\}$ is Cauchy, there is β in D such that $\gamma \geqq \beta$ implies $x_\gamma - x_\beta \in \langle \bigcup \{U_t : t \in A\} \rangle$.   Then, for each t, $x_\gamma(t) - x_\beta(t) \in U_t$.   Since U_t is closed, $x(t) - x_\beta(t) \in U_t$.   Hence, if $x(t) \neq 0$ then $x_\beta(t) \neq 0$.   It follows that x belongs to the direct sum and the net $\{x_\alpha\}$ converges to x relative to the topology $\mathscr{P}$ of coordinate-wise convergence.   In order to prove that $\{x_\alpha\}$ converges to x relative to the direct sum topology, it is sufficient to show that the net $x_\alpha - x$ is eventually in the closure of each neighborhood (of 0) of the form $U = \langle \bigcup \{U_t : t \in A\} \rangle$, where U_t is a convex neighborhood of 0 in E_t. Because the net $\{x_\alpha - x_\beta\}$ is eventually in U, the proof is complete if it is shown that the closure U^- relative to the direct sum topology is also $\mathscr{P}$-closed.

The rest of the proof is devoted to showing that U^- is $\mathscr{P}$-closed. Let y be a point of E belonging to the $\mathscr{P}$-closure of U.   Let $B =$

$\{t: y(t) \neq 0\}$; then y belongs to the $\mathscr{P}$-closure of $U \cap (\sum \{E_t: t \in B\})$. (This is a consequence of the fact that, if $z \in U$, then a point obtained from z by replacing one or more coordinates of z by 0 is still in U.) The set B is finite, and the direct sum topology relativized to $\sum \{E_t: t \in B\}$ is identical to the relativized $\mathscr{P}$; hence $y \in U^-$. Since the direct sum topology is stronger than $\mathscr{P}$, U^- is $\mathscr{P}$-closed.|||

Finally, the following description of the bounded and the totally bounded subsets of a product and of a direct sum is given for reference; its proof is left to the reader.

14.8 Bounded and Totally Bounded Subsets of Products and Sums

(i) *The family of all products of totally bounded (or bounded) subsets of the factors is a co-base for the family of totally bounded (or bounded) subsets of the product of linear topological spaces.*

Equivalently, a subset of a product is totally bounded (or bounded) if and only if its projection into each coordinate space is totally bounded (or bounded).

(ii) *The family of all convex extensions of finite unions of images under injection of totally bounded (or bounded) subsets of the factors is a co-base for the family of totally bounded (or bounded) subsets of the direct sum of locally convex Hausdorff spaces.*

PROBLEMS

A EXERCISES

(1) There is a non-trivial continuous linear functional on a linear topological space E if and only if E properly contains a convex body.

(2) If for each x in a linear topological space E there is a convex neighborhood of zero not containing x, then there are enough continuous linear functionals on E to distinguish points.

(3) A linear manifold in a linear topological space is contained in a closed hyperplane if and only if its complement contains a convex body.

(4) If A is a closed convex subset of a locally convex space E and B is a subset of E, then $B \subset A$ if and only if $f[B] \subset f[A]$ for every continuous linear functional f on E.

(5) A closed linear manifold in a locally convex space is the intersection of all the closed hyperplanes containing it.

B FURTHER SEPARATION THEOREMS

Let E be a linear topological space and A and B non-void subsets of E.

(a) If A and B are strongly separated by a continuous linear functional on the real restriction of E, there is a continuous linear functional on E such that the distance between $f[A]$ and $f[B]$ is positive.

(b) If $B - A$ is midpoint convex (see 13A) and somewhere dense, the following are equivalent:
  (i) A and B are strongly separated by a continuous linear functional on the real restriction of E;
  (ii) for some continuous linear functional f on E the distance between $f[A]$ and $f[B]$ is positive;
  (iii) 0 is not in the closure of $B - A$.

(c) Suppose that A and B are disjoint and convex and that E is locally convex. Then there is a continuous linear functional f such that the distance between $f[A]$ and $f[B]$ is positive if and only if 0 is not a member of the closure of $B - A$.

C   A FIXED POINT THEOREM

Lemma Let A be a compact convex subset of a linear topological space E. If x_0 is a point such that for each y in A there is a continuous linear functional strongly separating x_0 and y, then there is a continuous linear functional on the real restriction of E strongly separating x_0 and A.

Theorem Let A be a non-void compact convex subset of a linear topological space E and suppose that for each non-zero member x of $A - A$ there is a continuous linear functional f on E with $f(x) \neq 0$. Let T be a continuous mapping of A into itself such that $T(\sum \{a_i x_i \colon 1 \leq i \leq n\}) = \sum (a_i T(x_i) \colon 1 \leq i \leq n\}$ wherever $x_i \in A$ and $a_i \geq 0$ for $1 \leq i \leq n$ and $\sum \{a_i \colon 1 \leq i \leq n\} = 1$. Then there is a point x_0 in A such that $T(x_0) = x_0$. (Suppose not, and put $S(x) = x - T(x)$. Then $S[A]$ is convex and compact and $0 \notin S[A]$. By the lemma there is a continuous linear functional f on the real restriction of E positive (say) on $S[A]$. Consider the point at which f attains its minimum.)

D   STRONGEST LOCALLY CONVEX TOPOLOGY III (see 6I, 12D, 20G)

Let E be a linear space of dimension α with its strongest locally convex topology. Then E is the direct sum of α copies of the scalar field and is complete (see 14.7). Every linear manifold in E is closed (see 6I) and the relative topology for any linear subspace F of E is the strongest locally convex topology for F.

E   STRONGEST VECTOR TOPOLOGY II (see 5E)

Let E be a vector space with its strongest vector topology. Regarding E as a direct sum of copies of the scalar field, relative to some fixed Hamel base $\{e_t \colon t \in A\}$, let $x(t)$ denote the t-th coordinate of x, and I_t the injection of the t-th coordinate space into E.

(a) For convenience, call x and y disjoint iff they take their non-zero values on disjoint subsets of A. Then E has a local base $\mathscr{U}$ of neighborhoods U for which x and y disjoint and $x + y \in U$ implies $x \in U$ and $y \in U$. (The topology of E can be defined by the family of all functions q with the properties (i) through (iv) of 6C. For such q, let

$$q'(x) = \sum \{q(I_t(x(t))) \colon t \in A\}.$$

Then q' satisfies the conditions (i) through (iv) of 6C and $q'(x + y) = q'(x) + q'(y)$ whenever x and y are disjoint.)

(b) The space E is complete.

(Let $\{x_\gamma : \gamma \in \Gamma\}$ be a Cauchy net in E. Then $\{x_\gamma\}$ converges in the strongest locally convex topology on E, to x_0 say. Put $y_\gamma = x_\gamma - x_0$, let $U \in \mathscr{U}$ and choose a neighborhood V of the origin with $V + V \subset U$. There is an α with $y_\beta - y_\gamma \in V$ wherever $\beta, \gamma \geq \alpha$. Let B be the finite set of those t with $y_\alpha(t) \neq 0$. There is a $\beta \geq \alpha$ with $\sum \{I_t(y_\beta(t)): t \in B\} \in V$; if $z_\beta = \sum \{I_t(y_\beta(t)): t \in A \sim B\}$, then $y_\alpha - z_\beta \in U$ and so $y_\alpha \in U$, and $x_\gamma \to x_0$.)

F   ALGEBRAIC CLOSURE OF CONVEX SETS II (see 5G)

If A is a convex circled subset of a linear space E, then $\bigcap \{rA : r > 1\}$ is the closure of A in the strongest locally convex topology on E. (In the linear subspace generated by A, with its strongest locally convex topology, A is a neighborhood of zero; use 5G.)

G   LINEARLY CLOSED CONVEX SETS II (see 5H)

A convex *circled* subset of a linear space with its strongest locally convex topology is closed if its intersection with every straight line is closed. (Use the previous problem.)

If the linear space has a countable base, convex subsets have the same property (see also 18H).

H   A FUNDAMENTAL THEOREM OF GAME THEORY

Let A and B be non-void compact convex subsets of the linear topological spaces E and F, respectively, and let ϕ be a real-valued function on $A \times B$ with the following properties:

(i) for each $y \in B$ the mapping $x \to \phi(x, y)$ is continuous on A,

(ii) if $a_i \geq 0$ for $1 \leq i \leq n$ and $\sum \{a_i : 1 \leq i \leq n\} = 1$, then

$$\phi(\sum \{a_i x_i : 1 \leq i \leq n\}, y) = \sum \{a_i \phi(x_i, y) : 1 \leq i \leq n\}$$

for all $x_i \in A$ and

$$\phi(x, \sum \{a_i y_i : 1 \leq i \leq n\}) = \sum \{a_i \phi(x, y_i) : 1 \leq i \leq n\}$$

for all $y_i \in B$.

Then $\sup_{x \in A} \inf_{y \in B} \phi(x, y) = \inf_{y \in B} \sup_{x \in A} \phi(x, y)$. (In this equality, $\leq$ is immediate; to prove the reverse inequality denote the right side by c. Then c may be assumed finite and it is enough to prove that, if $A_y = \{x : x \in A, \phi(x, y) \geq c\}$, then $\bigcap \{A_y : y \in B\}$ is non-void. Since each A_y is closed and A is compact, it is sufficient to show that any finite intersection of sets A_y is non-void. Suppose $\bigcap \{A_{y_i} : 1 \leq i \leq n\} = \phi$ and map A into R^n by f where $f(x) = (\phi(x, y_1) - c, \phi(x, y_2) - c, \cdots, \phi(x, y_n) - c)$. By (i) and (ii) $f[A]$ is a compact convex subset of R^n and $f[A] \cap P = \phi$, where P is the positive cone in R^n, that is, the set of points all of whose coordinates are non-negative. There is a linear functional g on R^n strongly separating $f[A]$ and P. All the coordinates g_i of g have the same sign; we may suppose them non-negative and that $\sum \{g_i : 1 \leqq i \leqq n\} = 1$. Then if $y_0 = \sum \{g_i y_i : 1 \leqq i \leqq n\}$,

$$\sup_{x \in A} \phi(x, y_0) = c + \sup_{x \in A} \sum \{g_i(\phi(x, y_i) - c) : 1 \leqq i \leqq n\} < c,$$

which contradicts the definition of c.)

I   COMPLEX MEASURES

The adjoints of spaces of continuous functions can be identified with spaces of measures. When the functions are complex valued, so are the measures. Some of the properties of complex valued measures which are not immediate extensions of those of real valued measures are collected in this problem; they are sometimes omitted from texts on measure theory. If μ_1 and μ_2 are two signed measures defined on the same σ-ring $\mathscr{S}$ of subsets of a set X, the set function $\mu = \mu_1 + i\mu_2$ will be called a *complex measure* on X.

(a) For each A in $\mathscr{S}$, let $|\mu|(A)$ denote the supremum, over all finite families $\{B_r : 1 \leqq r \leqq n\}$ of disjoint subsets of A, of the sums $\sum \{|\mu(B_r)| : 1 \leqq r \leqq n\}$. Then $|\mu|$ is a measure on X, called the *total variation* of μ. For a signed measure μ, $|\mu| = \mu^+ + \mu^-$, where μ^+ and μ^- are the upper and lower variations of μ, at least one of which is finite, and $\mu = \mu^+ - \mu^-$. The set of all finite complex measures on X, defined on the σ-ring $\mathscr{S}$ of subsets of X is a Banach space with the norm $\|\mu\| = \sup \{|\mu|(A) : A \in \mathscr{S}\}$.

(b) A complex valued function $f = f_1 + if_2$ on X is integrable with respect to the complex measure $\mu = \mu_1 + i\mu_2$ iff both f_1 and f_2 are integrable with respect to the upper and lower variations of both μ_1 and μ_2; when it is, $\int f d\mu$ is defined to be the obvious linear combination of eight terms. If f is measurable with respect to $\mathscr{S}$, f is integrable with respect to μ iff $|f|$ is integrable with respect to $|\mu|$, and then $|\int f d\mu| \leqq \int |f| d|\mu|$.

(c) If X is a topological space, the set belonging to the σ-ring generated by the compact subsets of X are called *Borel sets*. A (positive, signed or complex) measure μ on X defined on the Borel sets is called a *Borel measure* iff $|\mu|(A)$ is finite for every compact subset A of X. The Borel measure $\mu = \mu_1 + i\mu_2$ is said to be *regular* iff, for each Borel set A,

$$|\mu|(A) = \sup \{|\mu|(B) : B \subset A \text{ and } B \text{ compact}\}$$
$$= \inf \{|\mu|(C) : C \supset A \text{ and } C \text{ open Borel}\},$$

or equivalently, iff the upper and lower variations of μ_1 and μ_2 are all regular. For a finite regular Borel measure, the total variation, $\|\mu\|$ as defined in (a) is the supremum, over all finite families $\{B_r : 1 \leqq r \leqq n\}$ of disjoint compact subsets of X, of the sums $\sum \{|\mu(B_r)| : 1 \leqq r \leqq n\}$. Every continuous function vanishing at infinity on X is Borel measurable.

J   SPACES OF CONTINUOUS FUNCTIONS II (see 8I)

Let X be a locally compact Hausdorff space and $K(X)$ the space of continuous scalar valued functions of compact support on X. The starting point of this representation theory is the theorem that, when the scalars are real, every positive linear functional ϕ on $K(X)$ can be expressed in the form $\phi(f) = \int f(x) d\mu(x)$, where μ is a regular Borel measure on X, uniquely determined by ϕ. (See, for example, Halmos [4] pp. 247–248.)

(a) Assume that the scalars are real; then each positive linear functional ϕ on $K(X)$ is continuous with respect to the inductive limit topology $\mathcal{T}$ (see 8I). Conversely, if $\phi \in (K(X), \mathcal{T})^*$, then ϕ can be written as $\phi = \phi^+ - \phi^-$, where both ϕ^+ and ϕ^- are positive linear functionals on $K(X)$. Therefore there are two regular Borel measures μ^+ and μ^- on X such that $\phi(f) = \int f d\mu^+ - \int f d\mu^-$ for all f in $K(X)$. (Warning: if neither μ^+ nor μ^- is finite $\mu^+ - \mu^-$ is not a signed measure.) If the scalars are complex, then there is a similar representation of a continuous linear functional on $K(X)$ using four regular Borel measures on X.

(Suppose first that the scalars are real and let $\phi \in (K(X))^*$. For each $f \geq 0$, put $\phi^+(f) = \sup \{\phi(g) : 0 \leq g \leq f\}$; then if $f \in K(X)$, $\phi^+(f) < \infty$, by the continuity of ϕ. Both ϕ^+ and $\phi^+ - \phi$ can be extended to give positive linear functionals on $K(X)$. For complex scalars, consider the real and imaginary parts of ϕ acting on the real subspace of $K(X)$ consisting of the real valued functions.)

(b) *Riesz's Theorem* The adjoint of the Banach space $C_0(X)$, with the uniform norm, is the space of all finite regular Borel signed or complex measures on X, with the total variation as norm. (Since the inductive limit topology on $K(X)$ is stronger than the uniform topology, to each $\phi \in (C_0(X))^*$ corresponds a unique regular Borel measure μ with $\phi(f) = \int f d\mu$ for each $f \in K(X)$. If $\{B_r : 1 \leq r \leq n\}$ are disjoint compact subsets of X and $e > 0$ is given, there is an open relatively compact set C containing the sets B_r and such that $\mu(C \sim \bigcup \{B_r : 1 \leq r \leq n\}) < e$, since μ is regular. Then there is a function f with support contained in C such that $|f(x)| \leq 1$ for all x and $f(x)\mu(B_r) = |\mu(B_r)|$ for all x in $B_r (1 \leq r \leq n)$. Since $\sum \{|\mu(B_r)| : 1 \leq r \leq n\} \leq |\int f d\mu| + e$, μ is finite and $\|\mu\| \leq \|\phi\|$. The formula $\phi(f) = \int f d\mu$ now extends to $C_0(X)$ and the inequality of 14I(b) gives $\|\phi\| \leq \|\mu\|$.)

(c) The *support* (or carrier) of a regular Borel measure μ is the smallest closed set A such that $\mu(B) = 0$ for all Borel sets B disjoint from A. (Such a set exists; it is the intersection of all sets with the above property.) The adjoint of the space $C(X)$ of all continuous functions on X with the topology of uniform convergence on compact subsets of X is the space of regular Borel measures with compact support on X. (If $\phi \in (C(X))^*$, there is a compact set A and a constant k such that $|\phi(f)| \leq k \sup \{|f(x)| : x \in A\}$. The restrictions to A of the functions of $C(X)$ form the space $C(A)$ and so ϕ defines a member ϕ_A of $(C(A))^*$. By (b), there is a regular Borel measure μ on A, which can be extended to X by putting $\mu(B) = \mu(B \cap A)$, with $\phi(f) = \int f d\mu$.)

K SPACE OF CONVERGENT SEQUENCES

Let X be any set. When X is assigned the discrete topology, the corresponding space $C_0(X)$ (see the previous problem) is usually denoted by $c_0(X)$. In particular, when X is the set of positive integers we write simply c_0; thus c_0 is the space of all scalar-valued sequences convergent to zero. Let Y be the one-point compactification of X. Then the space $C(Y)$ is isomorphic to the space of all functions on X convergent at infinity, and

$c_0(X)$ is then a closed hyperplane in $C(Y)$ under the uniform norm.   If X is the set of positive integers, $C(Y)$ is denoted simply by c; it is the space of all convergent sequences.

If μ is a regular Borel measure on the discrete space X, there is a function g on X such that $\mu(A) = \sum\{g(x) : x \in A\}$ for each subset A of X.   Thus the previous problem shows that the adjoint of $c_0(X)$ is $l^1(X)$ and the adjoint of $c(Y)$ is $l^1(Y)$, the general form for continuous linear functionals being $f \to \sum\{f(x)g(x) : x \in A\}$.   Naturally, this result is more easily obtained directly.   In particular the continuous linear functionals on c are of the form

$$\{x_n : n = 1, 2, \cdots\} \to \sum\{x_n y_n : n = 1, 2, \cdots\} + (\lim_{n \to \infty} x_n) \cdot y_0$$

with $\{y_n : n = 0, 1, 2, \cdots\}$ in l^1.

L HILBERT SPACES II (see 7H *et seq.*)

Every Hilbert space H is self-adjoint in the following sense.   To each continuous linear functional f on H corresponds a unique element y such that $f(x) = (x,y)$ for all x in H, and the mapping $f \to y$ is a conjugate isomorphism of H^* onto H.   (By a conjugate isomorphism T is meant a one-to-one mapping satisfying $T(ax + by) = \bar{a}T(x) + \bar{b}T(y)$ and $\|T(x)\| = \|x\|$.   Given $f \neq 0$, let $N = \{x : f(x) = 0\}$ and choose $z \in N^\perp$ with $z \neq 0$.   Then y is a scalar multiple of z.)

M SPACES OF INTEGRABLE FUNCTIONS IV (see 6K, 6N, 7M)

Let $(X, \mathscr{S}, \mu)$ be a measure space.   For each p in the range $0 < p < \infty$, $L^p(X,\mu)$ is the space of measurable functions f on X such that $|f|^p$ is integrable, with the topology determined by the local base of sets $\{f : \|f\|_p < 1/n\}$ for $n = 1, 2, \cdots$, where $\|f\|_p = (\int |f|^p d\mu)^{1/p}$.   In $L^p(X,\mu)$, $\|f\|_p = 0$ if and only if $f(x) = 0$ almost everywhere.   The subset of $\mathscr{S}$ consisting of measurable sets of finite measure is denoted below by $\mathscr{S}_0$.

(a) If $1 < p < \infty$, any continuous linear functional ϕ on $L^p(X,\mu)$ has the form $\phi(f) = \int fg d\mu$ with $g \in L^q(X,\mu)$, where q is the index conjugate to p, so that $q = p/(p - 1)$.   Conversely any linear functional of the above form is continuous, with $\|\phi\| = \|g\|_q$; two functions of $L^q(X,\mu)$ define the same linear functional if and only if they are equal almost everywhere. Thus the conjugate of $L^p(X,\mu)$ is isometrically isomorphic to the quotient of $L^q(X,\mu)$ modulo, the subspace of functions equal to zero almost everywhere.   It is usually more convenient to regard the spaces $L^p(X,\mu)$ as consisting of (equivalence classes of) functions specified only up to a set of measure zero; this also makes the pseudo-norm $f \to \|f\|_p$ a true norm. With this slight gloss, the adjoint of the space $L^p(X,\mu)$ becomes the space $L^q(X,\mu)$ for $1 < p < \infty$.

(Hölder's inequality shows that any ϕ of the above form is continuous and that $\|\phi\| \leq \|g\|_q$.   To show that any $\phi \in (L^p)^*$ has the required form, first assume that $X \in \mathscr{S}_0$.   Then, if $\nu(A) = \phi(\chi_A)$, where χ_A is the characteristic function of A, ν is a signed or complex measure on $\mathscr{S}$.   By the Radon-Nikodym theorem there is a function $g \in L^1$ with $\nu(A) = \int_A g d\mu$ and therefore, for all $f \in L^p$, $\phi(f) = \int f d\nu = \int fg d\mu$.   If now $\{g_n\}$ is a sequence of bounded non-negative μ-measurable functions increasing to $|g|$,

$$\|g_n\|_q^q \leq \int g_n^{q-1}|g|d\mu = \phi(g_n^{q-1}\overline{\text{sgn}}\,g) \leq \|\phi\| \, \|g_n\|_q^{q/p},$$

and so

$$\|g\|_q = \sup\{\|g_n\|_q: n = 1, 2, \cdots\} \leq \|\phi\|.$$

In the general case when $(X,\mathscr{S},\mu)$ is not totally finite, there is, for each $A \in \mathscr{S}_0$, a function $g_A \in L^q$ vanishing off A with $\phi(f\chi_A) = \int g_A d\mu$. If $k = \sup\{\|g_A\|_q: A \in \mathscr{S}_0\}$, $k \leq \|\phi\|$ and, for any increasing sequence $\{A_n\}$ of sets of $\mathscr{S}_0$ for which $\|g_{A_n}\|_q \to k$, $\{g_{A_n}\}$ converges in L^q. The limit g is independent of the particular sequence chosen and is the required function.)

(b) The case $p = 1$ requires more delicate treatment. A scalar valued function f on X is called *locally measurable* if $f \cdot \chi_A$ is measurable for each $A \in \mathscr{S}_0$. A property is said to hold *locally almost everywhere* iff the set on which it fails intersects each set of $\mathscr{S}_0$ in a set of measure zero. The functions on X which are locally measurable and bounded locally almost everywhere form a linear space, denoted by $L^\infty(X,\mu)$. For such a function f, if

$$\|f\|_\infty = \inf\{k: |f(x)| \leq k \text{ locally almost everywhere}\},$$

then $|f(x)| \leq \|f\|_\infty$ locally almost everywhere. The mapping $f \to \|f\|_\infty$ is a pseudo-norm on $L^\infty(X,\mu)$ with respect to which $L^\infty(X,\mu)$ is complete. When X is totally finite or totally σ-finite, the word "locally" may be omitted without changing the meaning above.

The measure space $(X,\mathscr{S},\mu)$ is called *localizable* iff to each family $\{g_A: A \in \mathscr{S}_0\}$ of scalar valued functions, such that g_A is defined and measurable on A and $g_A(x) = g_B(x)$ for almost all x in $A \cap B$, corresponds a locally measurable function g on X coinciding with g_A on A except for a set of measure zero for each A in $\mathscr{S}_0$. Every totally finite, or totally σ-finite, space is localizable.

(c) If $(X,\mathscr{S},\mu)$ is localizable, any continuous linear functional ϕ on $L^1(X,\mu)$ has the form $\phi(f) = \int fg d\mu$ with $g \in L^\infty(X,\mu)$, and any linear functional of the above form is continuous, with $\|\phi\| = \|g\|_\infty$. Two functions of $L^\infty(X,\mu)$ define the same linear functional if and only if they are equal locally almost everywhere. Thus, if we agree to identify functions of $L^\infty(X,\mu)$ which are equal locally almost everywhere, the adjoint of $L^1(X,\mu)$ is $L^\infty(X,\mu)$, provided that $(X,\mathscr{S},\mu)$ is localizable.

(A linear functional of the above form is continuous and $\|\phi\| \leq \|g\|_\infty$. If $\phi \in (L^1)^*$ and $X \in \mathscr{S}_0$, a proof similar to that in (a) shows that there is a function $g \in L^1$ with $\phi(f) = \int fg d\mu$ for all f in L^1. If B is the set on which $|g(x)| \geq k$, $k\mu(B) \leq \int_B |g|d\mu = \int (\chi_B \overline{\text{sgn}}\,g)g d\mu = \phi(\chi_B \overline{\text{sgn}}\,g) \leq \|\phi\|\mu(B)$. Hence $\mu(B) = 0$ for $k > \|\phi\|$, so that $g \in L^\infty$ and $\|g\|_\infty \leq \|\phi\|$. In the general case, for each $A \in \mathscr{S}_0$ there is a function g_A defined and measurable on A with $|g_A(x)| \leq \|\phi\|$ almost everywhere and $\phi(f\chi_A) = \int_A fg_A d\mu$. Since X is localizable there is a function $g \in L^\infty$ with $g(x) = g_A(x)$ almost everywhere in A and this function g is the one required.)

(d) Conversely, if the adjoint of $L^1(X,\mu)$ is $L^\infty(X,\mu)$, the space $(X,\mathscr{S},\mu)$ is localizable.

(First consider a uniformly bounded family $\{g_A: A \in \mathscr{S}_0\}$. If $\mathscr{S}_1$ is the subset of $\mathscr{S}$ consisting of sets of σ-finite measure, there is, for each B in

$\mathscr{S}_1$, a g_B extending g_A for each $A \in \mathscr{S}_0$ with $A \subset B$. Now each $f \in L^1$ vanishes off some $B \in \mathscr{S}_1$; put $\phi(f) = \int fg_B d\mu$. There is a $g \in L^\infty$ with $\phi(f) = \int fg d\mu$ and g is the required locally measurable function. The case of a general family $\{g_A\}$ reduces to this special one.)

(e) The adjoint of $l^p(X)$ is $l^q(X)$, where, for $1 < p < \infty$, $q = p/(p-1)$ and, for $p = 1$, $q = \infty$. Here the space X is always localizable. But if X is uncountable, $\mathscr{S}$ can be taken to be the set of all finite or countable subsets of X and then the functions of L^∞ are not all measurable, but only locally measurable. An example of a non-localizable space is given by Halmos ([4] §31, (9)).

(f) If $0 < p < 1$, the adjoint of $l^p(X)$ is $l^\infty(X)$. (Use 6N(d).) On the other hand, if X is an interval of the real line and μ is Lebesgue measure, the adjoint of $L^p(X,\mu)$ for $0 < p < 1$ consists of the zero functional only. (6N(c)).

15 EXTREME POINTS

The principal result of this section is the powerful theorem of Krein and Milman: each convex compact subset of a locally convex linear topological Hausdorff space has extreme points, and is in fact the closed convex extension of the set of its extreme points.

A point x of a convex set A is an **extreme point** of A if and only if x is not an interior point of any line segment whose endpoints belong to A. Thus, in the Euclidean plane, the "corner" points of the set $\{(x,y) \colon \max \{|x|, |y|\} \leqq 1\}$ are extreme points, and every point of the circumference of $\{(x,y) \colon x^2 + y^2 \leqq 1\}$ is an extreme point of this set. An open convex set clearly has no extreme points; however, the existence of extreme points of a bounded closed convex set is a likely sounding conjecture. Unfortunately it is false even for Banach spaces (see problem A of this section). However, if a convex set is compact, and the containing linear topological space is locally convex and Hausdorff, then extreme points do exist. This existence theorem, which has far reaching consequences, is the principal result of this section.

A set A is a **support** of a convex set B in a linear space if and only if A satisfies the following conditions:

(i) A is a non-void convex subset of B;

(ii) if an interior point of a segment in B belongs to A, then the segment is a subset of A; in other words, if $x \in B$, $y \in B$, and $tx + (1 - t)y \in A$ for some t such that $0 < t < 1$, then $x \in A$ and $y \in A$.

The supports enjoy the following properties. If A is a support of a convex set B and B is a support of a convex set C, then A is a support of C. If, for each t in a set S, B_t is a support of a convex set C, then the intersection $\bigcap \{B_t : t \in S\}$ is either void or a support of C also. Finally, suppose f is a linear transformation of a linear space E into F. If B is a support of $f[C]$, where C is a convex set in E, then $f^{-1}[B] \cap C$ is a support of C. A point x is an extreme point of a convex set A in a linear space E if and only if $\{x\}$ is a support of A.

The following theorem is the fundamental result of the section.

15.1   EXISTENCE OF EXTREME POINTS (KREIN-MILMAN THEOREM)
Let A be a convex compact subset of a locally convex linear topological Hausdorff space E. Then each closed support of A contains an extreme point of A, and A is the closed convex extension of the set of all its extreme points.

PROOF   Let R be any closed (hence compact) support of A; then by the maximal principle there exists a family $\mathscr{A}$ of closed supports of A which is maximal with respect to the following properties: (i) $R \in \mathscr{A}$; and (ii) $\mathscr{A}$ has the finite intersection property. Let $S = \bigcap \{B : B \in \mathscr{A}\}$. Then S is a non-void closed support of A since A is compact. Furthermore, S is a minimal closed support of A in the sense that it contains as a proper subset no closed support of A. Hence each closed support of A contains a minimal closed support of A.

Let S be any minimal closed support of A, and assume that a line segment $[x{:}y]$, where $x \neq y$, is contained in S. Then there exists a continuous linear functional f on the real restriction of E such that $f(x) \neq f(y)$. Since S is compact, f is bounded on S, and $C = \{z : z \in S, f(z) = \sup \{f(y) : y \in S\}\}$ is a non-void closed proper subset of S. Then C is a support of S because it is the inverse under f of a support of $f[S]$, and hence C is a support of A. But S was presumed to be minimal, and this is a contradiction. Therefore, S consists of only one point, which is an extreme point of A.

The foregoing paragraphs show that each closed support of A contains an extreme point of A. Since A itself is a closed support of A, A has at least one extreme point.

Let D be the closed convex extension of the set of all extreme points of A. Then clearly $D \subset A$. Assume that $A \sim D$ is non-void, and take x_0 in $A \sim D$. Then, by 14.4, there exists a continuous linear functional g on the real restriction of E such that $g(x_0) > \sup \{g(y) : y \in D\}$. Let $B = \{x : x \in A, g(x) = \sup \{g(z) : z \in A\}\}$; as before, B is a closed support of A, and by the choice of g, $B \subset A \sim D$. Then, as shown in the first paragraph of the proof, B contains an extreme point of A contrary to the definition of D. $|||$

The following theorem gives information on the set of extreme points of certain convex compact sets.   In general, little is known of the structure of the set of extreme points.

15.2   THEOREM   *Let A be a compact subset of a locally convex Hausdorff space E, and suppose that the closed convex extension $\langle A \rangle^-$ of A is compact.   Then all of the extreme points of $\langle A \rangle^-$ belong to A.*

PROOF   Let U be a convex circled closed neighborhood of 0 in E. Then, since A is compact, there exists a finite number of points $x_1, \cdots, x_n$ in A such that $\bigcup \{(U + x_i): i = 1, \cdots, n\} \supset A$.   Let $A_i = A \cap (U + x_i)$; then $\langle A \rangle^-$ is contained in the convex extension of $\bigcup \{\langle A_i \rangle^- : i = 1, \cdots, n\}$, since the latter is closed by theorem 13.1. Then each point y of $\langle A \rangle^-$ can be expressed as $y = \sum \{a_i y_i: i = 1, \cdots, n\}$, where the a_i's are non-negative real numbers such that $\sum \{a_i: i = 1, \cdots, n\} = 1$, and $y_i \in \langle A_i \rangle^- \subset U + x_i$. If y is an extreme point of $\langle A \rangle^-$, then for some i, $y = y_i$.   Therefore, $y \in U + x_i \subset A + U$; since U is an arbitrary convex circled closed neighborhood of 0 and A is closed, it follows that $y \in A.|||$

PROBLEMS

A   A BOUNDED SET WITH NO EXTREME POINT

The unit sphere in the space c_0 (see 14K) has no extreme point.

B   EXISTENCE OF EXTREME POINTS

The requirement in 15.1 that the space E be locally convex may be replaced by the hypothesis that any two points in A may be separated by a linear functional continuous on E.

C   EXTREME IMAGE POINTS

Let f be a continuous linear map of E into F where E and F are locally convex and Hausdorff.   Let A be a compact convex subset of E.   Then every extreme point of $f[A]$ is the image of an extreme point of A.

D   MAXIMUM OF A LINEAR FUNCTIONAL

Each real continuous linear functional on a convex compact subset of a Hausdorff locally convex space assumes its maximum at an extreme point.

E   SUBSETS OF A COMPACT CONVEX SET

If C is a compact convex subset of a locally convex Hausdorff space E, then for every subset S of C the following are equivalent.
   (i) $\sup \{f(x): x \in S\} = \sup \{f(x): x \in C\}$ for all continuous linear functionals f on the real restriction of E;
   (ii) $C = \langle S \rangle^-$;
   (iii) S^- includes all extreme points of C.

F   TWO COUNTER-EXAMPLES

(1) In a finite dimensional space, a convex compact set is the convex extension of the set of its extreme points, this extension being closed. This fails in general, even in a Banach space. In the space l^∞ (6K), let e_n be the sequence with n-th term 1 and other terms 0, and let A be the closed convex extension of the set $\{e_n/n: n = 1, 2, \cdots\}$ together with 0. Then A is compact but is not the convex extension of the set of its extreme points.

(2) In theorem 15.2, the hypothesis that the closed convex extension of the compact set A should be compact is essential: failing this, there may be an extreme point of $\langle A \rangle^-$ which is not in A. Let $C(I)$ be the space of real valued continuous functions on the interval $I = [0,1]$ and let E be the space of real valued functions on $C(I)$ with the topology of pointwise convergence. Let e be the evaluation mapping of I into E: for each $x \in I$, $e(x)$ is the element of E defined by $e(x)(f) = f(x)$ for all $f \in C(I)$.

    (a) $e[I]$ is a compact subset of E.

    (b) If $g(f) = \int_0^1 f(x)dx$ for $f \in C(I)$, then $g \in \langle e[I] \rangle^-$ but $g \notin e[I]$.

    (c) If F is the subspace of E generated by $e[I] \cup \{g\}$, then g is an extreme point of $F \cap \langle e[I] \rangle^-$.

G   EXTREME HALF LINES

Let C be a convex cone such that $C \cap (-C) = \{0\}$. The only extreme point of C is the vertex, zero. A half-line L from zero is an *extreme half-line* of C if it is contained in C and if every open line segment contained in C and intersecting L is contained in L. Then a half-line L is extreme if and only if for every hyperplane H not containing 0 and intersecting C the one point of $H \cap L$ is an extreme point of $H \cap C$. Let $\geqq$ be the order induced by C. Then the point x in C is in an extreme half-line if and only if $x \geqq y$ for $y \in C$ implies $y = tx$ for some t.

H   LIMITS AND EXTREME POINTS

Let $\{A_\alpha\}$ be a directed family of subsets of a topological space. Then $\lim \sup A_\alpha$ is defined to be the set of points x such that every neighborhood of x intersects some A_α for arbitrarily large α.

    (a) If $A = \lim \sup A_\alpha$, then A is closed and contains $\bigcap \{A_\alpha^-\}$. If the family is decreasing, that is, $A_\alpha \supset A_\beta$ whenever $\beta \geqq \alpha$, then $A = \bigcap \{A_\alpha^-\}$. If $B_\alpha = (\bigcup \{A_\beta : \beta \geqq \alpha\})^-$ then $\lim \sup A_\alpha = \bigcap \{B_\alpha\}$.

    (b) Let $\{C_\alpha\}$ be a decreasing family of compact convex subsets of a locally convex Hausdorff space E and let A_α be the set of extreme points of C_α. Then $\lim \sup C_\alpha = \langle \lim \sup A_\alpha \rangle^-$.

(Clearly $\lim \sup A_\alpha \subset \lim \sup C_\alpha$, and $\lim \sup C_\alpha$ is a convex compact set. Let f be a continuous linear functional on the real restriction of E; then, for each α, there is x_α in A_α such that $f(x_\alpha) = \sup \{f(x): x \in C_\alpha\}$. Let x_0 be a cluster point of the net $\{x_\alpha\}$; then $x_0 \in \lim \sup A_\alpha$ and $f(x_0) \geqq \sup \{f(x): x \in \lim \sup C_\alpha\}$.

    (c) Let $\{C_\alpha\}$ be a directed family of compact convex subsets of a locally convex Hausdorff space E, whose union is contained in a compact convex subset of E, and let A_α be the set of extreme points of C_α. Then $\lim \sup C_\alpha \subset \langle \lim \sup A_\alpha \rangle^-$; if $\lim \sup C_\alpha$ is convex, then $\lim \sup C_\alpha =$

$\langle \lim \sup A_\alpha \rangle$, that is, $\lim \sup A_\alpha$ contains all the extreme points of $\lim \sup C_\alpha$.

(Let $B_\alpha = (\bigcup \{A_\beta : \beta \geq \alpha\})^-$ and use (b).)

I   EXTREME POINTS IN L^1 AND L^∞ (see 6K and 14M)

Let X be the real line and let μ be the Lebesgue measure on X.

(a) The unit sphere of $L^1(X,\mu)$ has no extreme points.

(b) A member f of $L^\infty(X,\mu)$ is an extreme point of the unit sphere if and only if $f^2 = 1$.

(In these problems regard $f = g$ to mean that $f(x) = g(x)$ almost everywhere.)

J   EXTREME POINTS IN $C(X)$ AND ITS ADJOINT

Let $C(X)$ be the space of all real valued continuous functions on a compact Hausdorff space X, with the supremum norm. Then f is an extreme point of the unit sphere of $C(X)$ if and only if $f^2 = 1$.

A member F of the unit sphere of the adjoint of $C(X)$ is an extreme point if and only if F is represented by "$\pm$ a point measure"; that is, if and only if there is x in X and $F(f) = f(x)$ for all f in $C(X)$, or $F(f) = -f(x)$ for all f in $C(X)$.   (See 14J.)

Chapter 5

DUALITY

This chapter is devoted to the duality which is the central part of the theory of linear topological spaces. The pattern of investigation is simple: we seek to find, for each proposition about a linear topological space E, an equivalent proposition which is stated in terms of the adjoint space E^*. Of course, it is necessary that E^*, in some sense, describe E rather closely, and consequently our results, with minor exceptions, are for locally convex spaces.

The first section is devoted to a number of propositions, primarily geometric in character, which are intended for application both to a linear topological space E and to its adjoint E^*. These results are consequently framed in terms which are immediately applicable to both cases: we consider two arbitrary linear spaces E and F and a bilinear functional (the pairing functional) on their product $E \times F$. The weak topologies for E and for F are the weakest topologies which make the pairing functional continuous in each variable separately. The pairing is (essentially) completely determined by either one of the spaces with its weak topology, since the weakly continuous linear functionals on E are precisely those functionals which are represented by members of F. The geometry of a pairing is investigated by means of polars, where the polar of a subset A of E is the set of all f in F such that $|\langle x,f \rangle| \leqq 1$ for all x in A. Among the geometric propositions which have important linear space consequences we note: a weakly closed convex circled subset A of E is weakly compact if and only if each linear functional which is bounded on the polar of A is represented by a member of E, and each linear functional which is weakly continuous on a convex circled subset A of E can be approximated, uniformly on A, by a functional which is weakly continuous on E.

The results on weak topologies have immediate consequences for the natural pairing of a space and its adjoint.   In particular, closed convex sets are weakly closed, weakly bounded sets are bounded, and the last result mentioned in the preceding paragraph yields a characterization of completeness.   The most striking result concerning the weak topology for a linear topological space is the Eberlein-Krein-Šmulian theorem: the weakly closed convex extension of a weakly countably compact subset of a complete locally convex space is weakly compact.

The structure of an arbitrary locally convex space E is considerably more complicated than the theory of Banach spaces would indicate. We use four families of subsets of E^* to classify useful properties of E. Consider the classes $\mathscr{E}$, $\mathscr{C}$, $\mathscr{S}$, and $\mathscr{W}$ of all convex subsets of E^* which are, respectively, equicontinuous, with weak* compact closure, strongly bounded, and weak* bounded.   If E is a Banach space, these classes are identical, but in general we have $\mathscr{E} \subset \mathscr{C} \subset \mathscr{S} \subset \mathscr{W}$, and each inclusion may be proper.   The importance of these classes, as a gauge of the properties of E, is the following: $\mathscr{E} = \mathscr{C}$ if and only if the topology for $\mathscr{E}$ is the maximal locally convex topology which yields E^* as adjoint, $\mathscr{E} = \mathscr{S}$ if and only if the evaluation map of E into the second adjoint E^{**} is continuous (this map is always relatively open), and $\mathscr{E} = \mathscr{W}$ if and only if the Banach-Steinhaus theorem for E is true (the space is then called a barrelled space, or tonnelé).   The complexity in structure is further displayed by a lack of permanence properties for "good" characteristics.   Thus, although quotients, products, and direct sums of spaces for which the Banach-Steinhaus theorem is true again have this property, closed subspaces and adjoints may fail to retain the property.   On the other hand, the property of being semi-reflexive (the evaluation map carries the space onto the second adjoint) is retained by products, direct sums, and closed subspaces, but not by quotient spaces or by adjoints.   In general, little can be said about the adjoint of a linear topological space, and even the adjoint of a complete metrizable space exhibits features which are quite pathological from the point of view of classical Banach space theory.

The theory of dual mappings is a natural part of the study of duality. We characterize continuity and openness of a linear-transformation T in terms of the adjoint map T^*.   Finally, the last section of the chapter is devoted to the theory of metrizable spaces.   As might have been expected, there are substantial results here which apparently have no generalization; however, even in this case, the adjoint space may have a relatively complex structure.

16 PAIRINGS

This section is concerned with a bilinear pairing of spaces E and F. The relationship between the spaces is studied by means of the topology $w(E,F)$ that F induces on E, and by means of polars. A number of properties of a subset of E, most notably $w(E,F)$-compactness, can be described in terms of polars. The completeness of E relative to the topology of uniform convergence on members of a family of subsets of F is characterized. There are also a number of "tool" theorems concerning subspaces, quotients, etc.

The study of a linear topological space and its adjoint is facilitated by a further abstraction of certain essentials. If E is a vector space and F is a linear space of linear functionals on E, then the relation in which E stands to F is not dissimilar to that in which F stands to E. In particular, each point x of E corresponds to a linear functional ϕ on F by means of the definition: $\phi(f) = f(x)$ for all f in F. It is convenient to choose a setting in which the roles of the two spaces are entirely symmetric since theorems are obtained thereby which have consequences both for the space E and for the space F of functionals on E. It turns out that the abstraction which is made also yields a highly efficient mechanism for many linear space computations.

Let E and F be two linear spaces over the same scalar field. A **bilinear functional** on the product $E \times F$ is a functional B such that

$$B(ax + by,f) = aB(x,f) + bB(y,f)$$

and

$$B(x,af + bg) = aB(x,f) + bB(x,g)$$

for all x and y in E, all f and g in F, and all scalars a and b. A **pairing** is an ordered pair $\langle E,F \rangle$ of linear spaces together with a fixed bilinear functional on their product. The fixed bilinear functional in a pairing is usually written in the common inner product notation; that is, $B(x,y)$ is written $\langle x,y \rangle$. Mention of the functional is sometimes omitted, and it is said that E and F are **paired spaces**. In contexts where an ordered pair of spaces has already been specified, the word "pairing", or the words "pairing of the spaces", may be used to signify the bilinear functional alone.

Each pairing of two linear spaces E and F defines a mapping from either of the two spaces into the space of all scalar functions on the other. The **canonical map** T on F carries a member f of F into the function $T(f)$ on E such that $T(f)(x) = \langle x,f \rangle$ for all x in E. The space P of all scalar functions on E is simply the product $\mathsf{X}\ \{K: x \in E\}$, where K is the scalar field, obviously, P is a linear space.

Because the pairing functional is bilinear, the canonical map T is linear, and the image of each member of F is a linear functional on E; consequently $T[F]$ is a linear subspace of the algebraic dual E' of E, which in turn is a linear subspace of the product P.   On the other hand, if F is an arbitrary linear subspace of the algebraic dual of E, then the **natural pairing** of E and F is the bilinear functional on $E \times F$ defined by $\langle x, f \rangle = f(x)$ for all x in E and f in F.   In this case the canonical map of F into E' is the identity.   In the general case, when F is not a subspace of the algebraic dual E', it is not unusual to identify (that is, to fail to distinguish between) a point f of F and its canonical image $T(f)$, and to say that f is a linear functional on E. Nevertheless, it must be remembered that $T(f)$ and $T(g)$ may be equal for distinct members f and g of F.   It is obvious that $T(f)$ and $T(g)$ are distinct for $f \neq g$ if and only if $T(h) \neq 0$ for every non-zero member h of F; and this condition is satisfied if and only if for each member f of F other than 0, there is some x in E such that $\langle x, f \rangle \neq 0$. This condition is described by saying that E **distinguishes points in F**.

Clearly the roles of E and F are interchangeable here.   There is a canonical linear mapping of E into the subspace F' of $\times \{K : f \in F\}$ such that the image of a point x of E is the functional which carries f of F into $\langle x, f \rangle$, and the mapping is one-to-one if and only if F distinguishes points in E.   When E distinguishes points in F and F distinguishes points in E, the pairing is said to be **separated**.   For the natural pairing of a linear space E and its algebraic dual E' it is obvious that E distinguishes points of E' and that E' distinguishes points of E.   Hence this pairing is separated.

By making use of the canonical map a very useful topology can be defined for a linear space E which is paired with a space F.   The canonical map T carries E into the space $P = \times \{K : f \in F\}$ of all functionals on F, and the product P may be assigned the product topology.   This topology is also called the topology of pointwise convergence.   The space P is complete, locally convex and Hausdorff, and the algebraic dual of E is a closed subspace.   The space E is now topologized by defining a set U to be $w(E,F)$-open if and only if U is the inverse image under T of a set which is open in P relative to the product topology.   Formally, the **weak** (E,F) topology, denoted by $w(E,F)$, is the family of all subsets U of E such that for some subset V of P, V is open relative to the topology of pointwise convergence and $T^{-1}[V] = U$.   Equivalently, a subset B of E is $w(E,F)$-closed if and only if B is the inverse image under T of a closed subset of P.   A word of caution is necessary: if B is $w(E,F)$-closed in E, it does not follow that $T[B]$ is closed in P relative to the topology of pointwise convergence. It is true, however, that if B is $w(E,F)$-closed, then $T[B]$ is closed in $T[E]$, where $T[E]$ has the relativized product topology. The space P is locally convex and Hausdorff, and it follows that E with the topology $w(E,F)$ is locally convex, and that it is Hausdorff if and only if T is one-to-one; equivalently, E with the topology $w(E,F)$ is Hausdorff if and only if F distinguishes members of E.

The situation with respect to boundedness, total boundedness, and compactness relative to the $w(E,F)$ topology is quite simple. If B is a bounded subset of $P = \underset{}{\times} \{K : f \in F\}$, then the projection of B into each coordinate space is bounded, and hence $a_f = \sup \{|\langle x, f\rangle| : x \in B\} < \infty$ for each f in F. Then B is a subset of the set $\underset{}{\times} \{A_f : f \in F\}$, where $A_f = \{t : t \in K \text{ and } |t| \leq a_f\}$. By the Tychonoff theorem 4.1, this product is compact. It follows that each bounded closed subset of P is compact, and that each bounded subset is totally bounded (it is also possible to derive total boundedness directly from boundedness without appealing to the Tychonoff theorem). In view of the definition of the topology $w(E,F)$, it is clear that a subset of E is $w(E,F)$-bounded or totally bounded if and only if the image under the canonical map T of the subset has the same property. Consequently a subset B of E is bounded if and only if it is totally bounded; B is totally bounded if and only if $\sup \{|\langle x, f\rangle| : x \in B\} < \infty$ for each f in F.

The following theorem summarizes the preceding results, and states a few other simple propositions about the topology $w(E,F)$ which follow from known properties of the product topology (see Section 4).

16.1 Elementary Properties of the $w(E,F)$-Topology   *Let E and F be paired linear spaces.   Then:*

(i) *the space E with the topology $w(E,F)$ is a locally convex linear topological space, and E is a Hausdorff space if and only if F distinguishes points of E;*

(ii) *the family of all sets of the form $\{x : |\langle x, f_i\rangle| \leq 1 \text{ for } i = 1, \cdots, n\}$, where $\{f_1, f_2, \cdots, f_n\}$ is an arbitrary finite subset of F, is a local base for $w(E,F)$;*

(iii) *a net $\{x_\alpha, \alpha \in A\}$ in E converges to an element x in E relative to the topology $w(E,F)$ if and only if $\{\langle x_\alpha, f\rangle, \alpha \in A\}$ converges to $\langle x, f\rangle$ for each f in F;*

(iv) *a function R on a topological space G to E is continuous relative to the topology $w(E,F)$ if and only if g_f is continuous on G for each f in F, where $g_f(z) = \langle R(z),f \rangle$ for every z in G;*

(v) *If $\mathcal{T}$ is a topology for E such that the map $x \rightarrow \langle x,f \rangle$ is $\mathcal{T}$-continuous for each f in F, then $\mathcal{T}$ is stronger than $w(E,F)$; and*

(vi) *a subset B of E is $w(E,F)$-totally bounded if and only if it is $w(E,F)$-bounded, and this is the case if and only if $\sup \{|\langle x,f \rangle| : x \in B\} < \infty$ for each f in F.*

It is clear that there is a complete duality between the roles of E and F in the foregoing discussion and that the argument which leads to theorem 16.1 can be carried out with the roles of the two spaces E and F interchanged. Alternatively, a pairing of F and E can be defined by setting $\langle\langle f, x \rangle\rangle = \langle x, f \rangle$, and the results on the topology $w(E,F)$ which have just been obtained yield theorems on a topology for F, the **weak** (F,E) **topology** for F, denoted by $w(F,E)$. Explicitly, a subset U of F is a member of the topology $w(F,E)$ if and only if U is the inverse image, under the canonical map, of a subset V of the space of all functionals on E, such that V is open under the topology of pointwise convergence. Clearly each theorem on the topology $w(E,F)$ has a dual which applies to $w(F,E)$. These dual results will not be stated.

If E and F are paired and g is a linear functional on E, then g is **represented by** an element f of F if and only if $g(x) = \langle x,f \rangle$ for all x in E. The $w(E,F)$-topology yields a precise description of those linear functionals which can be represented by members of F.

16.2 REPRESENTATION OF $w(E,F)$-CONTINUOUS LINEAR FUNCTIONALS
A functional g on E is represented by some member of F if and only if g is a $w(E,F)$-continuous linear functional. The member of F which represents a linear functional g is unique if and only if E distinguishes the members of F.

PROOF   If there is some element f in F such that $g(x) = \langle x,f \rangle$ for every x in E, it is obvious from the properties of the bilinear functional that g is linear, and from (iii) of theorem 16.1 that g is continuous with respect to the topology $w(E,F)$. If, conversely, g is linear and continuous on E, then from (ii) of theorem 16.1 there is a finite set $f_1, \cdots, f_n$ in F such that $|g(x)| \leq 1$ whenever x is in E and $|\langle x,f_i \rangle| \leq 1$ for $i = 1, \cdots, n$. In particular, if x is in E and $\langle x,f_i \rangle = 0$ for $i = 1, \cdots, n$, then $f_i(ax) = af_i(x) = 0$ for each scalar a and all i. Thus, $|g(ax)| \leq 1$ for every a, and hence $g(x) = 0$. Thus the null space of g contains the intersection of the null spaces of $T(f_1), \cdots,$ $T(f_n)$, where T is the canonical map of F into the algebraic dual of E, and hence g is a finite linear combination of $T(f_1), \cdots, T(f_n)$, by 1.3. Since T is linear, it follows that g is the canonical image of an element f of F. The final statement of the theorem is obvious.$|||$

The most economical device for defining locally convex topologies in linear spaces, and for performing calculations concerning these topologies, is that of polars. Let $\langle E,F \rangle$ be a pairing. The **polar** in F of a subset A of E, written A°, is the set of all f in F such that for each x in A, $|\langle x,f \rangle| \leq 1$; the polar in E of a subset B of F, written $B_\circ$, is the set of all x in E such that for each f in B, $|\langle x,f \rangle| \leq 1$. In the next section several topologies will be described by means of polars. Here it suffices to remark that the family of all polars of finite subsets of F is a local base for $w(E,F)$. Since polars will be used extensively throughout the rest of the book, several rules for computation with them are collected in the following theorem. Of course, each statement in the following list has a dual form, obtained by interchanging the roles of E and F.

16.3 COMPUTATION RULES FOR POLARS *Let $\langle E,F \rangle$ be a pairing. Then:*

(i) *for each subset A of E, A° is convex, circled, and $w(F,E)$-closed;*

(ii) *if $A \subset B \subset E$, then $A^\circ \supset B^\circ$;*

(iii) *if $A \subset E$ and if a is a non-zero scalar, then $(aA)^\circ = (1/a)A^\circ$;*

(iv) *if A is a non-void subset of E, then $(A^\circ)_\circ$, which will be written $A^\circ{}_\circ$ henceforth, is the smallest convex circled $w(E,F)$-closed set in E which contains A;*

(v) *if A is a non-void subset of E, then $A^\circ{}_\circ{}^\circ = A^\circ$;*

(vi) *if $\{A_t : t \in C\}$ is a family of subsets of E, then $(\bigcup \{A_t : t \in C\})^\circ = \bigcap \{A_t{}^\circ : t \in C\}$; and*

(vii) *if $\{A_t : t \in C\}$ is a family of convex circled $w(E,F)$-closed sets in E, then $(\bigcap \{A_t : t \in C\})^\circ$ is the smallest convex circled $w(F,E)$-closed set in F which contains $A_t{}^\circ$ for every t in C.*

PROOF Statements (i), (ii), and (iii) are clear from the definition of polar. As to (iv), it is evident that if C is the smallest convex circled $w(E,F)$-closed set which contains A, then $C \subset A^\circ{}_\circ$; on the other hand, if x is a point not in C, then by virtue of 14.4 there is a $w(E,F)$-continuous linear functional g on E which in absolute value is at most 1 on C and is greater than 1 at x. By the representation theorem there is an f in F such that $g(y) = \langle y,f \rangle$ for all y in E. Clearly $f \in A^\circ$; hence, $x \notin A^\circ{}_\circ$. Conclusion (v) follows immediately from (i) and the dual of (iv). As to (vi), it is obvious that $|\langle x,f\rangle| \leqq 1$ for each x in the union of the members of a family of sets if and only if it holds for each x in each member. From (iv) and the dual of (vi) it follows that

$$(\bigcap \{A_t : t \in C\})^\circ = (\bigcap \{A_t{}^\circ{}_\circ : t \in C\})^\circ = (\bigcup \{A_t{}^\circ : t \in C\})_\circ{}^\circ,$$

which yields (vii).|||

There are several direct consequences of the preceding proposition. First, recall (16.1) that a subset A of E is $w(E,F)$-bounded if and only if for each f in F it is true that sup $\{|\langle x,f\rangle| : x \in A\}$ is finite. But this supremum is finite if and only if $|\langle x,af\rangle| \leqq 1$ whenever a is a scalar which is sufficiently small in absolute value—that is, if and only if $af \in A^\circ$ for all sufficiently small a. This establishes the first statement of the following theorem, and the second follows in view of 16.3(iv).

16.4 POLARS OF BOUNDED SETS   *Let E and F be paired linear spaces. Then*

> (i) *a subset A of E is $w(E,F)$-bounded if and only if its polar A° is radial at 0; and*
> (ii) *if A is a subset of E, then A° is $w(F,E)$-bounded if and only if the $w(E,F)$-closed convex circled extension $A^\circ{}_\circ$ of A is radial at 0.*

The computation rules 16.3 yield an easy proof for a result concerning $w(E,F)$-dense subspaces of E. If G is a linear subspace of E, then $G^\circ{}_\circ$ is the $w(E,F)$-closure of G, in view of part (iv) of 16.3, and it follows from part (v) that $G^\circ{}_\circ = E$ if and only if $G^\circ = E^\circ$. This last condition can be restated: if $\langle x,f\rangle = 0$ for all x in G, then $\langle x,f\rangle = 0$ for all x in E. The following proposition is a trivial consequence.

16.5 $w(E,F)$-DENSE SUBSPACES   *The linear extension of a subset A of E is $w(E,F)$-dense in E if and only if each member of F which vanishes on A vanishes on E (more precisely, if and only if whenever $\langle x,f\rangle = 0$ for all x in A, then $\langle x,f\rangle = 0$ for all x in E).*

All of the compactness results of this chapter are derived from the following fundamental theorem.

16.6 ŠMULIAN'S CRITERION FOR $w(E,F)$-COMPACTNESS   *Let E and F be paired linear spaces and let B be a $w(E,F)$-closed convex circled subset of E. Then B is $w(E,F)$-compact if and only if B° is radial at 0 and each linear functional on F which is bounded on B° is represented by some member of E.*

PROOF   Let T be the canonical map of E into the algebraic dual F' of F, let E have the topology $w(E,F)$, and let F' have the topology $w(F',F)$. The set B is compact if and only if $T[B]$ is compact, in view of the definition of $w(E,F)$. But bounded subsets of F' are totally bounded by 16.1(vi), and F' is complete because it is a closed subset of a product of complete spaces. Consequently B is compact if and only if $T[B]$ is bounded and closed in F'.

The set B° is radial at 0 if and only if B is bounded, by 16.4, and B is bounded if and only if $T[B]$ is bounded because of the definition of $w(E,F)$.

Finally, each linear functional bounded on B° is represented by a member of E if and only if the polar of B° in F' is contained in $T[E]$. But B° is identical with the polar of $T[B]$ in F, by a direct verification, and the second polar of $T[B]$ is the closure of $T[B]$ in F' by 16.3(iv). Hence each linear functional bounded on B° is representable by a member of E if and only if the closure of $T[B]$ is contained in $T[E]$. But B was supposed to be closed; hence $T[B]^- \subset T[E]$ if and only if $T[B]$ is closed. The results of the last three paragraphs yield the theorem.|||

The topology $w(E,F)$ is seldom metrizable; however, there is a result on metrizability which is frequently useful. Assume that B is a $w(E,F)$-compact subset of E, and that A is a subset of F which distinguishes points of B in the sense that, if x and y are distinct members of B, then there is a member f of A such that $\langle x,f \rangle \neq \langle y,f \rangle$. This assumption implies that the topology of pointwise convergence on A is Hausdorff on B; it is weaker than the topology $w(E,F)$, with respect to which B is compact. It follows that the topology $w(F,E)$ and the topology of pointwise convergence on A coincide on B (the topological theorem used here is given in problem 4A). If A is countable, then the topology of pointwise convergence on A is the relativized product topology, where B is considered as embedded in $\mathsf{X} \{K: f \in A\}$, where K is the scalar field, and the product is countable. Hence B is metrizable. These considerations prove the following proposition.

16.7   Theorem   *If B is a $w(E,F)$-compact subset of E, and if A is a subset of F which distinguishes members of B, then the topology $w(E,F)$ for B is identical with the topology of pointwise convergence on A; if A is countable, the topology $w(E,F)$ for B is metrizable.*

The last theorems of the section concern completeness of E relative to the topology of uniform convergence on members of a family of subsets of F, where E and F are paired linear spaces. Suppose that $\mathscr{A}$ is a family of $w(F,E)$-bounded subsets of F and that T is the canonical map of E into the algebraic dual F' of F. We define the **topology of uniform convergence on members of $\mathscr{A}$ for E** (again denoted by $\mathscr{T}_\mathscr{A}$) as follows: a subset of E is open relative to $\mathscr{T}_\mathscr{A}$ if and only if the set is the inverse image under T of a subset of F' which is open relative to the topology of uniform convergence on members of $\mathscr{A}$. It is easy to verify that $(E,\mathscr{T}_\mathscr{A})$ is a locally convex space and non-zero scalar multiples of finite intersections of polars of members of $\mathscr{A}$ form a local base. A net in E converges to a limit relative to $\mathscr{T}_\mathscr{A}$ if and only if its image under T converges to the image of the limit uniformly on members of $\mathscr{A}$.

It was shown earlier (in Section 8 on function spaces) that if $\mathscr{A}$ is a family of subsets of a topological space X and E is the family of all those functions f on X to a complete linear topological space such that f is bounded and continuous on each member of $\mathscr{A}$, then E is complete relative to the topology $\mathscr{T}_\mathscr{A}$ of uniform convergence on members of $\mathscr{A}$. The principal remaining theorem can be regarded as a sort of converse to this proposition. Under certain circumstances completeness relative to $\mathscr{T}_\mathscr{A}$ of a family G of linear functionals implies that G is the family of *all* linear functionals which are bounded and continuous on each member of $\mathscr{A}$. The critical step in the proof of this converse is furnished by the following approximation theorem.

16.8 Approximation Theorem *Let E and F be paired linear spaces, and let A be a $w(E,F)$-closed convex circled subset of E. If f is any linear functional on E which is $w(E,F)$-continuous on A, then for each $e > 0$ there is a point g in F such that $|f(x) - \langle x,g \rangle| \leqq e$ for all x in A.*

proof It is enough to prove the theorem for the case in which F is a subspace of the algebraic dual E' of E; for, in general, if T is the canonical map of F into E', then the topologies $w(E,T[F])$ and $w(E,F)$ are identical.

If f is a linear functional on E which is $w(E,F)$-continuous on A, then for each positive real number e there is a $w(E,F)$-neighborhood U of 0 such that $|f|$ is at most e on $U \cap A$; that is, $(f/e) \in (A \cap U)^\circ$ where the polar is taken in E'. The set U may be supposed to be the polar of a finite subset B of F, and A is the polar of A° by 16.3. Hence, using 16.3 again, $(A \cap U)^\circ = (A^\circ_\circ \cap B_\circ)^\circ = (A^\circ \cup B)_\circ{}^\circ$, and the latter set is the $w(E',E)$-closed convex circled extension of $A^\circ \cup B$. But the $w(E',E)$-closed convex circled extension $B_\circ{}^\circ$ of the finite set B is $w(E',E)$-compact, and hence $A^\circ + B_0{}^\circ$ is $w(E',E)$-closed by 5.2(vii). Consequently $(A^\circ \cup B)_0{}^\circ \subset A^\circ + B_0{}^\circ$. Then f/e is a member of $A^\circ + B_0{}^\circ$, and therefore there is a member g of F (in fact a member of $eB_0{}^\circ$) such that $f - g \in eA^\circ$; restated, there is a $w(E,F)$-continuous linear functional g on E which approximates f within e on A.|||

The completeness theorem is a direct consequence of the foregoing. Recall that a family $\mathscr{A}$ of sets is directed by $\supset$ if and only if for A and B in $\mathscr{A}$ there is C in $\mathscr{A}$ such that $C \supset A \cup B$.

16.9 GROTHENDIECK'S COMPLETENESS THEOREM *Let E and F be paired linear spaces, and let $\mathscr{A}$ be a non-void family of $w(F,E)$-bounded, $w(F,E)$-closed, convex and circled subsets of F such that $\mathscr{A}$ is directed by $\supset$.*

- (i) *The canonical image $T[E]$ of E in F' is dense relative to $\mathscr{T}_{\mathscr{A}}$ in the space of all linear functionals on F which are $w(F,E)$-continuous on each member of $\mathscr{A}$.*
- (ii) *If the linear extension of the union of the members of $\mathscr{A}$ is F, then E is complete relative to $\mathscr{T}_{\mathscr{A}}$ if and only if each linear functional on F which is $w(F,E)$-continuous on each member of $\mathscr{A}$ is $w(F,E)$-continuous on F; equivalently, this is the case if and only if each hyperplane which intersects every member of $\mathscr{A}$ in a $w(F,E)$-closed set is $w(F,E)$-closed.*

PROOF   Let G be the space of all linear functionals on F which are $w(F,E)$-continuous on each member of $\mathscr{A}$. First observe that each f in G is bounded on each A in $\mathscr{A}$, because f is $w(F,E)$-uniformly continuous on A and A is totally bounded relative to $w(F,E)$. Therefore the space G is a locally convex space with the topology $\mathscr{T}_{\mathscr{A}}$, and the family of all the sets of the form $\{f: |f(x)| < e \text{ for all } x \text{ in } A\}$ is a local base for $\mathscr{T}_{\mathscr{A}}$. In view of the approximation theorem, part (i) is clear. If the linear extension of the union of $\mathscr{A}$ is F, then $(G,\mathscr{T}_{\mathscr{A}})$ is a complete Hausdorff space. Since E is complete relative to $\mathscr{T}_{\mathscr{A}}$ if and only if the canonical image of E is a complete subspace of G, E is $\mathscr{T}_{\mathscr{A}}$-complete if and only if $T[E] = G$, which is exactly the content of part (ii).|||

The section is concluded with a computation of polars in sums and products and a discussion of subspaces and quotient spaces. If E_t is paired to F_t for each t in an index set A, then the **natural pairing** of $\sum\{E_t: t \in A\}$ and $\bigtimes\{F_t: t \in A\}$ is defined by letting $\langle x,f \rangle = \sum\{\langle x(t),f(t)\rangle: t \in A\}$ for each x in the direct sum and each f in the product. It will be convenient in the statement and proof of the following proposition to consider the spaces E_t as subspaces of the direct sum, thus avoiding explicit mention of the injection maps.

16.10   LEMMA ON SUMS AND PRODUCTS   *Let E_t and F_t be paired linear spaces for each t in an index set A and let $\sum \{E_t: t \in A\}$ and $\times \{F_t: t \in A\}$ have the natural pairing.   Then:*

(i) *if B_t is a subset of E_t for each t in A, then the polar of the union of the sets B_t is the product $\times \{B_t{}^\circ: t \in A\}$ of the polars $B_t{}^\circ$ in F_t of B_t; and*

(ii) *if C_t is a circled subset of F_t for each t in A, then the polar of the product $\times \{C_t: t \in A\}$ contains the convex extension B of the union of the polars $(C_t)_0$ of C_t in E_t and is contained in aB for each real number a, $a > 1$.*

PROOF   To prove (i), observe that if $f \in \times \{B_t{}^\circ: t \in A\}$ and x belongs to B_t, then $|\langle x, f\rangle| = |\langle x(t), f(t)\rangle|$, which is at most one because $f(t) \in B_t{}^\circ$.   On the other hand, if f belongs to the polar of the union of the sets B_t, then for each t and each x in B_t, $|\langle x(t), f(t)\rangle| = |\langle x, f\rangle| \leq 1$; hence, $f(t) \in B_t{}^\circ$.   To prove (ii), note that if $x \in (C_t)_0$, then $|\langle x, f\rangle| = |\langle x(t), f(t)\rangle| \leq 1$ for each f in $\times \{C_t: t \in A\}$ and hence $\bigcup \{(C_t)_0: t \in A\} \subset (\times \{C_t: t \in A\})_0$.   Since the polar of $\times \{C_t: t \in A\}$ is convex, it contains B.   To prove the other inclusion, let x belong to the polar in E of the product $\times \{C_t: t \in A\}$, and let a_t be the supremum of $|\langle x(t), f(t)\rangle|$ for f in $\times \{C_t: t \in A\}$.   Because C_t is circled, if $f(t) \in C_t$ and $|\langle x(t), f(t)\rangle| = a$, then $\langle x(t), bf(t)\rangle = a$ for some suitable b with $|b| = 1$; that is, with $bf(t) \in C_t$.   Since $|\langle x, f\rangle| = |\sum \{\langle x(t), f(t)\rangle: t \in A\}| \leq 1$ for f in $\times \{C_t: t \in A\}$, it follows that $\sum \{a_t: t \in A\} \leq 1$.   By the definition of a_t, $x(t)/a_t$ belongs to the polar of C_t when $a_t \neq 0$, and hence, if $y = \sum \{a_t(x(t)/a_t): a_t \neq 0\}$, y belongs to the convex extension of the union of the sets $(C_t)_0$.   Finally, $x = (x - y) + y$, and $\langle x - y, f\rangle = 0$ for all f in $\times \{C_t: t \in A\}$; hence, $b(x - y)$ belongs to the convex extension of the union of $(C_t)_0$ for all scalars b.   Therefore, for $a > 1$,

$$x = a \left[\frac{1}{a} y + \left(1 - \frac{1}{a} \right) \cdot \frac{(x - y)}{(a - 1)} \right] \in aB.|||$$

If E and F are paired linear spaces and H is a subspace of E, then there is an **induced pairing** of E/H and H° which is defined by letting $\langle A, f\rangle = \langle x, f\rangle$, where A is an arbitrary member of E/H, x is an arbitrary member of A, and f belongs to H°.   (Notice that the polar of H coincides with the annihilator $H^\perp$.)   The next proposition identifies the topology $w(E/H, H^\circ)$, and the topology $w(H, F/H^\circ)$ which is derived from the induced pairings of H and F/H°.

16.11 TOPOLOGIES FOR SUBSPACES AND QUOTIENTS *Let E and F be paired linear spaces, let H be a subspace of E, and let H and F/H° and E/H and H° have the induced pairings. Then*

(i) *the topology $w(E, F)$ relativized to H is identical with the topology $w(H, F/H^\circ)$, and*

(ii) *the quotient topology for E/H which is derived from $w(E, F)$ is identical with $w(E/H, H^\circ)$.*

PROOF   The family of polars in H of finite subsets $f_1 + H^\circ, \cdots, f_n + H^\circ$ of F/H° is a local base for the topology $w(H, F/H^\circ)$. But the polar in H of such a subset is simply the set $\{x: x \in H \text{ and } |\langle x, f_i \rangle| \leq 1 \text{ for } i = 1, \cdots, n\}$, and this is the intersection with H of the polar of $f_1, \cdots, f_n$ in E. It follows that $w(H, F/H^\circ)$ is the relativization of $w(E, F)$, and (i) is established.

To establish (ii), observe that the quotient topology for E/H has a local base consisting of all collections of the form $\{x + H: x \in X_0\}$, where X is a finite subset of F and X_0 is the polar of X in E. On the other hand, the topology $w(E/H, H^\circ)$ has a local base consisting of collections $\{y + H: y \in Y_0\}$ where Y is a finite subset of H°. Clearly the latter topology is weaker, and identity of the two is established if it is shown that for each finite subset X of F there is a finite subset Y of H° such that $Y_0 + H \subset X_0 + H$. To prove this fact first note that given a finite subset X of F there is a finite subset Y of H° such that $X_0^\circ \cap H^\circ \subset Y_0^\circ$. (In case F is a Hausdorff space the closed convex circled extension X_0° of X is a finite dimensional compact set, and there is clearly a finite subset Y of H° whose closed convex extension contains $X_0^\circ \cap H^\circ$. If F is not Hausdorff, then a simple argument using the description 5.11 of non-Hausdorff spaces serves to establish the same result.) Since $X_0^\circ \cap H^\circ \subset Y_0^\circ$, it follows from the computation rules 16.3 and from 5.2(v) that

$$Y_0 \subset (X_0 \cup H)_0^\circ \subset (X_0 + H)^- \subset X_0 + H + X_0 \subset (\tfrac{1}{2}X)_0 + H.$$

Hence $Y_0 + H \subset (\tfrac{1}{2}X)_0 + H$ and the theorem follows.|||

There is an asymmetry in the preceding result which stems from the fact that H° is necessarily $w(F, E)$-closed. This difference is exhibited in the following proposition, which is an immediate corollary.

16.12 COROLLARY *If E and F are paired linear spaces and G is a subspace of F the quotient topology for E/G_0 is the topology $w(E/G_0, G^-)$, where E/G_0 and the closure G^- of G have the induced pairing.*

A particular consequence of this corollary is that the quotient topology for E/G_0 is certainly not identical to the topology $w(E/G_0,G)$ if E separates points of F and G is not closed, for in this case the two topologies yield distinct classes of continuous linear functionals.

PROBLEMS

A DUALITY BETWEEN TOTALLY BOUNDED SETS

Suppose that E and F are paired linear spaces, that A is a $w(E,F)$-bounded subset of E and that B is a $w(F,E)$-bounded subset of F. Then A is totally bounded in the topology on E of uniform convergence on B if and only if B is totally bounded in the topology on F of uniform convergence on A. (Use 8.17 and 16.1(vi).)

If $\mathscr{A}$ and $\mathscr{B}$ are families of $w(E,F)$- and $w(F,E)$-bounded subsets of E and F, then each $A \in \mathscr{A}$ is totally bounded in the topology of uniform convergence on members of $\mathscr{B}$ if and only if each $B \in \mathscr{B}$ is totally bounded in the topology of uniform convergence on members of $\mathscr{A}$.

B POLAR OF A SUM

Let $\langle E,F \rangle$ be a pairing and let A and B be non-void subsets of E. If A and B are convex and circled, then $\frac{1}{2}(A^\circ \cap B^\circ) \subset (A + B)^\circ \subset A^\circ \cap B^\circ$. To obtain a precise expression for $(A + B)^\circ$, valid more generally, define the following operation $\triangle$: if X and Y are subsets of F, let $X \triangle Y = \bigcup \{rX \cap sY : r \geq 0, s \geq 0, r + s \leq 1\}$. Then $X \cap Y \subset 2(X \triangle Y)$ and if X and Y are circled, $X \triangle Y \subset X \cap Y$. Now if A and B are circled, and e is a number such that $0 < e < 1$, then $e(A + B)^\circ \subset A^\circ \triangle B^\circ \subset (A+B)^\circ$.

C INDUCTIVE LIMITS II (see 1I, 17G, 19A, 22C)

Let E be a linear space and $\{E_t, f_t : t \in A\}$ be a family of locally convex spaces E_t and linear mappings f_t of E_t into E. The strongest locally convex topology for E which makes each f_t continuous is called the *inductive topology* for E determined by $\{E_t, f_t : t \in A\}$.

(a) Let E_0 be the linear span of $\bigcup \{f_t[E_t] : t \in A\}$. If M is a complement of E_0 in E, then the inductive topology is the direct sum of its relativization to E_0 and the strongest locally convex topology for M.

This means, roughly speaking, that any parts of E outside E_0 have a topology independent of the determining family $\{E_t, f_t : t \in A\}$: this contributes very little to the utility of the notion of inductive topology. It is therefore reasonable to assume, whenever this is convenient, that the linear span of $\bigcup \{f_t[E_t] : t \in A\}$ is the whole given linear space E.

(b) Let $\{E_t : t \in A\}$ be an inductive system (see Section 1) and let $E = \lim \text{ind} \{E_t : t \in A\}$. The space E is (algebraically) isomorphic to a quotient of the direct sum $\sum \{E_t : t \in A\}$. If each E_t is locally convex, there is a natural topology for E, namely, the quotient of the direct sum topology. If Q denotes the quotient mapping and I_t the injection of E_t into the direct sum, then this topology for E is the strongest locally convex topology which makes each $Q \circ I_t$ continuous. Thus the natural topology of an inductive limit is an inductive topology.

Suppose also that, for $t \geqq s$, the canonical mappings Q_{ts} of E_s into E_t are all continuous.   The natural topology of E is then the same as the natural topology defined by any cofinal subset B of A (cf. 11).

In these circumstances, this topology will be called the *inductive limit topology* for lim ind $\{E_t : t \in A\}$.

Suppose now that the space E has the inductive topology determined by $\{E_t, f_t : t \in A\}$, and let E_0 be the linear span of $\bigcup \{f_t[E_t] : t \in A\}$.

(c) A local base is formed by the family of all convex circled subsets U of E radial at 0 for which $f_t^{-1}[U]$ is a neighborhood of zero in E_t for each $t \in A$.   Let $\mathscr{U}_t$ be a local base in E_t for each $t \in A$ and assume that $E_0 = E$. Then the family of all convex circled extensions of sets $\bigcup \{f_t[U_t] : t \in A\}$, as U_t runs through $\mathscr{U}_t$ for each $t \in A$, is also a local base.

(d) A linear mapping T of E into a locally convex space is continuous if and only if each $T \circ f_t$ is continuous.

(e) The quotient topology of any locally convex space by a subspace is a special case of an inductive topology; so also is a direct sum topology.

(f) When $E_0 = E$, the inductive topology determined by $\{E_t, f_t : t \in A\}$ is a quotient of the direct sum topology for $\sum \{E_t : t \in A\}$.   (The kernel of the quotient mapping is the null space of the mapping $\sum \{x(t) : t \in A\} \to \sum \{f_t(x(t)) : t \in A\}$.)

(g) An inductive topology may fail to be Hausdorff, even if each E_t is a Hausdorff space; it may fail to be complete, even if each E_t is complete. (For the last part, see the counter-example in 20D.)

(h) Let F be a locally convex space such that, for each t, any linear mapping T of E_t into F is continuous provided that its graph is closed. Then any linear mapping of E into F has the same property.   (Use (e), and exhibit the graph $T \circ f_t$ as an inverse image of the graph of T.)

(i) Let $K(X)$ be the set of continuous real or complex valued functions of compact support on a topological space X.   For each compact subset B of X, $K_B(X)$ is the subspace of $K(X)$ of functions whose supports lie in B, with the topology of uniform convergence on B; the topology of $K(X)$ is defined by the local base consisting of all convex circled sets U radial at 0 such that, for each compact subset B of X, $U \cap K_B(X)$ is a neighborhood of 0 in $K_B(X)$ (see 8I).   This is the inductive topology determined by the spaces $K_B(X)$ and their injections into $K(X)$.   For, ordering the compact subsets of X by inclusion, the spaces $K_B(X)$ form an inductive system, and if $B \subset C$, the injection of $K_B(X)$ into $K_C(X)$ is continuous (in fact a homeomorphism).   Then $K(X) = $ lim ind $\{K_B(X) : B$ compact, $B \subset X\}$ and the inductive topology for $K(X)$, being the strongest locally convex topology making the injection of each $K_B(X)$ into $K(X)$ continuous, has the local base described.   In view of (b), it is sufficient to determine the topology of $K(X)$ by means of any co-base for the compact subsets of X; that is, any family of compact subsets $\{B_\alpha\}$ of X with the property that each compact subset of X is contained in some $\{B_\alpha\}$.

(j) Let $\mathscr{D}$ be the set of infinitely differentiable real or complex valued functions on R^n of compact support.   For each compact subset K of R^n, $\mathscr{D}_K$ is the subspace of $\mathscr{D}$ of functions with supports in K, with the metrizable topology defined by the sequence $\{q_m\}$ of pseudonorms, where $q_m(f) = \sup \{|D^p f(x)| : x \in K, 0 \leqq |p| \leqq m\}$ (see 8J).   A local base for the topology of $\mathscr{D}$ is formed by all convex circled sets U radial at 0 such that, for each compact subset K of R^n, $U \cap \mathscr{D}_K$ is a neighborhood of 0 in $\mathscr{D}_K$. This is the inductive topology determined by the spaces $\mathscr{D}_K$ and their injections into $\mathscr{D}$. For each positive integer r, let $K(r) = \{x : |x_i| \leqq r, 1 \leqq i \leqq n\}$; then $\{K(r) : r = 1, 2, \cdots\}$ is a co-base for the compact subsets of R^n, and, again by (b), the family $\{\mathscr{D}_{K(r)} : r = 1, 2, \cdots\}$ is sufficient to determine the inductive topology for $\mathscr{D}$.

D   PROJECTIVE LIMITS

Let F be a linear space and $\{F_t, g_t : t \in A\}$ a family of locally convex spaces F_t and linear mappings g_t of F into F_t. The weakest topology for F which makes each g_t continuous is called the *projective topology* for F determined by $\{F_t, g_t : t \in A\}$.

(a) Let $N = \bigcap \{g_t^{-1}(0) : t \in A\}$. Then the projective topology for F is the product of its quotient for F/N and the trivial topology for N.

Thus the assumption that $N = \{0\}$ involves no loss; clearly it is essential if F with the projective topology is to be a Hausdorff space.

(b) A projective limit (see Section 1) has a natural topology, the relativization of the product; this is a projective topology.

(c) If F has the projective topology determined by $\{F_t, g_t : t \in A\}$ and if $N = \{0\}$, then F may be exhibited as lim proj $\{g_t[F] : t \in A\}$ with its natural topology.

Suppose now that F has the projective topology determined by $\{F_t, g_t : t \in A\}$, and let $N = \bigcap \{g_t^{-1}(0) : t \in A\}$.

(d) The projective topology is locally convex; a local subbase is formed by the family of all sets $g_t^{-1}[U_t]$, as U_t runs through a local base in F_t.

(e) A linear mapping T of any linear topological space into F is continuous if and only if each $g_t \circ T$ is continuous.

(f) Relativization and product are special cases of projective topologies.

(g) If $N = \{0\}$, the projective topology is a relativization of the product topology for $\mathsf{X} \{F_t : t \in A\}$.

(h) Any locally convex topology is the projective topology determined by a family of normed spaces or Banach spaces. (For each convex circled neighborhood U of 0 in F, let $F_U = F/\bigcap \{eU : e > 0\}$. Compare (g) and (h) with 6.4.)

(i) If $N = \{0\}$ and each F_t is a Hausdorff space, then so is F; even if each F_t is complete, F may fail to be complete.

(j) A subset B of F is [totally] bounded if and only if each $g_t[B]$ is [totally] bounded.

(k) The space $\mathscr{E}$ is the projective limit of the spaces $\mathscr{E}^{(m)}$ (see 8J).

E   DUALITY BETWEEN INDUCTIVE AND PROJECTIVE LIMITS

Let $\{E_t : t \in A, \geqq, Q_{ts}(t \geqq s)\}$ be an inductive system. Suppose also that each E_t is a locally convex Hausdorff space and that each $Q_{ts}(t \geqq s)$ is continuous. Let F_t be the adjoint of E_t and define P_{st} for $t \geqq s$ by $\langle x_s, P_{st}(y_t) \rangle = \langle Q_{ts}(x_s), y_t \rangle$ for all $x_s \in E_s$ and $y_t \in F_t$. Then $\{F_t : t \in A, \geqq, P_{st}(t \geqq s)\}$ is a projective system. The adjoint of lim ind $\{E_t : t \in A\}$ with the inductive topology is lim proj $\{F_t : t \in A\}$.

There is an exactly similar dual result.

Topological conjectures generally fail: see 20D and 22G.

F SEQUENTIAL CONVERGENCE IN $L^1(X,\mu)$ II (see 9I)

(a) The space $L^1(X,\mu)$ is weakly sequentially complete: that is, if $\{f_n\}$ is a sequence of elements of $L^1(X,\mu)$ Cauchy in the weak topology on $L^1(X,\mu)$, then there is an element f in $L^1(X,\mu)$ such that $f_n \to f$ weakly. (Without loss of generality, X may be supposed to be totally σ-finite, so that 14M applies. Use 9I(a) to show the existence of $f \in L^1(X,\mu)$ with $\lim \int_A f_n d\mu = \int_A f d\mu$ for each measurable $A \subset X$. It follows that $\int f_n g d\mu \to \int f g d\mu$ for every simple function g. The uniform boundedness theorem shows that $\{f_n\}$ is bounded in norm; since the set of simple functions is dense in the adjoint of $L^1(X,\mu)$, it follows that $f_n \to f$ weakly.)

(b) If $\{f_n\}$ is a sequence in $L^1(X,\mu)$ weakly convergent and pointwise convergent a.e., then it is convergent in norm. (Use 9I(b).)

(c) In the space l^1, a weakly convergent sequence is convergent in norm. (Use (b): weak convergence implies pointwise convergence on a discrete measure space.)

G DENSE SUBSPACES

Let E and F be paired linear spaces, and let $\mathscr{A}$ be a family of $w(F,E)$-bounded, $w(F,E)$-closed, convex and circled subsets of F, directed by $\supset$, whose union spans F. If G is a subspace of E and $\mathscr{T}_{\mathscr{A}}$ the topology of uniform convergence on the members of $\mathscr{A}$, the following are equivalent:

(i) G is $\mathscr{T}_{\mathscr{A}}$-dense in E;
(ii) $w(F,G)$ and $w(F,E)$ agree on members of $\mathscr{A}$;
(iii) $w(F,E)$-closed convex subsets of members of $\mathscr{A}$ are $w(F,G)$-closed;
(iv) whenever B is a $w(F,E)$-closed convex subset of some member of $\mathscr{A}$ and $x \notin B$, there is an element of G which strongly separates B and x.

(The implications (i) $\Rightarrow$ (ii) $\Rightarrow$ (iii) and the equivalence of (iii) and (iv) are straightforward; to show that (iii) implies (i) use Grothendieck's completeness theorem 16.9.)

H HELLY'S CONDITION

(a) Let E be a linear space and B a convex circled subset of E. Suppose that $\{f_i : 1 \leq i \leq n\}$ is a finite set of linear functionals on E and $\{c_i : 1 \leq i \leq n\}$ a set of scalars. Then for each $e > 0$ there is a point $x_e \in (1 + e)B$ with $f_i(x_e) = c_i$ if and only if

$$(*) \quad |\sum \{a_i c_i : 1 \leq i \leq n\}| \leq \sup \{|\sum \{a_i f_i(x) : 1 \leq i \leq n\}| : x \in B\}$$

for all choices of the scalars $\{a_i : 1 \leq i \leq n\}$. (Map each z in E into the point $(f_1(x), \cdots, f_n(x))$ of K^n, and use a Hahn-Banach theorem in K^n.)

(b) It can be deduced from (a) that if $\langle E,F \rangle$ is a pairing and A a nonvoid subset of E, the $w(E,F)$-closed convex circled extension of A is $A^{\circ}{}_{\circ}$ (16.3(iv)). (Let B be the convex circled extension of A; then $B^{\circ} = A^{\circ}{}_{\circ}$. If $z \in A^{\circ}{}_{\circ}$ but $z \notin B^-$, there are linear functionals $f_i \in F$ with $|f_i(z - x)| > 1$ for all $x \in B$. Put $c_i = f_i(z)$ and use (a).)

(c) Let E be a linear topological space and B a convex circled compact subset of E. Suppose that F is a family of continuous linear functionals on E and $\{c_f : f \in F\}$ a family of scalars. Then there is a point $x_0 \in B$ with

$f(x_0) = c_f$ for all $f \in F$ if and only if the condition $(*)$ holds for every choice of $f_1, f_2, \cdots, f_n$ in F and $c_i = c_{f_i}$ for $1 \leq i \leq n$.

(d) It follows from (c) that if $\langle E, F \rangle$ is a pairing and B a convex circled $w(E,F)$-compact subset of E, then any linear functional on F bounded on B° is represented by a member of E (cf. Šmulian's weak compactness criterion 16.6).

[When E is the adjoint of a Banach space, or a reflexive Banach space, and B is its unit sphere, (c) becomes a theorem of Hahn on the existence of solutions of an infinite system of linear equations.]

I   Tensor Products I

Let E and F be linear spaces, with algebraic duals E' and F'. For each x in E and y in F, define the bilinear form $x \otimes y$ on $E' \times F'$ by

$$(x \otimes y)(x', y') = \langle x, x' \rangle \langle y, y' \rangle.$$

Let k be the canonical mapping of $E \times F$ into the space $B(E', F')$ of bilinear forms on $E' \times F'$ defined by $k(x, y) = x \otimes y$, and let $E \otimes F$, the *tensor product* of E and F, be the linear subspace of $B(E', F')$ spanned by $k(E, F)$.

(a) Any element z of $E \otimes F$ may be written in the form $\sum \{ x_i \otimes y_i : x_i \in E, \; y_i \in F, \; 1 \leq i \leq n \}$; z is the bilinear form defined by $z(x', y') = \sum \{ \langle x_i, x' \rangle \langle y_i, y' \rangle : 1 \leq i \leq n \}$.

(b) The set of linear forms on $E \otimes F$ is isomorphic to the set of bilinear forms on $E \times F$: to $g \in (E \otimes F)'$ corresponds the bilinear form $g \circ k$ on $E \times F$. (To show that each bilinear form b is the image of some g, put $g(x \otimes y) = b(x, y)$ and $g(\sum x_i \otimes y_i) = \sum b(x_i, y_i)$. For this to be meaningful, it is necessary that $\sum x_i \otimes y_i = 0$ should imply $\sum b(x_i, y_i) = 0$. To prove this, express each x_i and y_i in terms of linearly independent elements, and choose $x' \in E'$ and $y' \in F'$ suitably.)

(c) If G is any third linear space, the set of linear mappings of $E \otimes F$ into G is isomorphic to the set of bilinear mappings of $E \times F$ into G.

Suppose that E and F are locally convex spaces.

(d) There is one and only one locally convex topology $\mathscr{T}$ for $E \otimes F$ such that, for every locally convex space G, the set of continuous linear mappings of $E \otimes F$ into G corresponds to the set of continuous bilinear mappings of $E \times F$ into G, that is, the isomorphism in (c) preserves continuity. If $\mathscr{U}$ and $\mathscr{V}$ are local bases in E and F, a local base for $\mathscr{T}$ is formed by the convex circled extensions of all the sets $U \otimes V = \{ x \otimes y : x \in U, y \in V \}$, as U and V run through $\mathscr{U}$ and $\mathscr{V}$. (Show that the topology with this local base produces the correct continuous linear mappings; to prove uniqueness, take $G = E \otimes F$ and consider the identity mapping.)

(e) If E and F are Hausdorff spaces, so is $E \otimes F$ with the topology $\mathscr{T}$. (If $z_0 \neq 0$, z_0 may be written $\sum \{ x_i \otimes y_i : 1 \leq i \leq n \}$ with the x_i and the y_i linearly independent. There are continuous linear forms f and g with $f(x_1) > 1$, $g(y_1) > 1$, and $f(x_i) = g(y_i) = 0$ for $2 \leq i \leq n$. If $U = \{ x : |f(x)| \leq 1 \}$ and $V = \{ y : |g(y)| \leq 1 \}$, then $|z(f,g)| \leq 1$ for $z \in U \otimes V$ and $z_0(f,g) > 1$.)

The completion of $E \otimes F$ with the *projective tensor product topology* $\mathcal{T}$ is denoted by $E \,\hat{\otimes}\, F$.

(f) If E and F are metrizable, $E \,\hat{\otimes}\, F$ is a Fréchet space.   If also E and F are barrelled, so is $E \otimes F$.   (See 12G(b).)

17   THE WEAK TOPOLOGIES

The results on pairing are applied here to the case of a space E with locally convex topology $\mathcal{T}$ and adjoint $(E,\mathcal{T})^*$.   Many of the standard theorems on weak and strong closure, weak boundedness, weak* compactness, and continuity of functionals follow easily from earlier work.   The most profound theorem of the section gives conditions ensuring the weak compactness of the closed convex extension of a set.   Finally, we identify the weak and weak* topologies for subspaces, quotients, and so on.

Throughout this section E will be a linear space with a vector topology $\mathcal{T}$, and E^* (or $(E,\mathcal{T})^*$) will be the space of all $\mathcal{T}$-continuous linear functionals.   This section is devoted primarily to application of earlier results to the natural pairing of E and E^*.   Other applications—notably to the relationship between $w(E^*,E)$-compact, $w(E^*,E)$-bounded, and $\mathcal{T}$-equicontinuous subsets of E^*—are postponed to the following section.

The **weak topology** w for a linear space E with vector topology $\mathcal{T}$ is the topology $w(E,E^*)$ of the natural pairing of E and E^*.   We review a few of the properties of this topology.   (See 16.1.)   A net $\{x_\alpha, \alpha \in A\}$ in E converges weakly to a point x if and only if $\{f(x_\alpha), \alpha \in A\}$ converges to $f(x)$ for each continuous linear functional f.   The weak topology is locally convex, and the family of all polars of finite subsets of E^* is a local base.   The topology w is Hausdorff if and only if E^* distinguishes points of E, and this is always the case if the topology $\mathcal{T}$ is locally convex and Hausdorff.   A function R on a topological space to E is continuous relative to the weak topology if and only if the composition $f \circ R$ is continuous for every f in E^*.   A subset A of E is weakly bounded if and only if it is weakly totally bounded, and this is the case if and only if each member of E^* is bounded on A.

Each $w(E,E^*)$-continuous linear functional is a member of E^*, and consequently the class of weakly continuous linear functionals is identical with the class of $\mathcal{T}$-continuous linear functionals.   This last proposition may be phrased as a statement about hyperplanes as: a hyperplane is $\mathcal{T}$-closed if and only if it is weakly closed. This last formulation is also a direct corollary to the following theorem.

17.1 WEAKLY CLOSED CONVEX SETS   *A convex subset of a locally convex space $(E,\mathcal{T})$ is $\mathcal{T}$-closed if and only if it is weakly closed.*

PROOF   Let A be convex and $\mathcal{T}$-closed, and let x be an element which is not in A. Then there is by theorem 14.4 a $\mathcal{T}$-continuous linear functional f, which strongly separates A and x. Since f is weakly continuous, no net in A can converge weakly to x; hence, A is weakly closed. The converse is obvious since $\mathcal{T}$ is stronger than w.|||

17.2 COROLLARY   *The weak closure of a subset A of E is a subset of the $\mathcal{T}$-closure of the convex extension of A. If E is metrizable, each point of the weak closure of A is the $\mathcal{T}$-limit of a sequence $\{x_n\}$; each x_n is of the form $\sum \{t_i y_i : i = 1, \cdots, n\}$, where the t_i are positive real numbers such that $\sum \{t_i : i = 1, \cdots, n\} = 1$, and $y_i \in A$.*

17.3 COROLLARY   *If A is a closed convex subset of E, and if f is a linear functional on E, then f is weakly continuous on A if and only if f is $\mathcal{T}$-continuous on A.*

PROOF   It may be supposed without loss of generality that f is a real linear functional on a real space, for f is continuous if and only if its real part is continuous. By theorem 13.5, f is w-continuous on A if and only if $f^{-1}[a] \cap A$ is w-closed relative to A for all scalars a. Since $f^{-1}[a]$ and A both are convex, and since A is w-closed and $\mathcal{T}$-closed, $f^{-1}[a] \cap A$ is w-closed relative to A if and only if $f^{-1}[a] \cap A$ is w-closed; that is, if and only if $f^{-1}[a] \cap A$ is $\mathcal{T}$-closed relative to A. Thus, f is w-continuous on A if and only if f is $\mathcal{T}$-continuous on A.|||

It may be noted that the proof of theorem 17.1, on weakly closed convex sets, depends on the Hahn-Banach theorem, via the proposition on strong separation. The relationship between $\mathcal{T}$-bounded sets and weakly bounded sets will be established by means of a category argument (via the absorption theorem 10.1) and a compactness result which is the earliest of the compactness theorems of linear space theory. If U is a neighborhood of 0 in a linear topological space $(E,\mathcal{T})$, then each linear functional which is bounded on U is $\mathcal{T}$-continuous and is hence a member of E^*. The Šmulian compactness criterion 16.6 then shows that the polar of U in E^* is $w(E^*,E)$-

compact. This topology $w(E^*,E)$ is called the **weak* topology** and is denoted w^*.

17.4 BANACH-ALAOGLU THEOREM *If U is a neighborhood of 0 in a linear topological space E, then the polar of U in E^* is weak* compact.*

17.5 WEAKLY BOUNDED SETS *A subset of a locally convex space $(E,\mathscr{T})$ is weakly bounded if and only if it is $\mathscr{T}$-bounded.*

PROOF Each $\mathscr{T}$-bounded set is weakly bounded because $\mathscr{T}$ is stronger than w. Suppose that B is a weakly bounded set and that U is a $\mathscr{T}$-neighborhood of 0; it can be assumed that U is convex, circled and $\mathscr{T}$-closed, and hence that U is weakly closed. It follows that U is the polar of U°, where U° is the polar of U in E^* (see computation rules 16.3). The Banach-Alaoglu theorem 17.4 shows that the polar U° is $w(E^*,E)$-compact. The polar of B in E^* is $w(E^*,E)$-closed and radial at 0, and it follows from the absorption theorem 10.1 (or its corollary 10.2) that B° absorbs U°. Finally, taking polars in E, it is clear that U absorbs $B^\circ{}_0$ and therefore absorbs the subset B of $B^\circ{}_0$.|||

The elementary properties of the weak* topology are similar to those of the weak topology (see the paragraph prior to 17.1, or theorem 16.1); we shall not list these in detail, but merely note a few special properties. The space E^* is a linear subspace of the algebraic dual E' of E, and the weak* topology is the relativized topology of pointwise convergence. Consequently w^* is always a Hausdorff, as well as a locally convex, topology. A linear functional on E^* is w^*-continuous, according to 16.2, if and only if it is represented by a member of E. This yields the following proposition.

17.6 WEAK*-CONTINUOUS LINEAR FUNCTIONALS *A linear functional ϕ on the adjoint E^* of a linear topological space E is weak* continuous if and only if it is the evaluation at some point of E; that is, if and only if for some member x of E and all f in E^* it is true that $\phi(f) = f(x)$.*

We shall need, in the discussion of weak compactness, a description of a locally convex linear topological space in terms of its adjoint. If E is a linear space with locally convex topology $\mathscr{T}$, then the family $\mathscr{U}$ of all convex, circled $\mathscr{T}$-closed neighborhoods of 0 is a local base for the topology. But each member of $\mathscr{U}$ is also weakly closed, by 17.1, and is therefore the polar of a subset of the adjoint E^*. If $\mathscr{A}$ is the family of all polars of members of $\mathscr{U}$, then $\mathscr{T}$ can be described as the topology $\mathcal{T}_{\mathscr{A}}$ of uniform convergence on members of $\mathscr{A}$, since a local base for $\mathcal{T}_{\mathscr{A}}$ is the family of positive scalar multiples of polars of members of $\mathscr{A}$, and this is precisely the family $\mathscr{U}$. Finally, it is easy to see (using theorem 8.16) that the members of $\mathscr{A}$ are equicontinuous sets, and in fact $\mathscr{A}$ is a co-base for the family of all equicontinuous sets. These facts and the completeness theorem 16.9 yield:

17.7 UNIFORM CONVERGENCE ON EQUICONTINUOUS SETS, AND COMPLETENESS *Each locally convex topology is the topology of uniform convergence on equicontinuous subsets of the adjoint.*

Consequently, a locally convex linear topological space E is complete if and only if each linear functional on E^* which is weak* continuous on every equicontinuous set is weak* continuous on E^* (equivalently, if each such linear functional is evaluation at some member of E).

A few remarks may further clarify the situation. The adjoint of a non-complete locally convex space E and the adjoint of the completion $E^\wedge$ of E are (roughly speaking) the same, because each continuous linear functional on E is uniformly continuous and thus has a unique continuous extension to $E^\wedge$. But the topologies $w(E^*,E)$ and $w(E^*,E^\wedge)$ are surely not identical, since the class of continuous linear functionals consists in one case of evaluation at points of E and in the other case of evaluations at points of $E^\wedge$. Nevertheless, the topologies $w(E^*,E)$ and $w(E^*,E^\wedge)$ agree on each equicontinuous subset of E^*, for each such subset is contained in the polar of a neighborhood of 0, this polar is $w(E^*,E^\wedge)$-compact by 17.4, and the identity map of the polar, being a $w(E^*,E^\wedge)$-$w(E^*,E)$-continuous map of a compact space onto a Hausdorff space, is topological.

There is a result on weak compactness which is an immediate corollary to the preceding theorem.

17.8 COROLLARY *Each weakly compact subset of a locally convex space is complete.*

PROOF Let $\{x_\alpha, \alpha \in A\}$ be a Cauchy net in a weakly compact set B, and let x be a weak cluster point. Let ϕ be a linear functional on E^* such that the functionals on E^* corresponding to x_α converge uniformly on equicontinuous sets to ϕ, and note that $\phi(f) = f(x)$ for each f in E^* because x is a weak cluster point of the net $\{x_\alpha, \alpha \in A\}$. Thus the functionals corresponding to x_α converge uniformly on equicontinuous sets to the functional corresponding to x, and in view of 17.7, x_α converges to x.|||

The next three propositions are essentially preparation for the proof of the principal result on weak compactness. A non-void family $\mathscr{A}$ of sets is called an **algebra** if the union I of members of $\mathscr{A}$ belongs to $\mathscr{A}$ and, whenever A and B are members of $\mathscr{A}$, $A \cup B$ and $I \sim A$ (and hence $A \cap B$ and the null set) belong to $\mathscr{A}$. A real valued function m on $\mathscr{A}$ is called **additive** if $m(A \cup B) = m(A) + m(B)$ for any disjoint members A and B of $\mathscr{A}$. Given a countable family of subsets of I, the smallest algebra containing the family and I is countable (see Halmos [4] p. 23).

17.9 LEMMA *Let m be a non-negative real valued additive function defined on an algebra $\mathscr{A}$, and let $\{A_k\}$ be a sequence of members of $\mathscr{A}$ such that, for some positive number e, $m(A_k) > e$ for all k. Then there exists a subsequence $\{A_{k_i}\}$ of $\{A_k\}$ such that $m(\bigcap \{A_{k_i}: i = 1, \cdots, n\})$ is positive for all n.*

PROOF It is sufficient to prove that there is an integer s and a positive number d such that $m(A_s \cap A_k) \geq d$ for infinitely many integers k. Application of the diagonal process, beginning with the subsequence of all such A_k, then yields the desired subsequence.

Let $B_n = \bigcup \{A_k: k = 1, \cdots, n\}$; then $\{m(B_n)\}$ is a bounded non-decreasing sequence of numbers. Let $a = \lim_n m(B_n)$; then there is a positive integer r such that $a - m(B_r) < e/2$. Hence for any integer k greater than r, there is an integer s such that $1 \leq s \leq r$ and $m(A_s \cap A_k) \geq e/2r$. Therefore there is an s such that $m(A_s \cap A_k) \geq e/2r$ holds for an infinite number of k's, and the assertion of the previous paragraph is established.|||

In the following we use the notation $c(A)$ for the number of elements in a finite set A.

17.10 LEMMA *Let $\{I_i\}$ be a sequence of pairwise disjoint finite non-void sets and let $I = \bigcup \{I_i: i = 1, 2, \cdots\}$. If $\{A_k\}$ is a sequence of subsets of I such that, for some positive number e, $\liminf_i c(A_k \cap I_i)/c(I_i)$*
$> e$ for each k; then there is a subsequence $\{A_{k_i}\}$ of $\{A_k\}$ such that $\bigcap \{A_{k_i}: i = 1, 2, \cdots, n\}$ is not void for each n.

PROOF Let $\mathscr{A}$ be the smallest algebra of sets containing the sequence $\{A_k\}$ and I. The algebra $\mathscr{A}$ is countable; hence there is a sequence $\{i_j\}$ of integers such that $\lim_j c(A \cap I_{i_j})/c(I_{i_j})$ exists for each A in $\mathscr{A}$.

Denote this limit by $m(A)$. Then m is a non-negative real valued additive function on $\mathscr{A}$, and $m(A_k) > e$ for each k. In view of the previous lemma the present one is now obvious.|||

The proof of the next theorem for the case in which C is compact (the only case we need later) is easy if the representation of the dual of the space of all continuous scalar valued functions on a compact Hausdorff space, together with Lebesgue's bounded convergence theorem, is assumed to be known (see problem 17H). However, the proof which is given here avoids measure theory.

17.11 THEOREM *Let C be a compact (or countably compact) subset of a linear topological space E, and let $\{f_n\}$ be a sequence of continuous linear functionals on E which is uniformly bounded on C. If, for each x in C, $\lim_n f_n(x) = 0$, then the same equality holds for each x in the closed convex extension $\langle C \rangle^-$ of C.*

PROOF   Without loss of generality, one can assume that the sequence $\{f_n\}$ is in the polar C° in E^* of C. Assume that for some x in $\langle C \rangle^-$ it is false that $\lim_n f_n(x) = 0$. Then, if necessary by taking a subsequence, we can assume that, for some positive number e, $|f_n(x)| > e$ for all n. For each i, there is an x_i in $\langle C \rangle$ such that $|f_n(x_i)| > e$ for each $n \leq i$. It is possible to choose x_i in the form: $x_i = 1/c(I_i) \sum \{x_a : a \in I_i\}$, where I_i is a finite index set and $x_a \in C$ for each a in I_i. Let $I = \bigcup \{I_i : i = 1, 2, \cdots\}$ and $A_k = \{a : a \in I \text{ and } |f_k(x_a)| > e/2\}$. If $i \geq k$, then

$$
\begin{aligned}
e < |f_k(x_i)| &\leq 1/c(I_i) \sum \{|f_k(x_a)| : a \in I_i\} \\
&= 1/c(I_i)[\sum \{|f_k(x_a)| : a \in A_k \cap I_i\} \\
&\qquad + \sum \{|f_k(x_a)| : a \in I_i \sim A_k\}] \\
&\leq 1/c(I_i)[c(A_k \cap I_i) + (e/2)c(I_i \sim A_k)] \\
&\qquad\qquad \leq c(A_k \cap I_i)/c(I_i) + e/2.
\end{aligned}
$$

Hence, $i \geq k$ implies that $c(A_k \cap I_i)/c(I_i) > e/2$. In view of the lemma above, there is a subsequence $\{A_{k_i}\}$ of $\{A_k\}$ such that, for each n, the intersection $\bigcap \{A_{k_i} : i = 1, \cdots, n\}$ contains at least one element, say a_n. Let $y_n = x_{a_n}$; then y_n is in C, and for each i such that $i \leq n$, it holds that $|f_{k_i}(y_n)| > e/2$. Let y_0 be a cluster point in C of the sequence $\{y_n\}$. Then, for each i, $|f_{k_i}(y_0)| \geq e/2$, but this contradicts $\lim_n f_n(y_0) = 0$.|||

The next theorem is the principal result on weak compactness.

17.12   THEOREM ON WEAK COMPACTNESS   *Let E be a Hausdorff locally convex linear topological space, and let A be a bounded subset of E such that the closure C ⁻ of the convex extension C of A is complete. Then the following are equivalent:*

(i)   *Each sequence in A has a weak cluster point in E.*

(ii)   *For each sequence $\{x_n\}$ in A and each equicontinuous sequence $\{f_n\}$ of linear functionals it is true that $\lim_m \lim_n f_n(x_m) = \lim_n \lim_m f_n(x_m)$ whenever each of the limits exist.*

(iii)   *The weak closure of A is weakly compact.*

(iv)   *The weak closure of C is weakly compact.*

PROOF   To establish that (i) implies (ii), let x_0 be a cluster point in E of $\{x_m\}$ and let f_0 be a weak* cluster point in E^* of $\{f_n\}$.   The latter exists because $\{f_n\}$ being equicontinuous, is contained in a weak* compact set (17.4).   If $\lim_m \lim_n f_n(x_m)$ exists, then $\lim_m \lim_n f_n(x_m) = \lim_m f_0(x_m) = f_0(x_0)$.   Similarly, if $\lim_n \lim_m f_n(x_m)$ exists, it is equal to $f_0(x_0)$.

In order to prove (iii) from (ii), first observe that it suffices to prove that the weak closure A^- of A in the completion $E^\wedge$ is weakly compact.   For, the closure C^- (in E) is closed in $E^\wedge$ and hence weakly closed in $E^\wedge$; therefore the weak closures of A in E and $E^\wedge$ are identical.   Consequently without a loss of generality it can be supposed that E is complete.   Now let $F = E^*$ and embed E in F' in the canonical way; then the $w(F',E^*)$-closure A^- of A is compact with respect to the $w(F',E^*)$-topology, because A is bounded.   Since the relativization of $w(F',E^*)$ to E is simply the weak topology, it remains to show that $A^- \subset E$.   Let ϕ be an element of A^- and let B be an equicontinuous weak*-closed (hence weak* compact) subset of E^*.   By applying theorem 8.18 to the members of A restricted to B, one sees that the iterated limit condition of (ii) implies that ϕ restricted to B is weak* continuous.   Hence by 17.7, $\phi \in E$.

It is clear that (iii) implies (i), and that (iv) implies (iii).   It remains to demonstrate (iv), assuming the first three conditions, and this will be done by establishing the iterated limit condition for an arbitrary sequence $\{x_m\}$ in C and an arbitrary equicontinuous sequence $\{f_n\}$ in $F(= E^*)$.   The sequence $\{f_n\}$ is contained in a weak* compact set, and in view of theorem 8.20 there is a subsequence which converges pointwise on the weakly compact set A^- (the weak closure of A) to some f_0 of F.   We may, therefore, assume that $\lim_n f_n(x) = f_0(x)$ for each x in A^-. As before embed E in F', and regard $\{f_n - f_0\}$ as a sequence of $w(F', E^*)$-continuous linear functionals of F' which are uniformly bounded on A^-. Since $\{f_n - f_0\}$ converges pointwise to zero on A^-, theorem 17.11 implies that $\lim_n \langle f_n, \phi \rangle = \langle f_0, \phi \rangle$ for each member ϕ of the $w(F', E^*)$-closure C^- of C in F'. Since C^- is $w(F', E^*)$-compact, there is a cluster point ϕ_0 of $\{x_m\}$ in C^-. Hence, if the iterated limit $\lim_n \lim_m f_n(x_m)$ exists, it is equal to $\lim_n \langle f_n, \phi_0 \rangle = \langle f_0, \phi_0 \rangle$. On the other hand, $\lim_m \lim_n f_n(x_m) = \lim_m f_0(x_m) = \langle f_0, \phi_0 \rangle$, provided that the iterated limit on the left-hand side exists. It follows that the convex extension C of A satisfies the iterated limit condition.|||

We conclude this section with a brief discussion of the weak topologies for subspaces, quotient spaces, products, and direct sums.

17.13 THE WEAK TOPOLOGY FOR SUBSPACES, QUOTIENTS, AND PRODUCTS

(i) *The weak topology for a subspace F of a locally convex space E is the relativization of the weak topology for E.*

(ii) *If E is a linear topological space and F is a subspace, then the weak topology for the quotient space E/F is the quotient topology derived from the weak topology for E.*

(iii) *The weak topology for a product of linear topological spaces is the product of the weak topologies for the coordinate spaces.*

PROOF   In view of the identification 14.5 of the continuous linear functionals on a subspace F of a locally convex space E, each member of F^* is represented by a member of E^*/F°, in the induced pairing. Theorem 16.11 then yields proposition (i), and proposition (ii) follows from the same two theorems in similar fashion. The result (iii) may be derived from 14.6 by a simple argument, which we leave to the reader.|||

The weak topology for a direct sum is not the direct sum of the weak topologies for the factors. This may be verified easily by consideration of the direct sum of infinitely many copies of the scalar field.

17.14 THE WEAK* TOPOLOGY FOR ADJOINTS OF SUBSPACES, QUOTIENTS, AND DIRECT SUMS

(i) *If F is a closed subspace of a locally convex space E and F° is its polar in E^*, then E^*/F°, with the quotient topology derived from* *the weak* topology for E*, is topologically isomorphic to F* with the weak* topology.*

(ii) *If F is a subspace of a linear topological space E, then the subspace F° of E* with the relativized weak* topology is topologically isomorphic to (E/F)* with the weak* topology.*

(iii) *The adjoint of a direct sum, with the weak* topology, is topologically isomorphic to the product of the adjoints with the product of the weak* topologies.*

The proof of the preceding theorem is left to the reader.   It may be accomplished by identifying the adjoint via 14.5 and 14.7, then applying 16.11 to get (i) and (ii), and making a simple direct verification to establish (iii).

PROBLEMS

A   EXERCISES

(1) An infinite dimensional pseudo-normed linear space is of the first category in its weak topology.

(2) If U is a weak neighborhood of zero in a linear topological space E and $N = \bigcap \{eU : e > 0\}$, then E/N is finite dimensional.

(3) The subsets of the adjoint E^* of a linear topological space E which are equicontinuous relative to the weak topology are the finite dimensional weak* bounded subsets.

B   TOTAL SUBSETS

(a) A subset A of a linear topological space is *total* if and only if each continuous linear functional vanishing on A is identically zero.   If $N_x = \{f : f \in E^*, f(x) = 0\}$, then A is total if and only if $\bigcap \{N_x : x \in A\} = \{0\}$. A subset A is total if and only if the subspace generated by A is $w(E,E^*)$-dense in E.

(b) A subset A of the adjoint of a Hausdorff linear topological space E is *total* if and only if $x = 0$ whenever $f(x) = 0$ for all $f \in A$.   A subset A is total if and only if the subspace generated by A is $w(E^*,E)$-dense in E^*. If there is a finite total subset of E^*, then E is finite dimensional.

(c) Let E be the space of all bounded real valued functions on an infinite set S with the topology of uniform convergence.   If A is a total subset of E, then $k(A) \geq 2^{k(S)}$; there is a total subset A with $k(A) = 2^{k(S)}$.   ($k(X)$ denotes the cardinal of X.)

C   UNIFORMLY CONVEX SPACES I (see 20L)

A normed linear space is *uniformly convex* iff for any $e > 0$ there is some $d(e) > 0$ such that $\|x - y\| < e$ whenever $\|\frac{1}{2}(x + y)\| > 1 - d(e)$ and $\|x\| = \|y\| = 1$.

(a) In a uniformly convex normed space, a net $\{x_\alpha\}$ converges to x_0 in the norm topology if and only if it converges to x_0 in the weak topology and $\{\|x_\alpha\|\}$ converges to $\|x_0\|$. (First prove the result when $\|x_\alpha\| = \|x_0\| = 1$ for all α. Let f be a continuous linear functional with $f(x_0) = 1$ and $\|f\| = 1$. Then for each $e > 0$, $1 - d(e) < |f(\frac{1}{2}(x_\alpha + x_0))| \leq \|\frac{1}{2}(x_\alpha + x_0)\|$ eventually.)

(b) The product of two uniformly convex spaces may be equivalently normed to be uniformly convex.

(c) Neither uniform convexity nor the convergence property in (a) are topological (that is, independent of the particular norm). (For uniform convexity, the plane provides a counter-example; for the convergence property, consider the product of a uniformly convex infinite dimensional normed space and the real numbers.)

(d) The convergence property in (a) does not imply uniform convexity. (Consider the space l^1 of sequences $x = \{\xi_n\}$ with $\|x\| = \sum \{|\xi_n| : n = 1, 2, \cdots\}$ Banach [1] Ch. IX, §3, p. 140.)

D   VECTOR-VALUED ANALYTIC FUNCTIONS

Let D be a domain in the complex plane and x a mapping of D into a sequentially complete locally convex Hausdorff space E. The function x is *analytic* in D iff, for each $z \in D$, $\lim \{[x(z + \zeta) - x(z)]/\zeta : \zeta \to 0\}$ exists. The function x is *weakly analytic* in D iff it is analytic in D relative to the weak topology, or, equivalently, iff for each $f \in E^*$ the function x_f defined by $x_f(z) = f(x(z))$ is an analytic function in D. Then every function weakly analytic in D is analytic in D. (Let $y(\zeta) = [x(z + \zeta) - x(z)]/\zeta$ for $z \in D$. There is a closed circle with center z and radius $2r$, say, lying in D. Show that $\{y(\zeta) : |\zeta| \leq 2r\}$ is a bounded set, by invoking 17.5. Deduce, using Cauchy's formula, that $(y(\zeta) - y(\zeta')) \to 0$ as $|\zeta - \zeta'| \to 0$, provided that $|\zeta| \leq r$ and $|\zeta'| \leq r$. It follows that $x'(z) = \lim \{y(\zeta) : \zeta \to 0\}$ exists uniquely, for each $z \in D$.)

E   STONE-WEIERSTRASS THEOREM

Let X be a locally compact Hausdorff space, let $C_0(X)$ be the Banach space of all real valued continuous functions on X vanishing at ∞, and let S be the unit sphere $\{f : \|f\| \leq 1\}$. The adjoint of $C_0(X)$ is the set of all finite regular Borel measures on X with total variation as the norm (14J).

Lemma   Let F be a subspace of $C_0(X)$ and let μ be a non-zero regular Borel measure which is an extreme point of the weak* compact set $F^\circ \cap S^\circ$. Let g be a member of $C_0(X)$ with the property that $\int fg d\mu = 0$ for all f in F. Then g is a constant function almost everywhere relative to $|\mu|$. (By adding a constant and by multiplying by a scalar, one can assume that $g \geq 0$ and $\int g d|\mu| = 1$. If $\|g\| \leq 1$, then g is 1 a.e. If $\|g\| > 1$, put $t = \|g\|^{-1}$; then $\mu = (1 - t)\mu_1 + t\mu_2$, where $\mu_1(A) = \int_A [(1 - tg)/(1 - t)] d\mu$ and $\mu_2(A) = \int_A g d\mu$.)

Theorem   Let F be a subspace of $C_0(X)$ satisfying:
(a) for each x in X, there is an f in F such that $f(x) \neq 0$;
(b) for each distinct pair x and y in X, there is an f in F such that $f(x) \neq f(y)$;
(c) F is closed under multiplication.

Then F is dense in $C_0(X)$. (If F is not dense in $C_0(X)$, then by the Krein-Milman theorem there is a measure satisfying the condition of the lemma. Let C be the support of $|\mu|$, that is, $C = \{x\colon$ *each neighborhood of x has a positive $|\mu|$-measure*$\}$. Then each f in F is constant on C.)

F COMPLETENESS OF A DIRECT SUM

The form of Grothendieck's theorem given in 17.7 yields an alternative proof of the completeness of a direct sum of complete spaces (14.7(iv)). Let $E = \sum\{E_t\colon t \in A\}$, each E_t being complete; then $E^* = \bigtimes\{E_t{}^*\colon t \in A\}$. Let ϕ be a linear functional on E^*, weak* continuous on every equicontinuous set.

(a) Let I_t be the injection of $E_t{}^*$ into the product $\bigtimes\{E_t{}^*\colon t \in A\}$. Then there is a finite subset B of A such that $\phi \circ I_t = 0$ for each $t \in A \sim B$. (If $(\phi \circ I_{t_n})(f_{t_n}) = C_n \neq 0$ for $n = 1, 2, \cdots$, consider the set $\{nI_{t_n}(f_{t_n})/C_n\colon n = 1, 2, \cdots\}$).

(b) For each $f \in E^*$, $\phi(f) = \sum\{(\phi \circ I_t)(f_t)\colon t \in B\}$. (If $f \in E^*$, $(\phi \circ I_t)(f_t) = 0$ for all $t \in A \sim B$ implies $\phi(f) = 0$; use again the weak* continuity of ϕ.)

(c) There is a point $x \in \sum\{E_t\colon t \in B\}$ with $\phi(f) = f(x)$ for all $f \in E^*$. (A finite direct sum of complete spaces is complete.)

G INDUCTIVE LIMITS III (see 1I, 16C, 19A, 22C)

Let $\{(E_n,\mathcal{T}_n)\colon n = 1, 2, \cdots\}$ be a sequence of locally convex spaces such that whenever $n \geq m$, $E_m \subset E_n$ and the injection (identity mapping) of $(E_m,\mathcal{T}_m)$ into $(E_n,\mathcal{T}_n)$ is continuous. Let $E = \bigcup\{E_n\colon n = 1, 2, \cdots\}$ and let $\mathcal{T}$ be the inductive topology for E, determined by the spaces E_n and their injections into E; $\mathcal{T}$ is the strongest locally convex topology which makes each injection continuous, or, equivalently, whose relativization to each E_n is weaker than $\mathcal{T}_n$. The space $(E,\mathcal{T})$ is the inductive limit (see 16C) of the sequence of spaces $(E_n,\mathcal{T}_n)$. The local bases mentioned in problem 16C(d) become especially simple in this case: the family of all convex circled subsets U of E such that $U \cap E_n$ is a $\mathcal{T}_n$-neighborhood of 0 for each n is a local base for $\mathcal{T}$, and the convex circled extension of all sets $\bigcup\{U_n\colon n = 1, 2, \cdots\}$, as U_n runs through a local base for $\mathcal{T}_n$, constitute a local base for $\mathcal{T}$. Also a linear mapping of E into any locally convex space is continuous if and only if its restriction to each $(E_n,\mathcal{T}_n)$ is continuous.

(a) Assume also that $(E,\mathcal{T})$ is a Hausdorff space. Then $(E,\mathcal{T})$ is complete if and only if for each n, the closure of E_n in $(E,\mathcal{T})$ is $\mathcal{T}$-complete. (Sketch of proof: one way is immediate. Suppose that $E_n{}^-$ is complete, and that ϕ is a linear functional on E^* which is weak* continuous on every equicontinuous set. First, show that there is some n for which ϕ vanishes identically on $E_n{}^\circ = E_n{}^{-\circ}$. Next, let $(E_n{}^-)^*$ be the adjoint of $E_n{}^-$ with the relativization of $\mathcal{T}$. Then $\phi = \psi \circ Q$, where Q is the quotient mapping of E^* onto $(E_n{}^-)^* = E^*/E_n{}^\circ$. With the weak* topologies, Q is continuous; if A is an equicontinuous subset of $(E_n{}^-)^*$ then there is an equicontinuous weak* compact subset B of E^* such that $A \subset Q[B]$. It follows, from the hypothesis on ϕ, that ψ is weak* continuous on $Q[B]$. Hence (17.7) ψ is weak* continuous; therefore so also is ϕ.

(b) Assume that, for each n, E_n is a closed subspace of E_{n+1} and that $\mathcal{T}_n$ is the relativization of $\mathcal{T}_{n+1}$ to E_n. The space $(E,\mathcal{T})$ is then called the *strict* inductive limit of the sequence $\{(E_n,\mathcal{T}_n)\}$. Then

(i) each $\mathcal{T}_n$ is the relativization of $\mathcal{T}$, and E_n is a $\mathcal{T}$-closed subspace of E;

(ii) if each E_n is a Hausdorff space, then so is E;

(iii) each bounded subset of E is contained in some E_n.

(Let U_n be a convex circled $\mathcal{T}_n$-neighborhood of 0 in E_n; use 13E(b) to construct a sequence $\{U_{n+m}: m = 0, 1, \cdots\}$ of $\mathcal{T}_{n+m}$-neighborhoods of 0, with $E_{n+m} \cap U_{n+m+1} = U_{n+m}$, and put $U = \bigcup \{U_{n+m}: m = 0, 1, \cdots\}$. For (ii) and (iii), use 13E(c) and similar constructions.)

H INTEGRATION PROOF OF THEOREM 17.11

Let A be a compact subset of a linear topological space E.

(a) Let z be a point of the closed convex extension of A. Then there is a positive Borel measure μ on A with $\mu(A) = 1$ and $f(z) = \int_A f(x)d\mu(x)$ for each continuous linear functional f on E. (If μ is any Borel measure on A, the mapping $f \to \int_A f(x)d\mu(x)$ is a linear functional $T(\mu)$ on the adjoint E^* of E. Then T is a weak* continuous linear mapping of the adjoint $M(A)$ of $C(A)$ (14J) into $E^{*\prime}$. The set P of positive measures with $\mu(A) = 1$ is weak* compact in $M(A)$ and so $T[P]$ is weak* closed in $E^{*\prime}$. Being convex, it contains the canonical image of $\langle A \rangle^-$.)

(b) If $\{f_n\}$ is a sequence of continuous linear functionals on E bounded uniformly on A with $f_n(x) \to 0$ on A, then $f_n(x) \to 0$ on $\langle A \rangle^-$. (Use (a) and Lebesgue's theorem on bounded convergence.)

I WEAKLY COMPACT CONVEX EXTENSIONS

The weakly closed convex extension of a weakly compact set is not in general weakly compact: the set in 18F provides a simple counter-example. In fact, the hypothesis of theorem 17.12, that the weakly closed convex extension be complete, in some topology giving the same adjoint, is essential (by 17.8).

The following example shows considerably more. Let E be the space of sequences $x = \{\xi_n\}$ with at most a finite number of non-zero terms, with the norm $\|x\| = \sum |\xi_n|$. Then any complete convex set in E (and so certainly any convex weakly compact set) is finite dimensional, although there are infinite dimensional compact sets. (Suppose that K is a complete infinite dimensional convex set. Then there is an infinite subsequence $\{n(k)\}$ for which the set $G_k = \{x: x \in K, \xi_{n(k)} \neq 0\}$ is not empty. Use the convexity of K to show that each G_k is dense in K; since K is of the second category, $\bigcap \{G_k: k = 1, 2, \cdots\}$ is not empty, and this achieves a contradiction. On the other hand, if x_n is the sequence with all its terms zero except the n-th, which is $1/n$, then the set consisting of the points x_n together with 0 is compact and infinite dimensional.)

J WEAK* SEPARABILITY

Let E be a locally convex space.

(a) If E is separable, the equicontinuous subsets of its adjoint E^* are weak* metrizable.

(b) If E is separable and pseudo-metrizable, then E^* is weak* separable.

(c) If E is pseudo-metrizable, the converse of (a) is true.

K   HELLY'S CHOICE PRINCIPLE

Let X be a locally compact Hausdorff space whose topology has a countable base and let $\{\mu_n\}$ be a sequence of Borel measures on X for which $\{\mu_n(X): n = 1, 2, \cdots\}$ is bounded. Then there is a Borel measure μ on X and a subsequence $\{n_k\}$ with $\int f d\mu_{n_k} \to \int f d\mu$ for each continuous real valued function f on X vanishing at infinity. (Regard $\{\mu_n\}$ as a sequence of elements of a sphere in the adjoint of $C_0(X)$, the normed space of continuous real valued functions on X vanishing at infinity, and use 17J(a).)

L   EXISTENCE OF WEAKLY CONVERGENT SEQUENCE

Suppose that E is metrizable and that A is a subset of E with the property that every sequence of points of A has a weak cluster point in E. Then any point of the weak closure of A is the weak limit of a sequence of points of A. (Sketch of proof: Let $\{U_n: n = 1, 2, \cdots\}$ be a local base with $U_{n+1} \subset U_n$ for each n, and let x be a point of the weak closure of A. For each n, apply 8.21 to show the existence of a sequence in A pointwise convergent to x on U_n°. Enumerate the points of all these sequences (triangular-wise, for example) to form a sequence $\{x_k\}$. Then x is a cluster point of $\{x_k\}$. The closed subspace H spanned by $\{x_k\}$ is separable and metrizable; it follows from 17J(b) that there is a countable subset D of E^* which is $w(E^*, H)$-dense in E^*. By successive refinement, there is a subsequence $\{x_{k(i)}\}$ with $f(x_{k(i)}) \to f(x)$ for all $f \in D$. Now any weak cluster point of $\{x_{k(i)}\}$ must be x because of the density of D; hence $x_{k(i)} \to x$ weakly.)

This theorem may be regarded as an application to a metrizable linear topological space of a generalization of theorem 8.20 on function spaces: suppose, in that theorem, that the functions are defined on a set S which is the union of a sequence of compact sets and that they take their values in a metric space Z. The proof above, which takes advantage of the properties of the weak* topology, may be carried through by using in its place the weakest topology $\mathscr{T}$ for S making the elements of the sequence continuous; any cluster point of the sequence is then also continuous for $\mathscr{T}$.

18   TOPOLOGIES FOR E AND E^*

There are a number of more or less reasonable (= admissible) topologies which may be assigned to a linear space which is paired with another linear space, and this section begins with results on completeness, normability, and metrizability of these. The primary interest is in the pairing of a linear topological space E and its adjoint E^*, and these results are applied to this case. It turns out that the classes of equicontinuous, strongly bounded, and weak* bounded subsets of E^* are generally distinct, but these classes coincide in the important special case that E is barrelled (i.e., the Banach-Steinhaus theorem holds). The collection of topologies for E that yield the same adjoint E^* is explicitly described. Finally, the canonical sorts of calculation for subspaces, quotients, and products are made.

This section is devoted to a study of certain topologies for a linear topological space and its adjoint. Properties of the strong topology for the adjoint E^* of a locally convex space E are examined; we find that the notions of equicontinuity, of weak* compactness, and of strong boundedness, which are (roughly speaking) equivalent for the adjoint of a normed space, are in general quite distinct. For a certain class of linear topological spaces, called barrelled spaces, these three notions coincide. Barrelled spaces have other useful properties, the most noteworthy being that a form of the Banach-Steinhaus theorem holds. Finally, a complete solution is given for the problem: what class of locally convex topologies yields a given space of linear functionals as adjoint? It turns out that there is a strongest and a weakest topology in this class, and the class consists of all locally convex topologies which lie between. The theorems of this section are, for the most part, very direct consequences of earlier developments. The first two theorems in particular are a simple recasting, in a form which is convenient for the present investigation, of some results from Sections 8, 16, and 17.

Since several topologies for a space and its adjoint will be considered, we shall describe the class of topologies which will be of concern. Suppose that E and F are paired linear spaces, that $\mathscr{A}$ is a family of $w(F,E)$-bounded subsets of F, and that $\mathscr{T}_\mathscr{A}$ is the topology of uniform convergence on members of $\mathscr{A}$ (see Section 16). Then the non-zero scalar multiples of finite intersections of polars of members of $\mathscr{A}$ form a local base for $\mathscr{T}_\mathscr{A}$. A topology for E is called **admissible** (see Kelley [5], p. 112) (relative to the pairing) if and only if it satisfies the three conditions: $\mathscr{T}$ is a vector topology for E, (the canonical image of) each member of F is $\mathscr{T}$-continuous, and $\mathscr{T} = \mathscr{T}_\mathscr{A}$ for some family $\mathscr{A}$ of subsets of F. The elementary facts about admissible topologies are stated in the next theorem. Part (iii) of this theorem has already been demonstrated (17.7), and is included here for reference.

18.1　Admissible Topologies

(i) *If E and F are paired linear spaces and $\mathscr{A}$ is a family of $w(F,E)$-bounded subsets of F such that the linear extension of $\bigcup \{A: A \in \mathscr{A}\}$ is F, then $\mathscr{T}_\mathscr{A}$ is an admissible topology for E.*

(ii) *Let E and F be paired linear spaces. Then a topology for E is admissible if and only if it is stronger than $w(E,F)$ and has a local base whose members are $w(E,F)$-barrels.*

(iii) *If $\mathscr{T}$ is an arbitrary locally convex topology for a linear space E, then $\mathscr{T}$ is admissible relative to the natural pairing of E and E^*; in fact, $\mathscr{T}$ is the topology of uniform convergence on $\mathscr{T}$-equicontinuous subsets of E^*, or equivalently, on the family of all polars of $\mathscr{T}$-neighborhoods of 0.*

PROOF   (i) It is only necessary to establish that the canonical image $T(f)$ of each member f of F is $\mathscr{T}_{\mathscr{A}}$-continuous. But this is immediate, since, if f is a member of some A in $\mathscr{A}$, then the functional $T(f)$ is bounded on the neighborhood A_0 of 0.

(ii) Suppose $\mathscr{T}_{\mathscr{A}}$ is an admissible topology for E. Recall from 16.1(iv) that a function R from a topological space Z to E is continuous relative to the $w(E,F)$-topology for E if the map $z \rightarrow \langle R(z),f \rangle$ is continuous on Z for each f in F. Applying this theorem to the identity map of E, with $\mathscr{T}_{\mathscr{A}}$, into E, with $w(E,F)$, shows that $\mathscr{T}_{\mathscr{A}}$ is stronger than $w(E,F)$. A local base for $\mathscr{T}_{\mathscr{A}}$ is the family of finite intersections of non-zero scalar multiples of polars of members of $\mathscr{A}$. Since members of $\mathscr{A}$ are bounded, by 16.4 the polars of members of $\mathscr{A}$ are $w(E,F)$-barrels. The local base just described therefore consists of barrels.

Conversely, if $\mathscr{T}$ is a topology for E which has a local base $\mathscr{B}$ consisting of $w(E,F)$-barrels, then by a straightforward use of the computation rules for polars it can be shown that $\mathscr{T}$ is the topology of uniform convergence on polars of members of $\mathscr{B}$. If $\mathscr{T}$ is stronger than $w(E,F)$, then each member of F is $\mathscr{T}$-continuous because it is $w(E,F)$-continuous.|||

It is clear from the foregoing that an admissible topology $\mathscr{T}$ by no means specifies the family $\mathscr{A}$ such that $\mathscr{T} = \mathscr{T}_{\mathscr{A}}$; many different families may give the same topology. It is desirable to construct some sort of normalization procedure to help in describing the situation. Recall that a subfamily $\mathscr{D}$ of a family $\mathscr{C}$ of sets is called a co-base for $\mathscr{C}$ if and only if each member of $\mathscr{C}$ is contained in some member of $\mathscr{D}$. It is evident that if $\mathscr{D}$ is a co-base for $\mathscr{C}$, then $\mathscr{T}_{\mathscr{D}} = \mathscr{T}_{\mathscr{C}}$. If E and F are paired linear spaces and $\mathscr{B}$ is a family of subsets of F, then the family $\mathscr{B}$ is **admissible** if and only if the following conditions are satisfied: each member of $\mathscr{B}$ is $w(F,E)$-closed $w(F,E)$-bounded circled and convex, each non-zero scalar multiple of a member of $\mathscr{B}$ is a member of $\mathscr{B}$, the union of any two members of $\mathscr{B}$ is contained in a member of $\mathscr{B}$, and $\bigcup \{B : B \in \mathscr{B}\} = F$. Given an arbitrary family $\mathscr{A}$ of $w(F,E)$-bounded sets which covers F, an admissible family may be constructed in the following fashion, let $\mathscr{A}_1$ consist of all finite unions of members of $\mathscr{A}$, let $\mathscr{A}_2$ consist of all $w(F,E)$-closed circled convex extensions of members of $\mathscr{A}_1$, and let $\mathscr{B}$ consist of scalar multiples of members of $\mathscr{A}_2$. The admissible family $\mathscr{B}$ is said to be **generated** by $\mathscr{A}$.

The following theorem summarizes those properties of admissible families which will be used in the sequel.

18.2 ADMISSIBLE FAMILIES   *Let E and F be paired linear spaces.*

(i) *If $\mathscr{A}$ is an admissible family of subsets of F, then $\mathscr{T}_{\mathscr{A}}$ is an admissible topology for E, the family of polars of members of $\mathscr{A}$ is a local base for $\mathscr{T}_{\mathscr{A}}$, and $\mathscr{A}$ is a co-base for the family of $\mathscr{T}_{\mathscr{A}}$-equicontinuous sets.*

(ii) *If $\mathscr{A}$ is a family of $w(F,E)$-bounded subsets of F such that the linear extension of $\bigcup \{A : A \in \mathscr{A}\}$ is F, and if $\mathscr{B}$ is the admissible family generated by $\mathscr{A}$, then $\mathscr{T}_{\mathscr{A}} = \mathscr{T}_{\mathscr{B}}$.*

(iii) *If $\mathscr{A}$ is an admissible family of subsets of F, then the topology $\mathscr{T}_{\mathscr{A}}$ is pseudo-metrizable if and only if there is a countable co-base for $\mathscr{A}$, and $\mathscr{T}_{\mathscr{A}}$ is pseudo-normable if and only if the family of scalar multiples of some member of $\mathscr{A}$ is a co-base for $\mathscr{A}$.*

PROOF   Both (i) and (ii) result from straightforward computations with polars, using the results of this section and of Section 16, and these proofs are omitted. Proposition (iii) is derived from 6.7 and 6.1, using the elementary facts on pairings and the following simple fact: if there is a countable local base for a vector topology, then each local base contains a countable local base.|||

There is a noteworthy corollary to Grothendieck's completeness theorem 16.9.

18.3   THEOREM   *If E and F are paired linear spaces and E is complete relative to an admissible topology $\mathscr{T}$, then E is complete relative to each stronger admissible topology for E.*

PROOF   Each admissible topology $\mathscr{T}$ is the topology $\mathscr{T}_{\mathscr{A}}$ where $\mathscr{A}$ is the family of all polars of $\mathscr{T}$-neighborhoods of 0. If $\mathscr{U}$ is a stronger admissible topology and $\mathscr{B}$ is the family of polars of $\mathscr{U}$-neighborhoods of 0, then $\mathscr{A} \subset \mathscr{B}$ because each $\mathscr{T}$-neighborhood is a $\mathscr{U}$-neighborhood. Applying (ii) of theorem 16.9 then yields the theorem.|||

If E and F are paired linear spaces, then there is a strongest admissible topology for E. This topology is denoted by $s(E,F)$. In view of the definition of "admissible", $s(E,F)$ is the topology of uniform convergence on $w(F,E)$-bounded subsets of F: alternatively, $s(E,F)$ can be described by the requirement that the family of $w(E,F)$-barrels be a local base (see 18.1(ii)).   The topology $s(F,E)$ for F is described dually; the family of $w(F,E)$-barrels is a local base for $s(F,E)$.

If E is a space with a locally convex topology $\mathscr{T}$ and E^* is the adjoint, then the **strong topology** for E^* is the topology of uniform convergence on $\mathscr{T}$-bounded sets.   Since a subset of the locally convex space E is $\mathscr{T}$-bounded if and only if it is weakly bounded (see 17.5), the strong topology is identical with $s(E^*,E)$.   It is the strongest admissible topology for E^*, and the family of weak* barrels is a base for the strong topology.   The adjoint E^* is a subspace of the space E^b of all bounded linear functionals on E (see 6.2).   The term "strong" will also be applied to $s(E^b,E)$.

18.4   METRIZABILITY AND COMPLETENESS OF THE ADJOINT   *Let E be a locally convex linear topological space, and let the adjoint E^* have the strong topology.   Then:*

(i) *the space E^* is metrizable if and only if the family of all bounded subsets of E has a countable co-base;*

(ii) *E^* is dense, relative to the strong topology, in the space of all linear functionals on E which are continuous on each bounded subset of E; and*

(iii) *E^* is complete relative to the strong topology if and only if each linear functional which is continuous on bounded subsets of E is continuous on E.*

PROOF   Let E be a linear space with locally convex topology $\mathscr{T}$, and let $\mathscr{A}$ be the family of all bounded closed convex circled subsets of E. Then $\mathscr{A}$ is an admissible family of subsets of E, because a convex set is weakly closed (or weakly bounded) if and only if it is $\mathscr{T}$-closed ($\mathscr{T}$-bounded, respectively).   All of the assertions of the theorem are now easy consequences of theorems 18.2, 16.9, and the fact that continuity of a linear functional on a convex circled set is equivalent to closure of the intersection of the set and the null space of the functional.|||

It is worthwhile noticing that part (iii) of the foregoing theorem can be restated: E^* is strongly complete if and only if a maximal linear subspace of E is closed whenever its intersection with every closed bounded convex circled subset is closed (see 13.5).

If E is a pseudo-normed space, then the strong topology for E^* is identical with the norm topology. Moreover, in this case a subset of E^* is strongly bounded if and only if it is equicontinuous. Unfortunately, no such simple equivalence exists in the more general case. The following theorem gives relations which do subsist between strong and weak* boundedness, weak* compactness, and equicontinuity. The proofs lean very heavily on earlier results.

18.5 BOUNDEDNESS, COMPACTNESS, AND EQUICONTINUITY *Let E be a locally convex space and let B be a subset of the adjoint E^*.*

   (i) *If B is equicontinuous, then the weak* closed convex circled extension of B is both equicontinuous and weak* compact.*
   (ii) *If B is weak* compact and convex, then B is strongly bounded.*
   (iii) *If B is strongly bounded, then B is weak* bounded, and if B is weak* bounded and E is sequentially complete, then B is strongly bounded.*

PROOF   A subset of E^* is equicontinuous if and only if its polar in E is a neighborhood of 0.   If B is equicontinuous, then $B_0{}^\circ$ is the weak* closed convex circled extension of B and the polar of $B_0{}^\circ$ is B_0 (see the computation rules for polars 16.3).   Hence $B_0{}^\circ$ is equicontinuous, and $B_0{}^\circ$ is weak* compact in view of the Banach-Alaoglu theorem 17.4.   This establishes (i).   To establish (ii), suppose that B is weak* compact and convex and U is a strong neighborhood of 0 in E^*.   Then one can suppose U is a (weak*) barrel, and the absorption corollary 10.2 shows that U absorbs B.   Hence B is strongly bounded.   Finally, (iii) follows directly from the uniform boundedness theorem 12.4.|||

It is noteworthy that certain of the results of the preceding theorem cannot be improved.   A subset B of E^* may be weak* compact convex and circled, and may fail to be equicontinuous (the relation here is clarified by the discussion of Mackey spaces later in this section).   Without the requirement of convexity, (ii) may fail: there are weak* compact sets which are not strongly bounded (see problem 18F).   Since such a set is certainly weak* bounded, the classes of strongly and weak* bounded sets do not coincide.   There are strongly bounded weak* closed sets which are not weak* compact, so that the converse of (ii) does not hold (problem 20A).

If E is a pseudo-normed space, then strongly bounded sets are equicontinuous, and the preceding theorem yields the following corollary.

18.6 COROLLARY *If E^* is the adjoint of a pseudo-normed space E, then:*

(i) *a subset of E^* is equicontinuous if and only if it is strongly bounded;*

(ii) *the weak* closed circled convex extension of each strongly bounded subset of E^* is weak* compact.*

The notions of strong and weak* boundedness fail to coincide even when E is normable. However, there is an important class of spaces for which the situation is very simple.

A linear topological space E is a **barrelled space** (or **disk space**, or **tonnelé space**) if and only if the space E is locally convex and each barrel is a neighborhood of 0. The requirement that each barrel be a neighborhood of 0 can be translated into a requirement concerning the adjoint space by taking polars. Explicitly, if $\mathcal{T}$ is a locally convex topology for E, then $(E,\mathcal{T})$ is a barrelled space if and only if each $\mathcal{T}$-barrel is a $\mathcal{T}$-neighborhood of 0, which is the case if and only if each $w(E,E^*)$-barrel is a $\mathcal{T}$-neighborhood of 0, which is true if and only if the polar of each $w(E^*,E)$-bounded subset of E^* is a $\mathcal{T}$-neighborhood of 0, and this condition holds if and only if each $w(E^*,E)$-bounded set is $\mathcal{T}$-equicontinuous. Thus a necessary and sufficient condition that a locally convex space be a barrelled space is that each weak* bounded subset of the adjoint be equicontinuous.

The most important fact about barrelled spaces is that theorem 12.3, which is a form of the Banach-Steinhaus theorem, applies. Each locally convex space which is of the second category, and in particular each locally convex complete pseudo-metrizable space, is a barrelled space (see the discussion preceding 12.3). It will be shown that direct sums and products of barrelled spaces are barrelled spaces. Since each complete locally convex Hausdorff space is topologically isomorphic to a closed subspace of a product of complete normed spaces and since there are examples of complete locally convex spaces which are not barrelled (see problem 20A), a closed subspace of a barrelled space is not necessarily barrelled. However, the completion of a barrelled space and each quotient space of a barrelled space are of the same sort, as is proven below.

18.7 PROPERTIES OF BARRELLED SPACES

(i) *If F is a family of continuous linear functionals on a barrelled space E to a locally convex space H, and if F is bounded relative to the topology of pointwise convergence, then F is equicontinuous.*

(ii) *A locally convex space is a barrelled space if and only if each weak* bounded subset of the adjoint is equicontinuous.*

(iii) *The four following conditions on a subset B of the adjoint of a barrelled space are equivalent: B is equicontinuous, the weak* closure of B is weak* compact, B is strongly bounded, and B is weak* bounded.*

(iv) *If E is a barrelled space and F a subspace, then the quotient space E/F is a barrelled space.*

(v) *If F is a dense subspace of a linear topological space E, and if F is a barrelled space, then E is a barrelled space. In particular, a completion of a barrelled space is a barrelled space.*

PROOF   Part (i) is theorem 12.3, and parts (ii) and (iii) have already been established by 18.5 and the discussion preceding the theorem. To establish (iv), let E be a barrelled space, let F be a subspace, and let Q be the quotient map of E onto E/F. If A is a barrel in E/F, then the inverse of A under Q is a barrel in E, is hence a neighborhood of 0 in E, and its image in E/F is therefore a neighborhood of 0 because Q is open. Hence E/F is a barrelled space. Finally, suppose that F is a barrelled space which is a dense subspace of E. If A is a barrel in E, then $A \cap F$ is a barrel in F and is hence a neighborhood of 0 in F. But then the closure of $A \cap F$ is a neighborhood of 0 in E, because F is dense in E, and since A contains this closure, A is itself a neighborhood of 0 in E. Consequently, E is a barrelled space.|||

The remainder of this section is devoted to a discussion of the following problem: given a linear space E and a linear space F of linear functionals on E, for what locally convex topologies $\mathscr{T}$ is it true that $F = (E,\mathscr{T})^*$? In other words, given E^*, what are the possible topologies for E?

18.8   MAXIMAL TOPOLOGY FOR A SPACE WITH A GIVEN ADJOINT   *Let E and F be paired linear spaces, and let $\mathscr{T}$ be a locally convex topology for E. Then the adjoint $(E,\mathscr{T})^*$ is exactly the set of all linear functionals which are represented by members of F if and only if $w(E,F) \subset \mathscr{T} \subset \mathscr{T}_\mathscr{C}$, where $\mathscr{T}_\mathscr{C}$ is the topology of uniform convergence on members of the family $\mathscr{C}$ of all $w(F,E)$-compact convex circled subsets of F.*

PROOF   Assume that $(E,\mathscr{T})^*$ is represented by members of F or, equivalently, that $T[F] = (E,\mathscr{T})^*$, where T is the canonical map on F into E'. Let U be a $\mathscr{T}$-closed convex circled neighborhood of 0; then 17.1 implies that U is $w(E,T[F])$-closed, that is, $w(E,F)$-closed. But in view of theorem 17.4, the polar U° in F of U is a member of $\mathscr{C}$; hence $U = U^\circ{}_\circ$ is a $\mathscr{T}_\mathscr{C}$-neighborhood of 0 and $\mathscr{T} \subset \mathscr{T}_\mathscr{C}$. That

$w(E,F) \subset \mathcal{T}$ follows from the continuity of the identity map on $(E,\mathcal{T})$ onto $(E,w(E,F))$ (16.1(iv)).

Suppose that $w(E,F) \subset \mathcal{T} \subset \mathcal{T}_\mathscr{C}$; then each member of F represents a $w(E,F)$-continuous (hence $\mathcal{T}$-continuous) linear functional on E by theorem 16.2. On the other hand, if f is a $\mathcal{T}$-continuous linear functional on E, f is bounded on some neighborhood of 0. Since $\mathcal{T} \subset \mathcal{T}_\mathscr{C}$, one can assume that f is bounded on C_0 for some C in $\mathscr{C}$. The criterion 16.6 then implies that f is represented by a member of F.|||

If E and F are paired linear spaces, then $m(E,F)$, the **Mackey topology** for E (also called the **maximal** or **relatively strong topology**) has for a local base the family of all polars of $w(F,E)$-compact circled convex subsets of F. It is the topology $\mathcal{T}_\mathscr{C}$ of uniform convergence on $w(F,E)$-compact circled convex sets. A linear topological space $(E,\mathcal{T})$ is a **Mackey space** if and only if $\mathcal{T} = m(E,E^*)$. Thus a Mackey space is a locally convex space such that each weak* compact convex circled subset of E^* is equicontinuous. The foregoing theorem characterizes a Mackey space as a locally convex space E whose topology $\mathcal{T}$ is the strongest locally convex topology yielding $(E,\mathcal{T})^*$ as adjoint.

A normed space is always a Mackey space, and it will be shown (Section 22) that each locally convex pseudo-metrizable space is a Mackey space. It will presently be proved that the product (and the direct sum) of Mackey spaces is again a Mackey space, and since each locally convex space is topologically isomorphic to a subspace of a product of pseudo-normed spaces (theorem 6.4), it follows that a subspace of a Mackey space is generally not of the same sort. The completion, and the image under a continuous open linear mapping, of a Mackey space is again a Mackey space, as shown in the following.

18.9   THEOREM *Let $\mathcal{T}$ be a locally convex topology for a linear space E. Then:*

   (i) *if E is a Mackey space and F is a subspace, then E/F with the quotient topology is a Mackey space;*

   (ii) *if F is a dense subspace of E such that F with the relativized topology is a Mackey space, then E is a Mackey space—in particular, each completion of a Mackey space is a Mackey space;*

   (iii) *each barrelled space is a Mackey space.*

PROOF   (i) Let F be a subspace of a Mackey space E, and let Q be the quotient map of E onto E/F. By theorem 17.14, the polar F°

with the relativized topology $w(E^*,E)$ is topologically isomorphic to the space $(E/F)^*$ with the weak* topology. If A is a weak*-compact convex circled subset of F°, the polar of A in E/F is equal to the image $Q[A_0]$ of the polar A_0 of A in E. Because E is a Mackey space, A_0 is a neighborhood of 0 in E; hence $Q[A_0]$ is a neighborhood of 0 in E/F since Q is open. Therefore E/F is a Mackey space.

(ii) Suppose E is a locally convex space, F is a dense subspace which with the relativized topology is a Mackey space, and C is a $w(E^*,E)$-compact convex circled subset of E^*. It must be shown that C_0 is a neighborhood of 0 in E, and since F is a dense subspace, it is sufficient to show $C_0 \cap F$ is a neighborhood of 0 in F. But if R is the map of E^* into F^* which carries a linear functional f into its restriction to F, then R is continuous with respect to the weak* topologies, as may be seen directly. Hence $R[C]$ is $w(F^*,F)$-compact convex circled, and $R[C]_0 = C_0 \cap F$ is a neighborhood of 0 in F.

(iii) This is a direct consequence of 18.7.|||

The section is concluded with a description of the strong topologies for the adjoints of direct sums and products of linear topological spaces, and a verification that a direct sum or a product of barrelled spaces (or of Mackey spaces) is a space of the same sort. (It should be pointed out that the property of being a strong adjoint is not necessarily inherited by quotients or subspaces (see problems 22G and 20D).)

18.10 The Strong Topology for the Adjoint of a Direct Sum or a Product *Let E_t, for each member t of an index set A, be a locally convex Hausdorff space. Then $(\sum \{E_t : t \in A\})^*$, with the strong topology, is topologically isomorphic to $\times \{E_t^* : t \in A\}$, where each factor has the strong topology; dually, $(\times \{E_t : t \in A\})^*$, with the strong topology, is topologically isomorphic to $\sum \{E_t^* : t \in A\}$, where each summand has the strong topology.*

PROOF It will be convenient to treat each E_t as a subspace of $\sum \{E_t : t \in A\}$, thus omitting specific mention of the injection isomorphism of E_t into the direct sum. The adjoint of the direct sum is, by 14.7, canonically isomorphic to $\times \{E_t^* : t \in A\}$, and the proof of (i) depends on identifying the family $\mathscr{A}$ of polars in $\times \{E_t^* : t \in A\}$ of bounded subsets of $\sum \{E_t : t \in A\}$. According to 14.8, the family of convex extensions of finite unions of bounded subsets of the summands E_t is a co-base for the family of bounded subsets of a direct sum; such a finite union may be described as $\bigcup \{B_t : t \in A\}$, where each B_t is bounded and $B_t = \{0\}$ except for a finite number of members t of A.

The computation 16.10 of the polar of such a union then shows that each member of $\mathscr{A}$ contains a member of $\mathscr{A}$ which is of the form $\mathsf{X}\,\{B_t{}^\circ\colon t \in A\}$, where B_t is a bounded subset of E_t for each t and (since $B_t = \{0\}$ except for finitely many t) $B_t{}^\circ = E_t$ except for finitely many t. In brief, $\mathscr{A}$ is a local base for the product of the strong topologies, and (i) is established. The proof of (ii) is omitted; it proceeds along similar lines, via identification of the adjoint 14.6, description of bounded subsets 14.8, and the computation on polars, 16.10.|||

18.11   Direct Sums and Products of Barrelled Spaces and Mackey Spaces

  (i) *The direct sum of barrelled spaces is a barrelled space, and the direct sum of Mackey spaces is a Mackey space.*
  (ii) *The product of barrelled spaces is a barrelled space, and the product of Mackey spaces is a Mackey space.*

PROOF   It is again convenient to treat each space E_t as a subspace of $\sum\{E_t\colon t \in A\}$, ignoring the injection map of E_t into the sum. A barrel in $\sum\{E_t\colon t \in A\}$ intersects each subspace E_t in a barrel in E_t, and if each summand is a barrelled space, then it follows that $\sum\{E_t\colon t \in A\}$ is, with the direct sum topology, a barrelled space. To establish the fact that the direct sum of Mackey spaces is a Mackey space, we recall (theorem 17.14) that the adjoint of $\sum\{E_t\colon t \in A\}$, with the weak* topology, is topologically isomorphic to $\mathsf{X}\,\{E_t{}^*\colon t \in A\}$ with the product of the weak* topologies for the factors. Each weak* compact subset of the product is consequently contained in a product of weak* compact subsets of the factors (explicitly, a weak* compact subset C is contained in the product of the projections of C on the spaces $E_t{}^*$), and it follows that the family of polars in $\sum\{E_t\colon t \in A\}$ of such products is a base for the Mackey topology for the direct sum. Finally, the polar of a product $\mathsf{X}\,\{C_t\colon t \in A\}$, by 16.10, contains the convex extension of the union of the polars of C_t in E_t, and, if each E_t is a Mackey space, then this convex extension is a neighborhood of 0 relative to the direct sum topology. Consequently the direct sum of Mackey spaces is a Mackey space.

The adjoint of a product $P = \mathsf{X}\,\{E_t\colon t \in A\}$ is canonically isomorphic to the direct sum $S = \sum\{E_t{}^*\colon t \in A\}$ according to 14.6. As a preliminary to the proof of (ii), the weak* $(= w(S,P))$-bounded subsets of S will be identified. If $w_t{}^*$ denotes the weak* topology for $E_t{}^*$, then E_t is isomorphic to the adjoint of $(E_t{}^*,w_t{}^*)$ by 17.6, and if S

is given the direct sum of the topologies $w_t{}^*$, then P is canonically isomorphic to the adjoint of S by 14.7. The $w(S,P)$-bounded sets are therefore precisely the subsets of S which are bounded relative to the direct sum of the topologies $w_t{}^*$, by 17.5. Finally, the bounded subsets relative to a direct sum topology are identified in 14.8 and thus: the family of convex extensions of finite unions of weak* bounded subsets of the summands is a co-base for the family of all weak* bounded subsets of $\sum \{E_t{}^*: t \in A\}$.

The proof of (ii) is now a routine verification, using the facts noted above and the computation 16.10 on polars; the details are omitted.|||

PROBLEMS

A EXERCISES

(1) If (E,F) is a pairing and $\mathscr{T}$ an admissible topology for E, then every $w(E,F)$-compact convex subset of E is $\mathscr{T}$-bounded.

(2) On a normed space, the norm is weakly lower semicontinuous.

(3) The adjoint $(E, \mathscr{T})^*$ with the strong topology is normable if and only if E with $\mathscr{T}^b$ (see §19) is pseudo-normable.

(4) In the adjoint of a barrelled space, the weak* closed convex extension of a weak* compact set is weak* compact.

B CHARACTERIZATION OF BARRELLED SPACES

A locally convex space with topology $\mathscr{T}$ is barrelled if and only if the only locally convex topologies which have local bases consisting of $\mathscr{T}$-closed sets are those weaker than $\mathscr{T}$.

C EXTENSION OF THE BANACH-STEINHAUS THEOREM (12.2)

Let E be a linear topological space with topology $\mathscr{T}$, such that

(P): the only vector topologies which have local bases consisting of $\mathscr{T}$-closed sets are those weaker than $\mathscr{T}$ (that is, $\mathscr{T}$ satisfies the condition of the previous problem with the hypothesis of local convexity removed). Then any pointwise bounded set of continuous linear mappings on E to any linear topological space G is equicontinuous. (If F is the set of mappings and $\mathscr{V}$ a local base in G of closed sets, then the sets $\bigcap \{f^{-1}[V] : f \in F\}$, as V runs through $\mathscr{V}$, form a local base of $\mathscr{T}$-closed sets for a vector topology on E.)

When E is of the second category, it has the property (P). There are linear topological spaces possessing (P) but not of the second category (e.g., an $\aleph_0$-dimensional linear space with its strongest vector topology). A locally convex space with the property (P) is barrelled, but not every barrelled space possesses (P) (e.g., the space $l^{1/2}$ with the relativization of the norm topology of l^1).

D TOPOLOGIES ADMISSIBLE FOR THE SAME PAIRING

The locally convex topologies $\mathscr{T}$ and $\mathscr{U}$ are admissible relative to the same pairing if and only if there is a local base of each consisting of sets closed in the other.   (This, together with 7C, yields an alternative proof of 18.3.)

E   EXTENSION OF THE BANACH-ALAOGLU THEOREM (17.4)

If U is any neighborhood of 0 in a linear topological space E, then the polar of U in E^* is compact relative to the topology of uniform convergence on the totally ·bounded subsets of E.   This is the strongest admissible topology having this property.   (Use theorem 8.17 and 16A.)

F   COUNTER-EXAMPLE ON WEAK* COMPACT SETS

Let E be the set of sequences $x = \{\xi_n\}$ with at most a finite number of non-zero terms, with the norm $\|x\| = \sum |\xi_n|$.   Let $f_n(x) = \xi_n$.   Then the set consisting of the points $2^{2n} f_n$ together with 0 is weak* compact but is not strongly bounded.   Thus its weak* closed convex extension is not weak* compact.   In fact, if $g_n = \sum \{2^r f_r : 1 \leq r \leq n\}$, then each g_n lies in this extension, and $\{g_n\}$ is a weak* Cauchy sequence which is not weak* convergent.   The weak* topology coincides with the Mackey topology $m(E^*,E)$ (see 17I).   Thus E^* with the Mackey topology is not even sequentially complete.

G   KREIN-ŠMULIAN THEOREM

In 17.7 it is proved that a locally convex space E is complete if and only if every hyperplane in its adjoint is weak* closed whenever its intersection with every equicontinuous set C is weak* closed in C.   The Krein-Šmulian theorem asserts that the same is true with "convex set" in place of "hyperplane" if E is metrizable (see 22.6).   Hypercompleteness (see 13F) is equivalent to the same property, this time for convex circled sets: a locally convex space E is hypercomplete if and only if every convex circled subset in its adjoint is weak* closed whenever its intersection with every equicontinuous set C is weak* closed in C.   (Sketch of proof: Let $\mathscr{U}$ be a base of convex circled neighborhoods of 0.

(a) Suppose that E is hypercomplete and that A is a convex circled subset of E^* with $U^\circ \cap A$ weak* closed for each $U \in \mathscr{U}$.   The net $\{(U^\circ \cap A)_0 : U \in \mathscr{U}\}$ is Cauchy.   (For if $V \subset U$ and $W \subset U$,

$$(V^\circ \cap A)_0 + 2U \supset ((V^\circ \cap A)_0 + U)^- = ((V^\circ \cap A)_0 + U)^\circ_0 \supset ((V^\circ \cap A)_0 \cup U)^\circ_0 = ((V^\circ \cap A)_0^\circ \cap U^\circ)_0 = (V^\circ \cap A \cap U^\circ)_0 \supset (W^\circ \cap A)_0).$$

Hence the net, being decreasing, converges to $B = \bigcap \{(U^\circ \cap A)_0 : U \in \mathscr{U}\}$.   Then $A = B^\circ$.   For $A \subset B^\circ$ is immediate; if $U \in \mathscr{U}$ and $r > 1$, there is $V \in \mathscr{U}$ with $(V^\circ \cap A)_0 \subset B + (r-1)U \subset (B \cup U)^\circ_0 + (r-1)(B \cup U)^\circ_0 = r(B \cup U)^\circ_0$.   By taking polars, it follows that $U^\circ \cap B^\circ \subset rA$ and thus

$$U^\circ \cap B^\circ \subset \bigcap \{r(U^\circ \cap A) : r > 1\} \subset (U^\circ \cap A)^- = U^\circ \cap A \subset A.$$

(b) For the converse, suppose that $\{A_\gamma : \gamma \in \Gamma\}$ is a decreasing Cauchy net (cf. 13F).   Let $B = \bigcup \{A_\gamma^\circ : \gamma \in \Gamma\}$ and let $B^\sim$ be the algebraic closure $\bigcap \{rB : r > 1\}$ of B (cf. 14F).   Then $B^\sim$ is convex and circled, since the net decreases, and $U^\circ \cap B^\sim$ is weak* closed for each $U \in \mathscr{U}$.

For if $r > 1$ there is some $\alpha \in \Gamma$ with $A_\alpha \subset A_\gamma + (r - 1)U$ for all γ. Hence $A_\alpha \subset r(A_\gamma \cup U)^\circ{}_\circ$; it follows that $(U^\circ \cap B)^- \subset rA_\alpha{}^\circ$, and hence that $U^\circ \cap B^\sim = \bigcap \{r(U^\circ \cap B) : r > 1\} = (U^\circ \cap B)^-$, which is weak* closed.

By hypothesis $B^\sim$ is weak* closed. Thus $B^\sim = B_\circ{}^\circ$. Now there is some γ with $A_\gamma \subset A_\beta + U$ for all β. It is enough to show $(B_\circ + U)^\circ \subset A_\gamma{}^\circ$, for then $A_\gamma \subset (B_\circ + U)^\circ{}_\circ \subset B_\circ + 2U$, and $\{A_\gamma : \gamma \in \Gamma\}$ converges to $B_\circ = \bigcap \{A_\gamma : \gamma \in \Gamma\}$. If $f \in (B_\circ + U)^\circ$, let $m = \sup \{|f(x)| : x \in B_\circ\}$ and $n = \sup \{|f(x)| : x \in U\}$. Then $m + n \leq 1$; if $m > 0$, $f \in mB_\circ{}^\circ = mB^\sim$ and so for $r > 1, f \in mrA_\beta{}^\circ$ for some β. Thus on $A_\gamma, |f(x)| \leq mr + n$. If $m = 0$, then for $r > 1, f \in (r - 1)B^\sim$ and so $f \in r(r - 1)A_\beta{}^\circ$ for some β, so that on $A_\gamma, |f(x)| \leq r(r - 1) + 1$.

H EXAMPLE AND COUNTER-EXAMPLE ON HYPERCOMPLETE SPACES

In the adjoint of a hypercomplete space, there may be a convex subset which is not weak* closed but whose intersection with every weak* closed equicontinuous set is weak* closed. That is, the word "circled" is essential in the previous problem; thus hypercomplete spaces do not have as strong a property in this respect as do complete metrizable spaces.

Let E be the algebraic dual F' of a vector space F, with the topology $w(E,F)$, so that E is isomorphic to a product of copies of the scalar field. Then E is hypercomplete. (The equicontinuous subsets of $E^* = F$ are finite dimensional; use the previous problem and 14G.)

For the counter-example, take F to be the space $S(X,\mu)$, where $X = [0,1]$ and μ is Lebesgue measure, of measurable functions (see 6L). Then the positive cone satisfies the intersection condition, but cannot be closed in any locally convex topology. (If it were, there would exist a positive linear functional.)

I FULLY COMPLETE SPACES

A locally convex Hausdorff space is called *fully complete* if a subspace of its adjoint is weak* closed whenever its intersection with every equicontinuous set C is weak* closed in C. (Cf. 17.7, 18G, 22.6.)

A locally convex complete metrizable space is fully complete (by 22.6). The algebraic dual F' of any linear space F with the topology $w(F',F)$ is fully complete (14D). The adjoint F^* of a Fréchet (complete metrizable locally convex) space F is fully complete under any topology between the topology of uniform convergence on the totally bounded subsets of F and the Mackey topology $m(F^*,E)$. The image under an open continuous linear mapping of a fully complete space is fully complete (cf. 11.3); so also are closed subspaces and Hausdorff quotients of fully complete spaces. (These last statements, which can be proved directly from the definition by using the weak* topologies for adjoints of subspaces and quotients (17.13), can also be deduced from the following characteristization of fully complete spaces.)

A hypercomplete space is fully complete, by 18G. Let $\{A_\gamma : \gamma \in \Gamma\}$ be a net of convex circled closed non-void subsets of E. We call this a *scalar net* if for each $\gamma \in \Gamma$ and $e > 0$ there is some $\alpha \geq \gamma$ with $A_\alpha \subset e A_\gamma$. The limit of a scalar net which converges in the Hausdorff uniformity is a subspace of E. A proof analogous to that sketched in 18G shows that E is fully complete if and only if every decreasing Cauchy scalar net converges. If E is fully complete, the space of all closed subspaces is complete, with the Hausdorff uniformity; the converse is not true.

J CLOSED GRAPH THEOREM III (see 12E, 13G, 19B)

A linear mapping of a locally convex Hausdorff space E into a fully complete space F (see the previous problem) is continuous, provided that
  (i) the graph of T in $E \times F$ is closed;
  (ii) for each neighborhood V of 0 in F, the closure of $T^{-1}[V]$ is a neighborhood of 0 in E.
(This is a generalization of the theorem in 13G; the same proof holds, because the Cauchy net used there is a decreasing scalar net.)

Full completeness is, in a certain sense, also a necessary property of F if, in addition to the above closed graph theorem, the corresponding open mapping theorem (cf. 11.4) is also to hold. To make this precise, let us call T^{-1} *somewhere dense* iff condition (ii) is satisfied, and also call a linear mapping S of E into F *somewhere dense* iff the closure of $S[U]$ is a neighborhood of 0 in F for each neighborhood U of 0 in E. Then, if F is a locally convex Hausdorff space, the following statements are equivalent:
  (a) F is fully complete;
  (b)   any linear mapping of any locally convex Hausdorff space into a Hausdorff quotient of F is continuous, provided that its graph is closed and its inverse is somewhere dense;
  (c)   any continuous somewhere dense linear mapping of F onto any locally convex Hausdorff space is open.
(If F is fully complete, (b) follows from the theorem above and the fact that full completeness is inherited by Hausdorff quotients; this fact enables (c) to be deduced from (b) in much the same way as the open mapping theorem 11.4 is deduced from the closed graph theorem 11.1. Finally, suppose that (c) holds and that M is a subspace of F^* whose intersection with each equicontinuous set C is weak* closed in C. By applying (c) to the quotient mapping of F onto F/M°, with the topology of uniform convergence on the sets $U^\circ \cap M$ as U runs through a local base in F, it can be shown that this topology coincides with the quotient topology; it follows that their adjoints M and $M^\circ{}_\circ$ are equal.)

If (c) is replaced by
  (c′) Any continuous linear mapping of F onto any barrelled Hausdorff space is open
then, if also F is barrelled, (c′) implies (a). (The image by a continuous somewhere dense linear mapping of a barrelled space is barrelled; hence (c′) implies (c).)

K SPACES OF BILINEAR MAPPINGS

Let E, F, and G be locally convex Hausdorff spaces. Various spaces of bilinear mappings of $E \times F$ into G can be topologized and a theory developed similar to that of spaces of linear mappings. Let $\mathscr{A}$ and $\mathscr{B}$ be families of bounded subsets of E and F such that $E = \cup \mathscr{A}$ and $F = \cup \mathscr{B}$, and H a linear space of bilinear mappings of $E \times F$ into G.

(a) If for each $f \in H$, $f[A \times B]$ is bounded for all $A \in \mathscr{A}$ and $B \in \mathscr{B}$, then H may be given the topology $\mathscr{T}_{\mathscr{A},\mathscr{B}}$ of uniform convergence on the sets $A \times B$, as A and B run through $\mathscr{A}$ and $\mathscr{B}$, making it a locally convex linear topological space.

(b) If the sets of either $\mathscr{A}$ or $\mathscr{B}$ are convex, circled, and complete (in particular if either E or F is complete), the set of all separately continuous bilinear mappings of $E \times F$ into G can be topologized in this way (see 12G).

(c) If the condition in (b) holds, the space of all separately continuous bilinear functionals on $E \times F$ is $\mathscr{T}_{\mathscr{A},\mathscr{B}}$-complete if (and only if) the adjoint of E is $\mathscr{T}_{\mathscr{A}}$-complete, and the adjoint of F is $\mathscr{T}_{\mathscr{B}}$-complete. (If $\{f_\alpha\}$ is Cauchy, then $f_\alpha(x,y) \to f(x,y)$ for each $x \in E$ and $y \in F$, and clearly f is bilinear. If y is fixed, the linear functional $x \to f(x,y)$ is the $\mathscr{T}_{\mathscr{A}}$-limit of continuous linear functionals $x \to f_\alpha(x,y)$.)

(d) More generally, if also G is complete, then (c) holds with "functionals on $E \times F$" replaced by "mappings of $E \times F$ into G", provided that E and F are Mackey spaces. (Consider the convergence of $\phi(f_\alpha(x,y))$ for each continuous linear functional ϕ on G.)

L   Tensor products II

Let E and F be locally convex Hausdorff spaces with adjoints E^* and F^*. For each $x \in E$ and $y \in F$, the bilinear functional $x \otimes y$ on $E' \times F'$ defines by restriction a bilinear functional on $E^* \times F^*$, which is separately continuous when E^* and F^* have their weak* topologies. On the space of all separately continuous bilinear functionals on $E^* \times F^*$, the topology of uniform convergence on products of equicontinuous subsets of E^* and F^* is a vector topology (see (b) of the previous problem). The relative topology for $E \otimes F$ is the *topology of bi-equicontinuous convergence*.

(a) The space $E^* \otimes F^*$ is a subspace of the algebraic dual of $E \otimes F$ (see 16I(b)). The topology of bi-equicontinuous convergence is the topology of uniform convergence on the sets $U^\circ \otimes V^\circ$, as U and V run through local bases in E and F. With this topology, $E \otimes F$ is a Hausdorff space.

(b) The topology of bi-equicontinuous convergence is weaker than the projective tensor product topology.

The completion of $E \otimes F$ with the topology of biequicontinuous convergence will be denoted by $E \overset{\smile}{\otimes} F$.

(c) If E and F are complete, $E \overset{\smile}{\otimes} F$ is a subspace of the space of separately continuous bilinear functionals on $E^* \times F^*$ (see (c) of the previous problem).

19   BOUNDEDNESS

The family of all bounded sets characterizes the topology of a normed space, but in general the relationship between boundedness and the topology is less intimate. Beginning with the family $\mathscr{B}$ of bounded sets we may try to reconstruct the topology in either of two ways. We may make an external construction by considering the class E^b of bounded linear functionals and from this obtain the Mackey topology $m(E,E^b)$ of the pairing, or we may consider the topology with local base the class of all convex circled subsets of E which absorbs each member of $\mathscr{B}$. Both of these constructions yield the same topology $\mathscr{T}^b$, but $\mathscr{T}^b$ may be distinct from the original topology $\mathscr{T}$. Spaces for which $\mathscr{T}^b = \mathscr{T}$ are called bound (or bornivore). We investigate the relation between this notion and that of barrelled spaces, and discuss the permanence properties of bound spaces.

If E is a linear space with pseudo-normable topology $\mathscr{T}$, then $\mathscr{T}$ determines and is determined by either the adjoint E^* or by the family $\mathscr{B}$ of all $\mathscr{T}$-bounded subsets of E. In this case each of $\mathscr{T}$, $\mathscr{B}$, and E^* determines the other two, and this fact has important consequences. In the more general situation the relations between $\mathscr{T}$, $\mathscr{B}$, and E^* are less intimate, but those which do exist are still of considerable importance. The relation between the topology $\mathscr{T}$ and the adjoint E^* has been studied at some length in the preceding section, and the present one is devoted primarily to a discussion of the relations of these two with $\mathscr{B}$. Certain relations are clear. The adjoint space E^* is surely determined by $\mathscr{T}$, and, if $\mathscr{T}$ is locally convex, E^* determines the family $\mathscr{B}$, for the $\mathscr{T}$-bounded sets are precisely the weakly bounded sets. In attempting to reconstruct $\mathscr{T}$ from the family $\mathscr{B}$ of bounded sets, two approaches suggest themselves. Directly, the family of all convex circled sets which absorb each member of $\mathscr{B}$ may be used as a local base for a topology, or, indirectly, the natural pairing of E with the space E^b of all bounded linear functionals may be used to define a topology. It turns out that these two approaches yield the same topology $\mathscr{T}^b$ for E, a topology which in general is properly stronger than $\mathscr{T}$. This topology $\mathscr{T}^b$ is examined briefly in the present section. A few of the results of the section are valid for spaces which are not locally convex; however, for simplicity of statement, the development is confined to the locally convex case.

First, let us consider briefly the abstract process which leads to the notion of bounded set. Let $\mathscr{C}$ be the family of all non-void convex circled subsets of a linear space E. For each subfamily $\mathscr{A}$ of $\mathscr{C}$ let $\mathscr{A}^\sim$ be the family of all members of $\mathscr{C}$ which absorb each member of $\mathscr{A}$, and dually, let $\mathscr{A}_\sim$ be the family of all members of $\mathscr{C}$ which are absorbed by each member of $\mathscr{A}$. It is easy to establish the following proposition about $\mathscr{A}_\sim$ and $\mathscr{A}^\sim$ by straightforward computation.

19.1 COMPUTATION RULES FOR BOUNDEDNESS *Let $\mathscr{C}$ be the class of all convex circled subsets of a linear space E, let $_\sim$ and $^\sim$ be the operations* *defined above, and let $\mathscr{A}$ and $\mathscr{B}$ be subfamilies of $\mathscr{C}$. Then:*

  (i) *if $\mathscr{A} \supset \mathscr{B}$, then $\mathscr{A}_\sim \subset \mathscr{B}_\sim$ and $\mathscr{A}^\sim \subset \mathscr{B}^\sim$;*

  (ii) *$\mathscr{A} \subset \mathscr{A}_\sim{}^\sim$ and $\mathscr{A} \subset \mathscr{A}^\sim{}_\sim$;*

 (iii) *$\mathscr{A}_\sim{}^\sim{}_\sim = \mathscr{A}_\sim$ and $\mathscr{A}^\sim{}_\sim{}^\sim = \mathscr{A}^\sim$;*

 (iv) *if each member of $\mathscr{A}$ is radial at 0, then $\mathscr{A}_\sim$ covers E; if $\mathscr{A}$ covers E, then each member of $\mathscr{A}^\sim$ is radial at 0;*

  (v) *each member of $\mathscr{C}$ which is absorbed by a member of $\mathscr{A}_\sim$, each scalar multiple of a member of $\mathscr{A}_\sim$, and the convex extension of the union of any two members of $\mathscr{A}_\sim$, belongs to $\mathscr{A}_\sim$;*

 (vi) *each member of $\mathscr{C}$ which absorbs a member of $\mathscr{A}^\sim$, each non-zero scalar multiple of a member of $\mathscr{A}^\sim$, and the intersection of any two members of $\mathscr{A}^\sim$, belong to $\mathscr{A}^\sim$.*

It is evident that, if $\mathscr{A}$ is a local base for a topology $\mathscr{T}$, then $\mathscr{A}_\sim$ is the family of convex circled $\mathscr{T}$-bounded sets. If $\mathscr{A}$ is a family of subsets of E which covers E, then the foregoing rules show that $\mathscr{A}^\sim$ is a local base for a locally convex topology. This topology, which has a local base consisting of all convex circled sets which absorb each member of $\mathscr{A}$, is called the **$\mathscr{A}$-absorbing topology.** A familiar concept can be rephrased in terms of the $\mathscr{A}$-absorbing topology. Suppose that F is a family of linear functions, each on E to a locally convex space H, and suppose that A is a subset of E. Then, according to 12.1, the following conditions are equivalent: the family F is uniformly bounded on A (that is, $F[A]$ is bounded), F is bounded relative to the topology of uniform convergence on A, and, for each neighborhood V of 0 in H the set $\bigcap \{f^{-1}[V] : f \in F\}$ absorbs A. These equivalences yield the following result about a family of subsets of E (the details of the verification are omitted).

19.2   LEMMA ON $\mathscr{T}_{\mathscr{A}}$-BOUNDEDNESS AND ON EQUICONTINUITY *Let $\mathscr{A}$ be a family of convex circled sets which covers a linear space E and let F be a family of linear functions on E to a locally convex space H. Then the following are equivalent:*

  (i) *the family F is uniformly bounded on each member of $\mathscr{A}$;*

 (ii) *the family F is bounded relative to the topology $\mathscr{T}_{\mathscr{A}}$ of uniform convergence on members of $\mathscr{A}$; and*

 (iii) *the family F is equicontinuous relative to the $\mathscr{A}$-absorbing topology for E.*

In particular, a linear function f on E to H is bounded on each member of $\mathscr{A}$ if and only if f is continuous relative to the $\mathscr{A}$-absorbing topology.

The foregoing discussion will now be applied to the family $\mathscr{A}$ of all circled convex neighborhoods of 0 in a locally convex space $(E,\mathscr{T})$. The family $\mathscr{A}_{\sim}$, consisting of all convex circled sets which are absorbed by each member of $\mathscr{A}$, is then a co-base for the family of $\mathscr{T}$-bounded subsets of E. Hence the family $\mathscr{A}_{\sim}{}^{\sim}$ consists of all convex circled subsets of E which absorb every $\mathscr{T}$-bounded set. Members of $\mathscr{A}_{\sim}{}^{\sim}$ are called **bound absorbing (bornivore)**; explicitly, U is bound absorbing if and only if U is convex and circled and absorbs each $\mathscr{T}$-bounded subset of E. It follows from the computation rules 19.1 that the family of all bound absorbing sets is a local base for a topology for E; this topology is called the **bound extension of $\mathscr{T}$**, or the **bound topology derived from** $\mathscr{T}$, and is denoted by $\mathscr{T}^{b}$. If $\mathscr{T}$ and $\mathscr{T}^{b}$ are identical, then $\mathscr{T}$ is called a **bound topology** and E with the topology $\mathscr{T}$ is called a **bound space** (a **bornivore space**). In other words, $\mathscr{T}$ is a bound topology if and only if $\mathscr{T}$ is locally convex and each convex circled set which absorbs every $\mathscr{T}$-bounded set is a $\mathscr{T}$-neighborhood of 0. Several properties of the bound extension of a topology are evident from 19.1.

19.3   BOUND EXTENSION OF A TOPOLOGY   *Let $\mathscr{T}^{b}$ be the bound extension of a locally convex topology $\mathscr{T}$ for a linear space E.   Then:*

(i) *$\mathscr{T}^{b}$ is a bound topology, and the class $\mathscr{B}$ of $\mathscr{T}$-bounded sets is identical with the class of $\mathscr{T}^{b}$-bounded sets; moreover, $\mathscr{T}^{b}$ is the strongest locally convex topology relative to which each member of $\mathscr{B}$ is bounded;*

(ii) *a family F of linear functions on E to a locally convex space H is $\mathscr{T}^{b}$-equicontinuous if and only if F is bounded relative to the topology $\mathscr{T}_{\mathscr{B}}$ of uniform convergence on bounded sets.*

In particular, a linear function on E to H is $\mathscr{T}^{b}$-continuous if and only if it is bounded.

PROOF   Let $\mathscr{A}$ be the family of all convex circled $\mathscr{T}$-neighborhoods of 0.   Then $\mathscr{A}_{\sim}$ is a co-base for the $\mathscr{T}$-bounded sets, $\mathscr{A}_{\sim}{}^{\sim}$ is a local base for $\mathscr{T}^{b}$, and $\mathscr{A}_{\sim}{}^{\sim}{}_{\sim} = \mathscr{A}_{\sim}$ is a co-base for the $\mathscr{T}^{b}$-bounded sets (the notation, and the equality just cited, are from 19.1).   Clearly $\mathscr{T}^{b}$ is a bound topology, for the family of all convex circled sets absorbing each $\mathscr{T}^{b}$-bounded set is simply $\mathscr{A}_{\sim}{}^{\sim}$.   Moreover, from the definition, $\mathscr{T}^{b}$ is stronger than any locally convex topology relative to which each member of $\mathscr{A}_{\sim}$ is bounded.   This establishes (i) and (ii) is an immediate consequence of the preceding proposition, 19.2.|||

The bound extension $\mathscr{T}^b$ of a topology $\mathscr{T}$ for a linear space E is defined internally, in the sense that the definition does not involve the dual space E'.   Nevertheless, it is possible to give an external description of $\mathscr{T}^b$.

19.4   EXTERNAL DESCRIPTION OF THE BOUND EXTENSION OF A TOPOLOGY   *The bound extension $\mathscr{T}^b$ of a locally convex topology $\mathscr{T}$ is the Mackey topology $m(E,E^b)$, where E^b is the space of all bounded linear functionals on E.*

Consequently, a locally convex topology $\mathscr{T}$ for E is a bound topology if and only if each bounded linear functional is continuous and E, with the topology $\mathscr{T}$, is a Mackey space.

PROOF   Each $w(E^b,E)$-compact circled convex subset of E^b is strongly bounded, by 18.5, and is hence $\mathscr{T}^b$-equicontinuous by 19.3(ii). Consequently each $m(E,E^b)$-neighborhood of 0 is a $\mathscr{T}^b$-neighborhood of 0, and hence $\mathscr{T}^b$ is stronger than $m(E,E^b)$.   But E^b is the space of all $\mathscr{T}^b$-continuous linear functionals, and by 18.8, $m(E,E^b)$ is the strongest locally convex topology which yields E^b as adjoint.   Hence $\mathscr{T}^b = m(E,E^b)$.|||

The relation between the notion of barrelled space and that of a bound space is of some interest.   Recall that a barrelled space (a locally convex space such that each barrel is a neighborhood of 0) can be characterized as a locally convex space such that each weak* bounded subset of the adjoint is equicontinuous.   A space with a bound topology has, in view of 19.3, the property that each strongly bounded subset of the adjoint is equicontinuous.   It is then clear that if a space has a bound topology, and if weak* bounded sets are strongly bounded, then the space is a barrelled space.   In particular, this is the case if the space is sequentially complete (see 18.5), and consequently a sequentially complete bound space is a barrelled space. These facts are noted for reference.

19.5   THEOREM   *Let E be a linear space with a bound topology. Then:*

   (i) *each strongly bounded subset of E^*, and hence each weak* compact convex subset, is equicontinuous;*
   (ii) *if E is sequentially complete, then E is a barrelled space; and*
   (iii) *the space E^* with the strong topology is complete.*

PROOF   Because of the remarks preceding the theorem, only (iii) requires a proof.   However, this is immediate in view of Grothendieck's completeness theorem (16.9) and the fact that each linear functional on E which is weakly continuous on bounded subsets of E is bounded (hence a member of E^*).|||

It is not true that a subspace of a bound space is necessarily a bound space (see problem 20D), nor is it true that the strong topology for the adjoint of a bound space is necessarily bound (see problem 22G). However, there is no difficulty seeing that direct sums and quotients of bound spaces are bound.

19.6 DIRECT SUMS AND QUOTIENTS OF BOUND SPACES *If F is a subspace of a bound space E, then the quotient topology for E/F is bound.*

The direct sum of Hausdorff spaces, each of which has a bound topology, is a bound space.

PROOF If f is a bounded linear functional on E/F and Q is the quotient map of E onto E/F, then $f \circ Q$ is a bounded linear function on E. If the topology of E is bound, then $f \circ Q$ is continuous, and f is therefore continuous on E/F. Finally, if the topology of E is bound, then E is a Mackey space, by 19.4, and therefore E/F is a Mackey space, by 18.9. But this fact, together with the fact that each bounded linear functional is continuous, implies that the quotient topology for E/F is bound. The fact that the direct sum of bound spaces is a bound space follows directly from the definition of a bound space, the definition of the direct sum topology and the description 14.8 of the bounded subsets of a direct sum.|||

There is a simple proposition which exhibits a bound space as the quotient of a direct sum of pseudo-normed spaces. If A is a convex, circled subset of a linear space E, then E_A is the subspace generated by A (or the linear extension of A). The Minkowski functional p_A of A is a pseudo-norm for E_A, and E_A with this pseudo-norm is called the **pseudo-normed space generated by** A. If $\mathscr{A}$ is a family of convex circled subsets of E, then the natural map of the direct sum $\sum \{E_A : A \in \mathscr{A}\}$ into E is the mapping which sends each member of the direct sum into the sum of its coordinates.

19.7 DIRECT SUM CHARACTERIZATION OF BOUND SPACES *Let $\mathscr{T}$ be a bound topology for a linear space E, let $\mathscr{A}$ be a co-base for the family of convex circled bounded subsets of E, and for each A in $\mathscr{A}$ let E_A be the pseudo-normed space generated by A. Then $\mathscr{T}$ is the quotient topology derived from the natural map T of the direct sum $\sum \{E_A : A \in \mathscr{A}\}$ onto E.*

PROOF It will be convenient to treat each E_A as a subspace of the direct sum $\sum \{E_A : A \in \mathscr{A}\}$. If U is a convex circled $\mathscr{T}$-neighborhood of 0 in E, then for each member A of $\mathscr{A}$ there is a non-zero scalar a such that $aA \subset U$, because $\mathscr{A}$ consists of bounded sets. Hence $T^{-1}[U]$ contains a non-zero scalar multiple of A in E_A for every A in $\mathscr{A}$, and therefore $T^{-1}[U]$ is a neighborhood of 0 relative to the direct sum topology. Hence $\mathscr{T}$ is weaker than the quotient topology. On the other hand, if V is a neighborhood of 0 relative to the quotient topology, then $T^{-1}[V] \cap E_A$ is a neighborhood of 0 in the subspace E_A of the direct sum; hence $T^{-1}[V]$ contains a non-zero scalar multiple of A, and it follows that V contains a non-zero scalar multiple of A. Thus V absorbs bounded sets, for $\mathscr{A}$ was supposed to be a dual base for the family of bounded sets, and consequently V is a $\mathscr{T}$-neighborhood of 0. The quotient topology is therefore weaker than $\mathscr{T}$, and the two topologies coincide.|||

In conclusion, it will be shown that whether or not the product $\mathsf{X}\,\{E_t \colon t \in A\}$ of bound spaces is a bound space depends solely on the cardinal number of A. To be explicit, let us agree that m is a **simple measure** on A if m is a countably additive function, defined on the class of all subsets of A, which assumes only the values zero and one. An **Ulam measure** is a simple measure m which is not identically zero and has the property that $m(\{t\}) = 0$ for each member t of A. It is not known whether Ulam measures exist; Ulam has shown that if $\aleph'$ is the smallest cardinal of a set with an Ulam measure (provided there is such a set), then $\aleph'$ is strongly inaccessible and $2^{\aleph} < \aleph'$ whenever $\aleph < \aleph'$. We shall prove that the product of bound spaces is a bound space if and only if there is no Ulam measure on the index set.

The following lemma, which furnishes the most important part of the proof of the result just cited, uses the following notation. If f is a function on $\mathsf{X}\,\{E_t \colon t \in A\}$ and $B \subset A$, then f_*B is defined to be the function on the product such that $f_*B(x) = f(K_B \cdot x)$, where $K_B \cdot x$ is the product of x and the characteristic function K_B of B.

19.8 Lemma   *Each bounded linear functional on a cartesian product* $\mathsf{X}\,\{E_t \colon t \in A\}$ *is the sum of a finite number of bounded linear functionals* f *which have the property: there is a simple measure m on A such that* $f_*B = m(B)f$ *for every subset B of A.*

proof   Let f be a bounded linear functional on $\mathsf{X}\,\{E_t \colon t \in A\}$, and let $\mathscr{A}$ be the class of all subsets B of A such that f_*B is not identically zero. Then each disjoint subfamily of $\mathscr{A}$ is finite, as the following argument demonstrates. If $\{B_n\}$ is a disjoint sequence in $\mathscr{A}$ such that $f_*B_n \neq 0$ for every n, then there is a sequence $\{x_n\}$ in the product such that $x_n(t) = 0$ if $t \notin B_n$ and $f(x_n) \neq 0$. Then the sequence $\{nx_n/f(x_n)\}$ is a bounded subset of the product, since its projection on each coordinate space is bounded (see 14.8), but f is clearly not bounded on this sequence.

It follows from the fact that each disjoint subfamily of $\mathscr{A}$ is finite that there is a disjoint finite cover $B_1, B_2, \cdots, B_n$ of A such that for each i and for every subset C of B_i, either $f_*C = f_*B_i$ or $f_*C = 0$. Letting $f_i = f_*B_i$, it is clear that f is the sum of the functionals f_i and that $(f_i)_*C$ is either f_i or 0 for every subset C of A. The proof of the lemma then reduces to showing that if g is a bounded linear functional such that g_*C is either g or 0 for each subset C of A, and if we set $m(C) = 1$ in the first case and $m(C) = 0$ in the second, then m is countably additive. Since m is evidently finitely additive, this amounts to proving that if C_n is an increasing sequence of subsets of A such that $g_*C_n = 0$, and if C is the union of the sets C_n, then $g_*C = 0$. But if $g_*C(x) \neq 0$, then, letting $y_n(t)$ be $nx(t)$ for t in $C \sim C_n$ and zero otherwise, we have $g(y_n) = g_*C_n(y_n) + g_*(C \sim C_n)(y_n) = g_*(C \sim C_n)(y_n) = ng_*(C \sim C_n)(x) = ng_*C(x)$. This is a contradiction because the sequence $\{y_n\}$ is bounded (its projection on each coordinate space is a finite set), and the functional g was supposed to be bounded.$\|\|\|$

19.9 PRODUCTS OF BOUND SPACES *The product $\times \{E_t : t \in A\}$ of bound spaces E_t is bound if and only if there is no Ulam measure on A.*

PROOF If there is no Ulam measure on A, then each simple measure m on A consists of unit mass situated at some point t of A—that is, for some t and for all subsets B of A it is true that $m(B)$ is one or zero depending on whether $t \in B$ or $t \notin B$. The preceding lemma then implies that each bounded functional is a finite sum of functionals of the form $f(x) = g(x_t)$, where g is a linear functional on E_t. It is clear that g must be bounded and hence continuous on E_t, and consequently f is continuous on the product. Hence each bounded functional on the product is continuous.

Conversely, suppose that m is an Ulam measure on A. If f is a scalar valued function on A and A_n is the set of t in A such that $|f(t)| > n$, then $\lim_n m(A_n) = 0$, and, since the measure m assumes only the values one and zero, there is a positive integer n such that $m(A_n) = 0$. Thus each scalar valued function f is essentially bounded on A, and, since each such function is measurable, it follows that every scalar valued function is integrable (in fact, each such function is constant m-almost everywhere, but we shall not need this fact). For each t in A choose f_t in E_t^* such that f_t is not identically zero, and for x in $\bigtimes \{E_t: t \in A\}$ let $g(x) = \int_A f_t(x_t)dm(t)$. Then g is evidently linear, and it may be seen that g is bounded by means of the following argument. Each bounded subset of the product is contained in a bounded set of the form $B = \bigtimes \{B_t: t \in A\}$, where each B_t is a bounded subset of E_t. Then $\sup \{|g(x)|: x \in B\} \leq \int_A \sup \{|f_t(x_t)|: x_t \in B_t\}dm(t)$, and this integral of a scalar valued function exists. Hence g is bounded on bounded subsets of the product and it remains to be proved that g is not continuous. But if g were continuous there would be a finite subset C of A such that $g(x) = 0$ whenever $x_t = 0$ for all t in C, by 14.6. But if x is chosen so that $x_t = 0$ for t in C and $f_t(x_t) = 1$ for t in $A \sim C$, then $g(x) = \int_A f_t(x_t)dm(t) = m(A \sim C) = 1$, and it follows that g is not continuous.|||

PROBLEMS

A   INDUCTIVE LIMITS IV (see 1I, 16C, 17G, 22C)

(a) Let E be a linear space with an inductive topology, determined by a family $\{E_t, f_t: t \in A\}$, each E_t being a locally convex space. Then if each E_t is barrelled, so is E; if each E_t is a bound space, so is E; if each E_t is a Mackey space, so is E.

(b) The topology of a locally convex space E is bound if and only if it is the inductive topology determined by a family of pseudo-normed spaces; if also E is a Hausdorff space, "pseudo-normed" may be replaced by "normed"; if in addition E is sequentially complete, "normed" may be replaced by "Banach". (Compare 19.7, and use the same device.)

B   CLOSED GRAPH THEOREM IV (see 12E, 13G, 18J)

Let E be a linear space with the inductive topology determined by a family of locally convex spaces of the second category. Let F be a Hausdorff inductive limit of a sequence of fully complete spaces (see 17R, 18I).

(a) A linear mapping of E into F is continuous if its graph is closed. (Sketch of proof: by 16C(i) it is enough to prove the theorem when E is a locally convex space of the second category. Suppose that F is the inductive limit of the sequence $\{F_n\}$. Then $E = \bigcup \{T^{-1}[F_n]: n = 1, 2, \cdots\}$ and so there is some n with $H = T^{-1}[F_n]$ of the second category in E, and then $H^- = E$. The graph of the restriction of T to H is closed in $H \times F_n$; it follows from 18J that T is continuous on H. There is a continuous extension S of T mapping E into F_n; the graph of S in $E \times F_n$ is the closure of its graph in $H \times F_n$. It follows that the graph of S is contained in the graph of T, because the graph of T is closed. Thus $S = T$ and so T is continuous.) In particular, the theorem applies when E has an inductive topology determined by Banach spaces, and so whenever E is a sequentially complete bound space (see the previous problem).

(b) Assume also that E is a Hausdorff space. Then a linear mapping of F onto E is open if its graph is closed. (If the graph of T is closed, so is

$T^{-1}(0)$; $F/T^{-1}(0)$ is then an inductive limit of a sequence of fully complete spaces, and, if $T = S \circ Q$, (a) applies to S^{-1}.)

C   COMPLETENESS OF THE ADJOINT

The adjoint of a bound space is complete under its strong topology (19.5 (iii)). This result may be sharpened considerably. Let $\mathscr{A}$ be an admissible family of subsets of a bound space E such that every sequence in E convergent to 0 is contained in some set of $\mathscr{A}$. Then E^* is complete under the topology of uniform convergence on the sets of $\mathscr{A}$. (It is sufficient, by 16.9 and 19.4, to show that every linear functional continuous on each $A \in \mathscr{A}$ is bounded.)

Let E be a bound space which is complete (or has the property that the closed convex extension of every compact set is compact). Then E^* is complete relative to any admissible topology between the strong topology and the topology of uniform convergence on convex circled compact sets. In particular, E^* with the Mackey topology $m(E^*,E)$ is then complete. The example in 18F shows that it is not sufficient to assume only that E is bound.

20   THE EVALUATION MAP INTO THE SECOND ADJOINT

If E is a normed linear space, then the natural mapping, evaluation, of E into its second adjoint E^{**} is an isometry. In case E maps onto E^* (we say E is reflexive) the situation of E and E^* is entirely symmetric, and linear space methods are particularly effective.

In general, the situation is more intricate. Evaluation is always relatively open, but may fail to be continuous (if evaluation is continuous we say E is evaluable), and even in case evaluation is discontinuous E may map onto E^{**} (we say E is semi-reflexive). This section is concerned with the exploration of this situation and its relation to the concepts introduced earlier.

The second adjoint E^{**} of a linear topological space E is the space of all strongly continuous linear functionals on E^*. The evaluation mapping of the space E into E^{**} is defined by letting $I(x)$, for x in E, be the linear functional on E^* whose value at a member f of E^* is $f(x)$; that is, $I(x)(f) = f(x)$. The principal concern of this section is to describe conditions which imply that I is a topological isomorphism of E onto E^{**}, where E^{**} has the strong topology (the topology of uniform convergence on strongly bounded subsets of E^*). The discussion falls naturally into two parts: we first consider conditions which ensure that the evaluation maps E onto E^{**}, and then discuss the continuity of the evaluation mapping.

A linear topological space $(E,\mathscr{T})$ is **semi-reflexive** if and only if $I[E] = E^{**}$. That is, E is semi-reflexive if and only if each strongly continuous linear functional on E^* is the evaluation at some point of E. The requirement that a locally convex space E be semi-reflexive can be restated conveniently in terms of the natural pairing of E and E^*. The family of polars in E^* of bounded subsets of E is a local base for the strong topology for E^*, and hence E is semi-reflexive if and only if for each closed convex circled bounded subset B of E it is true that each functional on E^* which is bounded in absolute value by one on B^o is represented by some member of E. In view of Smulian's compactness criterion 16.6, this is the case if and only if each such set B is weakly compact. The following proposition follows without difficulty (it may also be obtained from theorem 18.8).

20.1   CRITERION FOR SEMI-REFLEXIVENESS   *A locally convex space E is semi-reflexive if and only if each bounded weakly closed set is weakly compact, and this is the case if and only if $m(E^*,E) = s(E^*,E)$.*

It should be observed that the criterion above can be rephrased in terms of the weak topology only, and consequently the notion of semi-reflexiveness of a locally convex space $(E,\mathcal{T})$ depends only on the pairing of E and E^*. It follows from 18.8 that if $(E,\mathcal{T})$ is semi-reflexive, then E with any topology which is between the weak and the Mackey topology $m(E,E^*)$ is also semi-reflexive. In view of this fact it is not surprising that a semi-reflexive space need not be complete, for a space is seldom complete relative to the weak topology; however, we will see that each closed bounded subset of a semi-reflexive space is complete. Quotient spaces and adjoints of semi-reflexive spaces are not, in general, of the same sort (see problems 22G and 20D); however, products and direct sums of semi-reflexive spaces are semi-reflexive.

20.2   PROPERTIES OF SEMI-REFLEXIVE SPACES

(i)   *Each closed bounded subset of a semi-reflexive space is complete.*

(ii) *If F is a closed subspace of a locally convex semi-reflexive space E, then F is semi-reflexive, and F^*, with the strong topology, is topologically isomorphic to E^*/F^o with the quotient topology derived from the strong topology.*

(iii) *Direct sums of Hausdorff semi-reflexive spaces and products of semi-reflexive spaces are semi-reflexive.*

PROOF   The first statement of the theorem is a direct corollary of theorem 17.8 and the criterion 20.1 for semi-reflexivity. If F is a closed subspace of a locally convex semi-reflexive space E, then F is weakly closed by 17.1, and the weak topology for F is the relativized weak topology for E by 17.13. The fact that F is semi-reflexive then follows from the criterion 20.1. If B is a bounded subset of F, B^o is the polar of B in F^*, and $f \in E^*$, then the restriction of f to F belongs to B° if and only if f belongs to the polar C of B in E. That is, the inverse of B° under the canonical map T of E^* onto F^* is the polar of B in E^*, and T is therefore continuous relative to the strong topologies. The proof of (ii) now reduces to showing that T is open, for continuity and openness of T implies that the induced map of E^*/F° onto F^* is topological. Let B be a bounded closed convex circled subset of E; then the polar C of $B \cap F$ in F^* is equal to $T[(B \cap F)^\circ]$, since any continuous functional on F can be extended on E (14.1). By the computation rule 16.3, $(B \cap F)^\circ = (B^\circ \cup F^\circ)_\circ{}^\circ \subset (B^\circ + F^\circ)_\circ{}^\circ$. The set $(B^\circ + F^\circ)_\circ{}^\circ$ is the $m(E^*,E)$-closure of $B^\circ + F^\circ$ which is also the strong closure because of the semi-reflexivity of E (20.1). Hence $(B^\circ + F^\circ)_\circ{}^\circ \subset B^\circ + F^\circ + B^\circ \subset F^\circ + 2B^\circ$, and it follows that $C \subset T[F^\circ + 2B^\circ] = 2T[B^\circ]$, which implies that T is open.

Part (iii) of the theorem is a direct consequence of the criterion 20.1 for semi-reflexivity, the description 14.8 of bounded subsets of a direct sum and product, the identification 17.13 of the weak topology of a product, and the Tychonoff theorem 4.1. We omit the details of the verification.|||

The second half of part (ii) of the previous theorem is also a direct consequence of theorem 21.4 of the next section and of the fact that $m(E^*,E)$ is the same as the strong topology on E^* if E is semi-reflexive.

The traditional term for a Banach space E for which $I[E] = E^{**}$ is "reflexive", not "semi-reflexive". However, Banach spaces enjoy a property not common to locally convex spaces in general; namely the evaluation map is a topological isomorphism. A linear topological space E is called **reflexive** if and only if I is a topological isomorphism of E onto E^{**} In general, I is relatively open (that is, $I[U]$ is open in $I[E]$ whenever U is open in E), but not continuous. The following proposition gives necessary and sufficient conditions that I be continuous.

20.3 Continuity and Openness of Evaluation *If E is a locally convex space, then the evaluation map I of E into E^{**} is relatively open; the mapping I is continuous if and only if each strongly bounded subset of E^* is equicontinuous, and this is the case if and only if each bound absorbing barrel in E is a neighborhood of 0.*

proof Let B be an arbitrary subset of E^*, let B_0 be its polar in E and let B° be its polar in E^{**}. It follows directly from the definition of polar and of I that $I[B_0] = I[E] \cap B^\circ$. In particular, if B is itself the polar of a closed convex circled neighborhood U of 0 in E, then B is equicontinuous and hence strongly bounded, B° is a neighborhood of 0 in E^{**}, and $I[U] = I[B_0] = I[E] \cap B^\circ$. Therefore I is relatively open.   It is also evident from the definition of polar that $I^{-1}[B^\circ] = B_0$, and letting B be an arbitrary strongly bounded subset of E^*, it follows that I is continuous if and only if the polar of each strongly bounded subset of E^* is a neighborhood of 0 in E.   Rephrased, I is continuous if and only if each strongly bounded set is equicontinuous.   The last assertion of the theorem follows from the fact that a subset of E is the polar of a strongly bounded set if and only if A is a bound absorbing barrel.   For if A is a bound absorbing barrel, then A absorbs each bounded set, therefore A° is absorbed by each polar of a bounded set, hence A° is strongly bounded and $A = A^\circ{}_0$ (see 16.3).   Conversely, if A is the polar of a strongly bounded set B, then B is absorbed by the polar of each bounded subset C of E, and hence $B_0 = A$ absorbs $C^\circ{}_0$.   It is then clear that A absorbs each bounded set, and A is a barrel because it is the polar of a weak* bounded set (see 16.4).|||

A linear topological space E is **evaluable (symmetric, infratonnelé, quasi-tonnelé, quasi-barrelled)** if and only if E is locally convex and the evaluation map I is continuous.   In view of the foregoing theorem this is the case if and only if each bound absorbing barrel in E is a neighborhood of 0, or equivalently, if each strongly bounded subset of E^* is equicontinuous.   It is evident from the preceding that a bound space, or a barrelled space, is necessarily evaluable. The precise strength of the evaluability requirement can be visualized as follows: Let $\mathscr{E}$, $\mathscr{C}$, $\mathscr{S}$, and $\mathscr{W}$ be the classes of all convex circled subsets of E^* which are, respectively, equicontinuous, with weak* compact closure, strongly bounded, weak* bounded.   Then, for an arbitrary locally convex space E it is true that $\mathscr{E} \subset \mathscr{C} \subset \mathscr{S} \subset \mathscr{W}$. In these terms: E is a barrelled space if and only if $\mathscr{E} = \mathscr{W}$; if E is a bound space, then $\mathscr{E} = \mathscr{S}$; E is evaluable if and only if $\mathscr{E} = \mathscr{S}$; and E is a Mackey space if and only if $\mathscr{E} = \mathscr{C}$.   (It may be noticed that semi-reflexivity can be characterized in terms of the analogous families of subsets of E if E and E^* are interchanged; E is semi-reflexive if and only if $\mathscr{C}(E) = \mathscr{W}(E)$.)

The following theorem summarizes properties of evaluable spaces. The converses of the assertions of part (ii) are all false (see problems 20B and 20A).

20.4   PROPERTIES OF EVALUABLE SPACES

(i) *A locally convex space E is evaluable if and only if the evaluation* *map of E into E** is continuous, and this is the case if and only if each strongly bounded subset of E* is equicontinuous, or, equivalently, if and only if each bound absorbing barrel in E is a neighborhood of 0.*

(ii) *Each barrelled space and each bound space is evaluable, and each evaluable space is a Mackey space.*

(iii) *Each quotient space of an evaluable space is evaluable.*

(iv) *Each sequential completion of an evaluable space is a barrelled space.*

(v) *Products and direct sums of evaluable spaces are evaluable.*

PROOF   Propositions (i) and (ii) have already been noted in the discussion preceding the theorem.   To prove (iii), notice that if Q is the quotient map of an evaluable space E onto E/F and D is a bound absorbing barrel in E/F, then $Q^{-1}[D]$ is a bound absorbing barrel in E and is hence a neighborhood of 0.   But Q is open, and consequently D is a neighborhood of 0 in E/F.

To prove (iv), suppose that an evaluable space F is a dense subspace of a sequentially complete space E, and that D is a barrel in E. In view of theorem 10.3, D absorbs each closed convex circled bounded subset of E, that is, D is bound absorbing.   Therefore $D \cap F$ is a bound absorbing barrel in F.   Since F is evaluable, $D \cap F$ is a neighborhood of 0 in F; hence D is a neighborhood of 0 in E.   Therefore E is a barrelled space.

If D is a bound absorbing barrel in a direct sum $\sum \{E_t : t \in A\}$ of evaluable spaces, then (considering each E_t as a subspace of the sum) the intersection $D \cap E_t$ is a bound absorbing barrel in E_t and is consequently a neighborhood of 0 in E_t.   It follows that D is a neighborhood of 0 in the direct sum, and the direct sum of evaluable spaces is therefore evaluable.

The simplest proof that a product $\times \{E_t : t \in A\}$ of evaluable spaces is a space of the same sort relies on proposition 18.10, which states that the adjoint of a direct sum (a product) is, with the strong topology, topologically isomorphic to the product (the direct sum, respectively) of the adjoints, where each factor is given the strong topology.   From this fact it follows that (to a topological isomorphism) the evaluation map I of $\times \{E_t : t \in A\}$ can be described by $I(x)_t = I_t(x_t)$, where I_t is the evaluation map of E_t into E_t^{**}.   If each I_t is continuous, then I is also continuous, which shows that the product is evaluable.|||

There is a useful corollary to part (i) of the preceding theorem. Suppose E is a linear topological space and E^*, with the strong topology, is semi-reflexive.   Then each strongly bounded closed convex circled subset of E^* is weakly $(w(E^*,E^{**}))$ compact, and hence weak* (that is, $w(E^*,E)$) compact. Such a subset is consequently equicontinuous relative to the Mackey topology $m(E,E^*)$.

20.5   COROLLARY   *If E is a linear topological space and E^*, with the strong topology, is semi-reflexive, then E, with the Mackey topology, is evaluable.*

On the other hand, it may happen that E^* is evaluable and yet E, with the Mackey topology, fails to be evaluable (see problem 20A). That the adjoint of an evaluable space need not be evaluable is demonstrated by an example in problem 22G.

A linear topological space E is reflexive if and only if the evaluation map of E into E^{**} is a topological isomorphism onto. We have already derived conditions under which evaluation is continuous, and conditions which ensure that evaluation carries E onto E^{**}. These facts (propositions 20.1, 20.4, and 20.5), and the observation that evaluation map of a locally convex space E is one-to-one if and only if E is Hausdorff, yield the following theorem.

20.6   CHARACTERIZATIONS OF REFLEXIVE SPACES   *A locally convex Hausdorff space is reflexive if and only if it is semi-reflexive (bounded weakly closed sets are weakly compact) and evaluable (strongly bounded subsets of the adjoint are equicontinuous).*

Equivalently, a locally convex Hausdorff space E is reflexive if and only if it is a Mackey space and both E and E^* are semi-reflexive.

20.7   PROPERTIES OF REFLEXIVE SPACES

(i) *If E is reflexive so is E^*.*
(ii) *Each reflexive space is a barrelled space, and each closed bounded subset of a reflexive space is complete.*
(iii) *Products and direct sums of reflexive spaces are reflexive.*

PROOF   Part (i) is a direct consequence of the definition of a reflexive space. A reflexive space is evaluable, and hence by 20.4 each bound absorbing barrel is a neighborhood of 0. But each closed convex circled bounded set is weakly compact (E is semi-reflexive) and hence, by corollary 10.2 of the absorption theorem, every barrel absorbs such a set. Consequently each barrel is a neighborhood of 0, and it follows that each reflexive space is a barrelled space. It has already been observed in 20.2 that a closed bounded subset of a semi-reflexive space is complete, and part (ii) of the theorem is thereby established. Part (iii) is an immediate consequence of 20.2 and 20.4.|||

Neither a closed subspace nor a quotient space of a reflexive space is necessarily reflexive. In fact, of the two parts which make up reflexivity—being evaluable and being semi-reflexive—one may fail for subspaces and the other for quotients (see problem 20D). Finally, the converse of 20.7(i) does not hold in general: it may be that the adjoint E^* is reflexive but that E is not reflexive, or even semi-reflexive (with its Mackey topology, the only one for which it is sensible to ask the question; it is necessarily evaluable by 20.5). An example, and a restricted form of converse, are displayed in problem 20C.

PROBLEMS

A   EXAMPLE OF A NON-EVALUABLE SPACE

Let F be a non-reflexive Banach space and let $E = F^*$ with the topology $m(F^*,F)$. Then $E^* = F$ and the strong topology on E^* is the norm topology (for the $w(F^*,F)$-bounded sets are norm bounded, by 18.5(iii)). Thus

(a) E is a Mackey space which is not evaluable, and so neither barrelled nor bound;

(b) in E^*, there is a convex weak* closed strongly bounded set which is not weak* compact (cf. 18.5(ii));

(c) E^* is evaluable;

(d) E is semi-reflexive and E^* is not semi-reflexive;

(e) E is complete (see 19C).

B   AN EVALUABLE PRODUCT

Let E be a barrelled space which is not bound (see †) and F a bound space which is not barrelled (see 18F). Then $E \times F$ is evaluable, but is neither barrelled nor bound.

C   CONVERSE OF 20.7(i)

The result 20.7(i) states that if E is reflexive, so is its adjoint E^* under the strong topology. The converse fails: let E be a dense proper subspace of an infinite dimensional reflexive Banach space with the relativized norm topology. Then E^* is reflexive but E is not semi-reflexive.

The lack of completeness is the vital point here. For suppose that E is a Hausdorff complete Mackey space. If E^*, with the strong topology, is semi-reflexive, then E is reflexive. (Use the fact that E is a closed subspace of E^{**}.)

D   COUNTER-EXAMPLE ON QUOTIENTS AND SUBSPACES

Let $(E,\mathscr{T})$ and $(F,\mathscr{T}^1)$ be complete linear topological spaces, so that F is a $\mathscr{T}$-dense proper subspace of E and $\mathscr{T}^1$ is stronger than the $\mathscr{T}$-induced

† Examples of barrelled spaces that are not bound are given by Nachbin [*Proc. Nat. Acad. Sci. U.S.A.*, *40* (1954), 471–474] and Shirota [*Proc. Japan Acad.*, *30* (1954), 294–298].

topology on F.   Let X^ω denote a countable product and $X^{(\omega)}$ a countable direct sum of copies of X.   Let $G = E^{(\omega)} \times F^\omega$, let T be the mapping of G into E^ω defined by $T(x,y) = x + y$, and let $H = T^{-1}(0)$.

(a) The space G/H is a quotient of a complete space by a closed subspace, but is not complete.   (Show that T is a continuous open mapping of G onto the dense subspace $E^{(\omega)} + F^\omega$ of E^ω.)

(b) Suppose also that E and F are reflexive.   Then G is reflexive, but G/H is not semi-reflexive (for $(G/H)^{**}$ is isomorphic to $(E^\omega)^{**} = E^\omega$). Also G^* is reflexive, but the subspace H° of G^* is not evaluable.   (For it is semi-reflexive but cannot be reflexive, since its adjoint G/H is not reflexive.)

(c) The strong topology $S(H^\circ, G/H)$ does not coincide with the topology induced on H° by $S(G^*, G)$ (the adjoints being different), so that H° is a subspace of a strong adjoint space which is not itself a strong adjoint space. This also illustrates the fact that the strong adjoint of an inductive limit may fail to be the projective limit of the corresponding strong adjoints.

(d) If E and F are Fréchet spaces, then G is a countable direct sum of Fréchet spaces, but it cannot be fully complete (see 18I), since a Hausdorff quotient of a fully complete space is fully complete.   Also G^* is a direct sum of Fréchet spaces and so is bound (see 22.3 and 19.6) but the subspace H° of G^* is not bound.

The above construction may be realized by taking $E = l^p$ and $F = l^q$ with $1 < q < p < \infty$.

E   PROBLEM

Is every reflexive space complete?

F   MONTEL SPACES

A Hausdorff barrelled space in which every closed bounded set is compact is a *Montel space*.

(a) A Montel space is reflexive; its adjoint with the strong topology is also a Montel space.

(b) A normed space can only be a Montel space if it is finite dimensional.

(c) A Montel space topology coincides with its weak topology on every bounded set.

(d) A product or a direct sum of Montel spaces is a Montel space; neither a closed subspace nor a Hausdorff quotient of a Montel space is necessarily a Montel space (see 22I).

(e) A strict inductive limit (see 17G(b)) of a sequence of Montel spaces is a Montel space.

G   STRONGEST LOCALLY CONVEX TOPOLOGY (see 6I, 12D, 14D)

Let E be a linear space of dimension α with its strongest locally convex topology.   The adjoint of E is the algebraic dual E', and the topology of E is the Mackey topology $m(E, E')$.   All admissible topologies on E' coincide (use 6I); E' is the product of α copies of the scalar field, is complete and barrelled, and is metrizable if and only if $\alpha \leq \aleph_0$.   Both E and E' are reflexive Montel spaces; E is a bound space, but whether E' is always a bound space is not known (see 19.9).

H   SPACES OF ANALYTIC FUNCTIONS I (see 22D, 22J)

Let U be an open proper subset of the Riemann sphere (the compactification of the complex plane by the one point ∞). Let $A(U)$ be the space of complex-valued functions, analytic on U, and vanishing at ∞ if $\infty \in U$, with the topology of uniform convergence on compact subsets of U. Then $A(U)$ is a complete metrizable locally convex space which is not normable. If $\infty \in U$, the topology of $A(U)$ coincides with the topology of uniform convergence on compact subsets of $U \sim \{\infty\}$. Each bounded closed subset of $A(U)$ is compact; thus $A(U)$ is a Montel space and so reflexive. (The last part can be proved by using Cauchy's formula for a derivative to show that each bounded subset of $A(U)$ is equicontinuous at each point of $U \sim \{\infty\}$.)

I   DISTRIBUTION SPACES II (see 8J)

(a) The space $\mathscr{E}$ is a Montel space. (Sketch of proof: first, it is sufficient to prove that any bounded subset B of $\mathscr{E}^{(m+1)}$ is totally bounded as a subset of $\mathscr{E}^{(m)}$, by 16D(j) and (k). Now this will follow, from 8J(a) and 16D(j) again, if each $D^p[B]$, for $|p| \leqq m$, is a totally bounded subset of $C(R^n)$. But this is a consequence of Ascoli's theorem, $D^p[B]$ being equicontinuous because B is a subset of $\mathscr{E}^{(m+1)}$.)

(b) The space $\mathscr{D}$ is the strict inductive limit (17G) of the sequence $\{\mathscr{D}_{K(r)} \colon r = 1, 2, \cdots\}$ (see 16C(k)); it follows in particular that $\mathscr{D}$ is complete and that each bounded subset of $\mathscr{D}$ is contained in some $\mathscr{D}_{K(r)}$. Each $\mathscr{D}_{K(r)}$ is a Montel space (this follows from (a)); therefore $\mathscr{D}$ is also a Montel space and reflexive. Since $\{\mathscr{D}_{K(r)} \colon r = 1, 2, \cdots\}$ is a strictly increasing sequence of Fréchet spaces, $\mathscr{D}$ is a non-metrizable LF-space: for the definition and consequent properties, see 22C.

(c) The continuous linear functionals on $\mathscr{D}$ are called *distributions*. Every regular Borel measure is a distribution (embed $\mathscr{D}$ in $K(R^n)$: see 14J); the converse is not true. The space of distributions, with its strong topology, is a complete non-metrizable Montel space.

(d) The *support* of a distribution T is the smallest closed subset K of R^n such that $T(f) = 0$ for all functions f whose supports do not meet K. Such a set must always exist. The adjoint of $\mathscr{E}$ is the space of distributions of compact support. (The adjoint of $\mathscr{E}$ is a subspace of the adjoint of $\mathscr{D}$. Suppose that T is continuous but not of compact support. Choose inductively a sequence of compact n-dimensional cubes K_r and a sequence of functions f_r of compact support S_r, in such a way that $S_r \cap K_r = \phi$, $T(f_r) = 1$ and so that S_r lies in the interior of K_{r+1}. This can be managed because $\mathscr{D}$ is dense in $\mathscr{E}$. Then put $f = \sum \{f_r \colon r = 1, 2, \cdots\}$.)

J   CLOSED GRAPH THEOREM FOR A REFLEXIVE BANACH SPACE

The closed graph theorem can be proved especially simply in a useful special case, by using the fact that the unit sphere of a reflexive Banach space is weakly compact. Let T be a linear mapping of a locally convex space E into a reflexive Banach space F. Let V be the unit sphere in F. If the graph of T is closed and if $(T^{-1}[V])^-$ is a neighborhood of 0 in E, then T is continuous. (If $x \in (T^{-1}[V])^-$ and $\mathscr{U}$ is a local base of convex circled neighborhoods of 0 in E, then for each $U \in \mathscr{U}$, $A_U = (x + U) \cap$

$T^{-1}[V] \neq \phi$. It follows from the weak compactness of V that there is a point y in V with $y \in (T[A_U])^-$ for each $U \in \mathcal{U}$. It follows that, since the graph of T is closed, $y = T(x)$.)

The condition that $(T^{-1}[V])^-$ be a neighborhood is satisfied wherever E is barrelled, so that this proof covers a large number of the cases used in applications.

K   EVALUATION OF A NORMED SPACE

The evaluation mapping of a normed space onto its canonical image in the second adjoint is an isometric isomorphism.

L   UNIFORMLY CONVEX SPACES II (see 17C)

(a) Let E be a uniformly convex normed space. Then E is reflexive. (Let U be the unit sphere in E; suppose that $x \in E^{**}$, $\|x\| = 1$, and $x \notin U$. Then for some $e > 0$, $\sup \{\|x - y\| : y \in U\} > e$. There is $f \in E^*$ with $f(x) > 1 - d(e)$; thus $V = \{y : |1 - f(y)| < d(e)\}$ is a $w(E^{**},E^*)$-neighborhood of x. Use the uniform convexity to show that the diameter of $U \cap V$ is less than e; thus the $w(E^{**},E^*)$-closure of $U \cap V$, to which x belongs, has diameter e at most.)

(b) There are reflexive normed spaces which cannot be given an equivalent uniformly convex norm. (See, for example, Bourbaki [3] Ch. V, §2, ex. 13.)

M   A NEARLY REFLEXIVE BANACH SPACE

The following example, given by R. C. James, shows that a Banach space may be isomorphic to its second adjoint without being reflexive; in fact here its canonical image is a subspace of co-dimension one.

Let E be the space of all sequences $x = \{\xi_n\}$ of complex numbers with $\xi_n \to 0$ and with

$$\|x\|^2 = \sup \{\sum \{|\xi_{p(2i-1)} - \xi_{p(2i)}|^2 : 1 \leq i \leq n\} + |\xi_{p(2n+1)}|^2\} < \infty,$$

the supremum (here and later) being taken over all strictly increasing finite sequences $\{p(i) : 1 \leq i \leq n\}$ and $n = 1, 2, \cdots$.

(a) Let e_n be the sequence with n-th term 1 and all others 0. Then $\|e_n\| = 1$ and if $x \in E$, $x = \sum \{\xi_n e_n : n = 1, 2, \cdots\}$, the series being convergent in norm.

(b) For each n, put $f_n(x) = \xi_n$. Then $f_n \in E^*$ and $\|f_n\| = 1$; if $f \in E^*$, there are scalars η_n with $f = \sum \{\eta_n f_n : n = 1, 2, \cdots\}$, the series being convergent in norm, and for each $x \in E$, $f(x) = \sum \{\xi_n \eta_n : n = 1, 2, \cdots\}$. (If $f \in E^*$, let $\eta_n = f(e_n)$. If $\sum \{\eta_r f_r : r = 1, 2, \cdots, n\}$ is not convergent to f, there is some $d > 0$ and a sequence of elements x_n of E, the terms of the sequence $\{x_n\}$ being non-zero on disjoint sets of indices, with $\|x_n\| = 1$, $f(x_n) > d$ and $\sum \{n^{-1}x_n : n = 1, 2, \cdots\}$ convergent.)

(c) For each $z \in E^{**}$, let $\zeta_n = z(f_n)$. Then

$$\|z\|^2 = \sup \{\sum \{|\zeta_{p(2i-1)} - \zeta_{p(2i)}|^2 : 1 \leq i \leq n\} + |\zeta_{p(2n+1)}|^2\}.$$

(First, prove the inequality $\leq$ by considering linear combinations of the elements f_n of E^*. Next, if $z_n = \sum \{\zeta_r e_r : 1 \leq r \leq n\} \in E$, there is some $f \in E^*$ with $\|f\| = 1$, $f(z_n) = \|z_n\|$ and $f(e_r) = 0$ for $r > n$.)

(d) For each $z \in E^{**}$ there are scalars $\zeta_n{}'$ such that $z = \zeta_0{}'e_0 + \{\zeta_n{}'e_n:$ $n = 1, 2, \cdots\}$, where e_0 is the sequence with every term 1 and the series is convergent in norm; the canonical image of E in E^{**} has codimension one. (Show that $\zeta_n \to \zeta_0{}'$, say, and take $\zeta_n{}' = \zeta_n - \zeta_0{}'$.)

(e) With the norm given by

$$|z|^2 = \sup\{\textstyle\sum \{|\zeta_{p(2i-1)}' - \zeta_{p(2i)}'|^2 : 1 \leqq i \leqq n\} + |\zeta_{p(2n+1)}'|^2\},$$

E^{**} is isomorphic to E. This is equivalent to the norm of E^{**}, given by

$$\|z\|^2 = \sup\{\textstyle\sum \{\zeta_{p(2i-1)}' - \zeta_{p(2i)}'|^2 : 1 \leqq i \leqq n\} + |\zeta_{p(2n+1)}' + \zeta_0{}'|^2\}.$$

21   DUAL TRANSFORMATIONS

A natural continuation of the program of describing a linear topological space in terms of its adjoint is the description of a linear transformation in terms of the transformation induced on the adjoint. This section is devoted to the relationship between continuity and openness of a mapping and of its adjoint, relative to the several possible topologies. The results for Fréchet spaces are especially sharp.

This section is devoted to the study of linear transformations by means of the fundamental duality. Each continuous linear transformation T of a linear topological space E into a linear topological space F induces, in a natural way, an adjoint transformation T^* of F^* into E^*. The properties of T are intimately related to those of T^*, and, in particular, it is possible to characterize continuity and openness of T in terms of T^*.

In order to preserve symmetry as far as possible, we begin with a discussion of paired linear spaces. Let E and F, and G and H, be paired linear spaces, let T be a linear transformation of E into G and let T' be a linear transformation of H into F. The transformations T and T' are **dual** if and only if $\langle T(x), h \rangle = \langle x, T'(h) \rangle$ for all members x of E and all members h of H. If T is dual to T', and if S and S' are dual linear transformations of E into G and H into F, then $aS' + bT'$ is dual to $aS + bT$ for all scalars a and b. Moreover, if U is a linear transformation of G into a linear space I, and U' is a dual mapping of a space J which is paired to I, then $T' \circ U'$ is dual to $U \circ T$.

The existence of a transformation which is dual to a given linear transformation T can be described in terms of the continuity properties of T.

21.1   Existence and Uniqueness of Dual Mappings *Let E and F, and G and H, be paired linear spaces, and let T be a linear transformation of E into G. Then T has a dual T' if and only if T is $w(E,F)$-$w(G,H)$*

continuous, and this is the case if and only if the linear functional $x \to$ $\langle T(x), h \rangle$ *is* $w(E,F)$-*continuous for each h in H.*

If T has a dual T', then (since T' has a dual), the transformation T' is necessarily $w(H,G)$-$w(F,E)$ *continuous. The dual T' of T is unique if and only if E distinguishes members of F.*

PROOF   If T is $w(E,F)$-$w(G,H)$ continuous, then for each f in H the map $x \to \langle T(x), f \rangle$ is a $w(E,F)$-continuous linear functional on E; hence, by the representation theorem 16.2 for $w(E,F)$-continuous linear functionals, there is an element f^* of F such that $\langle T(x), f \rangle = \langle x, f^* \rangle$ for all x in E. Moreover, the element f^* is unique if and only if E distinguishes points of F. A dual T' of T may then be constructed as follows: for each member f of a Hamel base B for H let $T'(f)$ be an element f^* of F such that $\langle T(x), f \rangle = \langle x, f^* \rangle$ for all x in E, and let T' be the linear transformation of H into F which has the values so determined at the members of B. It follows without difficulty that T' is a dual for T, and that T' is unique if and only if E distinguishes points of F.

On the other hand, if T has a dual T', then for each h in H, the linear functional $x \to \langle T(x), h \rangle = \langle x, T'(h) \rangle$ is $w(E,F)$-continuous on E, and if each such linear functional is $w(E,F)$-continuous then, by 16.1, T is $w(E,F)$-$w(G,H)$ continuous on E.|||

21.2   COMPUTATIONS WITH DUAL MAPPINGS   *Let E and F, and G and H, be paired linear spaces, and let T be a linear transformation of E into G with dual T'.   Then:*

(i) *for each subset A of E, $T[A]^\circ = (T')^{-1}[A^\circ]$;*

(ii) *if E distinguishes points of F, then the polar of the range $T[E]$ of T is the null space of T';*

(iii) *if C is a subset of $T[E]$, then $T'[C^\circ] = T'[H] \cap T^{-1}[C]^\circ$; and*

(iv) *the transformation T' is relatively open, continuous, and maps closed sets onto relatively closed sets in $T'[H]$ with respect to the topologies $w(H, T[E])$ and $w(F,E)$.*

PROOF   Clearly $f \in T[A]^\circ$ if and only if $|\langle T(x), f \rangle| = |\langle x, T'(f) \rangle| \leq 1$ for all x in A, and this is the case if and only if $T'(f) \in A^\circ$. This establishes (i), and part (ii) is an immediate corollary. To prove (iii), assume that C is a subset of $T[E]$ and observe that $C = T[T^{-1}[C]]$. Hence, applying (i), $C^\circ = (T')^{-1}[T^{-1}[C]^\circ]$, and therefore $T'[C^\circ] = T^{-1}[C]^\circ \cap T'[H]$. This establishes (iii). To prove (iv), it is necessary only to notice that a net $T'(h_\alpha)$ in $T'[H]$ converges to $T'(h)$ if and only if h_α converges to h relative to $w(H, T[E])$.|||

We now prove the principal theorem of the section, which gives necessary and sufficient conditions for a linear transformation T be continuous or relatively open with respect to admissible topologies. (Recall that an admissible topology for E, where E is paired to F, is a topology $\mathcal{T}_{\mathscr{A}}$ of uniform convergence on members of $\mathscr{A}$, where $\mathscr{A}$ is a family of $w(F,E)$-bounded, closed, convex circled subsets of F such that scalar multiples of members of $\mathscr{A}$ belong to $\mathscr{A}$, the union of any two members is contained in a member, and $\mathscr{A}$ covers F. Every locally convex topology is admissible relative to the natural pairing of the space and its adjoint.)

21.3 CONTINUITY AND OPENNESS RELATIVE TO ADMISSIBLE TOPOL-OGIES   *Let E and F, and G and H, be paired linear spaces, let $\mathscr{A}$ and $\mathscr{B}$ be admissible families of subsets of F and H, respectively, and let T be a linear transformation of E into G with dual T'.   Then:*

(i) *the transformation T is $\mathcal{T}_{\mathscr{A}}$-$\mathcal{T}_{\mathscr{B}}$ continuous if and only if for each B in $\mathscr{B}$ there is a member A of $\mathscr{A}$ such that A contains $T'[B]$;*

(ii) *if T is $\mathcal{T}_{\mathscr{A}}$-$\mathcal{T}_{\mathscr{B}}$ relatively open, then for each member A of $\mathscr{A}$ there is a member B of $\mathscr{B}$ such that $A \cap T'[H]^- \subset T'[B]^-$, where closures are relative to the topology $w(F,E)$; if, in addition E distinguishes points of F and if each member of $\mathscr{B}$ is $w(H,G)$-compact (equivalently, $\mathcal{T}_{\mathscr{B}}$ is weaker than the Mackey topology $m(G,H)$), then $T'[H]$ is $w(F,E)$-closed; and*

(iii) *if H distinguishes points of G, if each member of $\mathscr{A}$ is $w(F,E)$-compact (equivalently, $\mathcal{T}_{\mathscr{A}}$ is weaker than the Mackey topology $m(E,F)$), and if for A in $\mathscr{A}$ there is B in $\mathscr{B}$ such that $A \cap T'[H]^- \subset T'[B]^-$, then T is $\mathcal{T}_{\mathscr{A}}$-$\mathcal{T}_{\mathscr{B}}$ relatively open.*

PROOF   In view of the definition of $\mathcal{T}_{\mathscr{A}}$ and $\mathcal{T}_{\mathscr{B}}$, the transformation T is $\mathcal{T}_{\mathscr{A}}$-$\mathcal{T}_{\mathscr{B}}$ continuous if and only if for each B in $\mathscr{B}$ there is A in $\mathscr{A}$ such that $A_0 \subset T^{-1}[B_0]$. According to 21.2(i), $T^{-1}[B_0] = T'[B]_0$, and $A_0 \subset T'[B]_0$ if and only if $T'[B]_0{}^0 = T'[B]^- \subset A_0{}^0 = A$, by taking polars. Since A is $w(F,E)$-closed, we have that T is $\mathcal{T}_{\mathscr{A}}$-$\mathcal{T}_{\mathscr{B}}$ continuous if and only if for each B in $\mathscr{B}$ there is A in $\mathscr{A}$ such that $T'[B] \subset A$, and (i) is proved.

To prove (ii) and (iii), we first note that T is $\mathcal{T}_{\mathscr{A}}$-$\mathcal{T}_{\mathscr{B}}$ relatively open if and only if for A in $\mathscr{A}$ there is B in $\mathscr{B}$ such that $T[A_0] \supset B_0 \cap T[E]$. But, taking inverses under T, this is true if and only if $A_0 + T^{-1}[0] \supset T^{-1}[B_0]$.

Suppose T is relatively open; then, since $T'[H]_0 \supset T^{-1}[0]$, for each A in $\mathscr{A}$ there is B in $\mathscr{B}$ such that $A_0 + T'[H]_0 \supset T'[B]_0$, which implies that $(A_0 + T'[H]_0)^\circ \subset T'[B]_0^\circ = T'[B]^-$. The first half of (ii) now follows, because $A \cap T'[H]^- \subset (A_0 + T'[H]_0)^\circ$. Under the additional hypotheses of (ii), $w(F,E)$ is Hausdorff and, for each B in $\mathscr{B}$, $T'[B]$ is $w(F,E)$-compact, since T' is w (H,G)-$w(F,E)$ continuous (21.1); hence $T'[B]$ is $w(F,E)$-closed for each B in $\mathscr{B}$. Using the result of the first half, we see that, for each A in $\mathscr{A}$, $A \cap T'[H]^- \subset T'[H]$, and, since $\mathscr{A}$ covers F, $T'[H]^- \subset T'[H]$.

To establish (iii), first observe that for each member A of $\mathscr{A}$ we have $2A_0 + T'[H]_0 \supset A_0 + \langle T'[H]_0 \cup A_0 \rangle$, and the latter set contains the $\mathscr{T}_{\mathscr{A}}$-closure of $\langle T'[H]_0 \cup A_0 \rangle$. The topology $\mathscr{T}_{\mathscr{A}}$ is, under the hypothesis of the theorem, stronger than $w(E,F)$ and weaker than the Mackey topology $m(E,F)$; hence the members of F represent all of $(E,\mathscr{T}_{\mathscr{A}})^*$, by 18.8. It follows from 17.1 that the $\mathscr{T}_{\mathscr{A}}$-closure of $\langle T'[H]_0 \cup A_0 \rangle$ is $w(E,F)$-closed. Hence $2A_0 + T'[H]_0 \supset \langle T'[H]_0 \cup A_0 \rangle^- = (T'[H]^- \cap A)_0$ for each A in $\mathscr{A}$. Finally, if B is a member of $\mathscr{B}$ such that $A \cap T'[H]^- \subset T'[B]^-$, then $(A \cap T'[H]^-)_0 \supset T'[B]_0$, and hence for each A in $\mathscr{A}$ there is B in $\mathscr{B}$ such that $2A_0 + T'[H]_0 \supset T'[B]_0$. Since H distinguishes members of G, $T'[H]_0 = T^{-1}[0]$, and the inclusion above can be written as $2A_0 + T^{-1}[0] \supset T^{-1}[B_0]$. It follows, in view of the criterion of the second paragraph, that T is $\mathscr{T}_{\mathscr{A}}$-$\mathscr{T}_{\mathscr{B}}$ relatively open.|||

There are three especially important topologies which are defined by a pairing. Recall that, if E and F are paired linear spaces, the Mackey topology $m(E,F)$ for E has a local base consisting of the polars in E of $w(F,E)$-closed compact convex circled subsets of F, and the strong topology $s(F,E)$ for F has a local base consisting of all polars of $w(E,F)$-bounded subsets of E. The following proposition is an application of the preceding theorem to these topologies. (In the statement of the result the term "weak" refers to the topologies $w(E,F)$ and $w(F,E)$.)

21.4 Continuity and Openness of Transformations and Their Duals *Let E and F, and G and H, be paired linear spaces, and let T be a linear transformation of E into G. Then:*

   (i) *T is weakly continuous if and only if T is Mackey continuous;*

   (ii) *in case E distinguishes points of F and H distinguishes points of G, T is weakly continuous and weakly relatively open if and only if T has a dual T' and the range of T' is weakly closed;*

   (iii) *in case E distinguishes points of F and the pairing $\langle G, H \rangle$ is*

*separated, T is weakly continuous and open if and only if T' is a
weakly topological isomorphism of H onto a weakly closed
subspace of F;*

(iv) *in case E distinguishes points of F and the pairing $\langle G,H \rangle$ is
separated, T is Mackey continuous and open if and only if T
is weakly continuous and open; and*

(v) *If T' is a dual of T, then T' is strongly continuous.*

PROOF   If T is weakly continuous, then there is a dual T' of T, T' is
automatically weakly continuous, and the image under T' of a
weakly compact subset of H is therefore weakly compact.   Therefore, by the criterion 21.3(i), T is Mackey continuous.   Conversely,
if T is Mackey continuous, then for each h in H the functional $x \rightarrow$
$\langle T(x),h \rangle$ is Mackey continuous, and is hence represented by a
member of F (theorem 18.8).   Therefore $x \rightarrow \langle T(x),h \rangle$ is weakly
continuous for every h in H, and, by theorem 21.1, T is weakly continuous.

To prove part (ii), first assume that T' exists and the range of T' is
weakly closed.   In view of 21.3(iii), T is weakly relatively open if for
each finite subset A of F there is a finite subset B of $T'[H]$ such that
$A_0{}^\circ \cap T'[H] \subset B_0{}^\circ$, and the latter is easily seen to be the case (see
the proof of 16.11(ii)).   The existence of T' implies the weak continuity of T (21.1).   Conversely, if T is weakly continuous and
relatively open, then 21.3(ii) implies that the range of T' is weakly
closed.

If T is weakly continuous and open, then $T[E] = G$, and since
$T[E]$ is weakly closed, T' is relatively open.   The null space of T'
is $T[E]^\circ = \{0\}$, by 21.2, and hence T' is a weakly topological isomorphism.   Conversely, if T' is a weakly topological isomorphism
onto a weakly closed subspace of F, then T' is weakly continuous and
relatively open, and $T[E]$ is therefore weakly closed.   Then $T[E]$ is
identical with the polar of the null space of T', by 21.2, and hence
$T[E] = G$.   Thus T is weakly continuous and open, and (iii) is
established.

If T is weakly continuous and open, then T' is a topological isomorphism onto a weakly closed subspace of F, and the criterion
21.3(iii) for relative openness shows that T is Mackey relatively open
and (since $T[E] = G$) hence open.   Conversely, if T is Mackey
continuous and open, then by 21.3(ii) the range $T'[H]$ is weakly
closed.   It follows from part (iii) that T is weakly open, and (iv) is
established.

Proposition (v) is a direct application of 21.3(i) to the strong topologies of H and F.|||

We now apply the foregoing theorems to the study of a linear transformation T of a linear topological space E into a linear topological space F with the natural pairings $\langle E, E^* \rangle$ and $\langle F, F^* \rangle$. Since E always distinguishes elements of E^*, there is at most one transformation which is dual to T. This dual transformation, if it exists, is called the **adjoint** of T and is denoted by T^*. The condition for dual transformations becomes, in this case, $h(T(x)) = T^*(h)(x)$ for all x in E and all h in F. Rewriting, this is: $T^*(h) = h \circ T$ for all h in F. The uniqueness of T^* is self-evident in this formulation.

The following results are immediate consequences of 21.1. We recall that the weak topology for a linear topological space E is the topology $w(E, E^*)$, and the weak* topology for E^* is the topology $w(E^*, E)$.

21.5 Existence, Uniqueness, and Weak* Continuity of Adjoints; One-to-One Transformations   *Let E and F be locally convex, linear topological spaces, and let T be a linear transformation of E into F.*

(i) *If T is continuous, then T is weakly continuous, and T has an adjoint T^* if and only if T is weakly continuous. The adjoint, if it exists, is unique and is necessarily weak* continuous.*

(ii) *A linear transformation S of F^* into E^* is the adjoint of some linear transformation T of E into F if and only if S is weak* continuous. The transformation T, if it exists, is weakly continuous, and is unique if F is Hausdorff.*

(iii) *If T has an adjoint, T^*, then the polar of $T[E]$ is the null space of T^*, and, if in addition F is Hausdorff, the polar of $T^*[F^*]$ is the null space of T. Hence T^* is one-to-one if and only if $T[E]$ is weakly dense in F, and, if E and F are Hausdorff, T is one-to-one if and only if $T^*[F^*]$ is weak* dense in E^*.*

Proposition 21.4 yields the following results on linear transformations of a locally convex space.

21.6 Continuity and Openness of Transformations and Their Adjoints   *Let T be a continuous linear transformation of a locally convex space E into a locally convex space F, and let T^* be the adjoint transformation of F^* into E^*. Then:*

(i) *T is continuous relative to the Mackey topologies, and T^* is strongly continuous;*

(ii) *if F is Hausdorff, T is weakly relatively open if and only if*

$T^*[F^*]$ *is weak* closed, and* T^* *is weak* relatively open if and only if* $T[E]$ *is closed; and*

(iii) *if F is Hausdorff, T is open relative to the Mackey topologies if and only if T is weakly open.*

If T is a continuous linear transformation on a locally convex space E into a locally convex Hausdorff space F and if T is relatively open, then theorem 21.3(ii) implies that the range of T^* is weak* closed, and hence, using part (ii) of the theorem above, T is weakly relatively open. Therefore we have the following:

21.7 COROLLARY *Let T be a continuous linear transformation on a locally convex space into a locally convex Hausdorff space. Then, if T is relatively open, T is weakly relatively open, and, if T is open, T is weakly open.*

The adjoint T^* of a linear transformation T of E into F is always strongly continuous, in view of theorem 21.6, and consequently the second adjoint T^{**} (the adjoint of T^*) exists. This second adjoint is an extension of T, in a sense made precise by the following simple proposition.

21.8 EVALUATION AND THE SECOND ADJOINT OF A TRANSFORMATION *Let T be a continuous linear transformation of a locally convex space E into a locally convex space F, and let I_E and I_F be the evaluation mappings of E into E^{**} and F into F^{**}, respectively. Then $T^{**} \circ I_E = I_F \circ T$.*

PROOF   Using the definitions of adjoint and evaluation, we have, for each x in E and each f in F^*,

$$T^{**} \circ I_E(x)(f) = T^{**}(I_E(x))(f) = I_E(x) \circ T^*(f)$$
$$= I_E(x)(f \circ T) = f \circ T(x) = I_F \circ T(x)(f).|||$$

The results concerning open mappings can be very much improved in case the domain and range spaces are Fréchet spaces (locally convex, metrizable, and complete). We give the results here, assuming for the moment that it is known that each metrizable space is a Mackey space (this is proved in the next section, and the proof given there is independent of the following theorem).

21.9 OPEN MAPPINGS OF FRÉCHET SPACES *Let T be a continuous linear transformation of a Fréchet space E into a Fréchet space F. Then the following are equivalent.*

(i) *The map T is relatively open.*
(ii) *The map T is weakly relatively open.*

(iii) *The range of T is closed.*

(iv) *The adjoint map T* is weak* relatively open.*

(v) *The range of T* is weak* closed.*

PROOF The equivalences (i) → (ii) ↔ (v) and (iii) ↔ (iv) result immediately from 21.6, 21.7, and the fact that each metrizable space is a Mackey space. That (i) → (iii) follows from the fact that the image under a continuous open linear transformation of a complete metric space is complete, (11.3), and that (iii) → (i) follows from the open mapping theorem 11.4. It remains to show that (ii) → (i). Assume (ii) and let $T[E] = G$; then the weak topology $w(G,G^*)$ is the relativization of $w(F,F^*)$ (theorem 17.13 (i)). Therefore, T is a weakly open map of E onto G, and, in view of 21.6(iii), T is open relative to the Mackey topologies of E and G. Finally, since both E and G are metrizable, T is a relatively open map of E into F.|||

Certain of the implications of the above theorem can obviously be strengthened. However, the theorem is, in substance, the best of the known results of its type on openness of mappings.

PROBLEMS

A COMPLETELY CONTINUOUS MAPPINGS

Let E and F be Hausdorff linear topological spaces and let T be a linear mapping of E into F. Then T is called *completely continuous* iff there is some neighborhood U of zero which is mapped by T into a compact set. For Banach spaces, this coincides with the usual definition of a completely continuous mapping as one which maps bounded sets into totally bounded sets (cf. 8B and 21D). A completely continuous mapping is continuous. Let I be a topological isomorphism of E onto F and let $S = I + T$. Then

(a) the null space of S is finite dimensional;

(b) S is an open mapping of E onto $S[E]$;

(c) the range of S is closed.

(Sketch of proof: let N be the null space $S^{-1}[0]$. For (a), show that $N \cap U$ is totally bounded and use 7.8. For (b), suppose that S is not open, so that there is a neighborhood U_1 of zero, which may be taken to be a subset of U, with $S[U_1]$ not a neighborhood of zero in $S[E]$. If $\mathscr{V}$ is a base of circled neighborhoods of zero in F, for each $V \in \mathscr{V}$ there is a point y in V with $y \notin S[U_1]$ and then there is some λ with $0 \leq \lambda \leq 1$ so that $\lambda y \in V \cap S[2U_1]$ but $\lambda y \notin S[U_1]$. Thus for each V there is a point $x_v \in 2U_1$ but $x_v \notin U_1 + N$ with $S(x_v) \in V$. Now use the properties of T and I to show that a subnet converges to a point of N, thus contradicting $x_v \notin U_1 + N$. For (c), suppose that $\{S(x_\alpha) : \alpha \in A\}$ converges to $y \in F$. It follows from (b) that $S(x_\alpha) \in S[x_\beta + U]$ for some β and all $\alpha \geq \beta$, and hence that $(x_\beta + U) \cap S^{-1}[y + V] \neq \phi$ for each $V \in \mathscr{V}$. If z_v is a point of this set, then $S(z_v) \to y$ and, by an argument similar to that for (b), a subnet of $\{z_v : V \in \mathscr{V}\}$ is convergent.)

B RIESZ THEORY

The theorem of the previous problem forms the foundation for the Riesz theory of completely continuous linear mappings of a linear topological space into itself. The following sequence of results gives an outline of the theory.

Let E be a linear topological space and T a completely continuous linear mapping of E into itself. Suppose throughout that λ is a non-zero scalar, and write $\lambda I - T = S$.

(a) *Lemma*: Let G be a closed proper subspace of a subspace F of E such that $S[F] \subset G$. If U is a convex circled neighborhood of 0 mapped by T into a compact set, then there is a point $x \in F \cap (2U)$ but $x \notin G + U$.

(b) The sequence $\{S^{-r}[0]\}$ of subspaces increases; there is an integer $n \geq 0$ such that $S^{-r}[0] \subset S^{-(r+1)}[0]$ properly for $0 \leq r < n$ and $S^{-r}[0] = S^{-(r+1)}[0]$ for $r \geq n$. (Suppose not, and apply (a) with $G = S^{-r}[0]$ and $F = S^{-(r+1)}[0]$. Use the sequence $\{x_r\}$ thus provided to contradict the fact that $T[2U]$ is totally bounded.)

(c) The sequence $\{S^r[E]\}$ decreases; there is an integer $m \geq 0$ such that $S^r[E] \supset S^{r+1}[E]$ properly for $0 \leq r < m$ and $S^r[E] = S^{r+1}[E]$ for $r \geq m$. (Proof similar to that of (b); each $S^r[E]$ is closed by (c) of the previous problem.)

(d) The integers n and m of (b) and (c) are equal, and E is the direct sum of $S^{-n}[0]$ and $S^n[E]$.

By (a) and (c) of 21A and by 7E, E is then isomorphic to the product of $S^{-n}[0]$ and $S^n[E]$ (in fact, it is their topological direct sum when it is locally convex).

It follows that the existence of a completely continuous endomorphism on a linear topological space which is not locally convex ensures the existence of continuous linear functionals, if $S^{-1}[0] \neq \{0\}$ (that is, if λ is an eigenvalue of T): in fact the linear functionals in the subspace $(S^n[E])^\circ$ of the algebraic dual E' are continuous, and this subspace has the dimension of $S^{-n}[0]$.

(e) The following are equivalent: (i) λ is not an eigenvalue of T; (ii) the range of $\lambda I - T$ is the whole of E; (iii) $\lambda I - T$ is an isomorphism. (The integer n above is zero: use (d) and 21A(b).)

(f) E is isomorphic to the product of two closed subspaces M and N, N being finite dimensional; $S = \lambda I - T$ maps each of M and N into itself and is an isomorphism on M. For all positive integers r, $\dim S^{-r}[0] = \dim E/S^r[E]$. (Use 21A(b) applied to the restriction of S to M; for the last part, observe that since N is finite dimensional, $\dim S^{-r}[0] = \dim N/S^r[N]$.)

(g) The eigenvalues of T form a finite set or a sequence convergent to zero. (If $e > 0$, there are at most a finite number of eigenvalues with $|\lambda| \geq e$. Suppose not; choose a sequence $\{x_n\}$ of linearly independent eigenvectors, and let H_n be the subspace spanned by $x_1, \cdots, x_n$. Apply (a) with $G = H_{n-1}$, $F = H_n$ and $S = \lambda_n I - T$, λ_n being the eigenvalue corresponding to x_n, and so again contradict the fact that $T[U]$ is totally bounded.)

C COMPLETE CONTINUITY OF THE ADJOINT

Let E and F be locally convex Hausdorff spaces and T a completely continuous linear mapping of E into F. Then T^* is completely continuous if E^* and F^* have the topologies of uniform convergence on the convex circled compact subsets of E and F, respectively. (Use 18E.)

Let E and F coincide and $S = \lambda I - T$, as in the previous problem. Then T and T^* have the same eigenvalues and the spaces $S^{-r}[0]$ and $S^{*-r}[0]$ have the same dimension, for each r. (Use the results of the previous problem applied to T^*, with the aid of the relations $(S^r[E])^\circ = S^{*-r}[0]$ and $(S^{*r}[E^*])^\circ = S^{-r}[0]$.)

D SCHAUDER'S THEOREM

Let E and F be locally convex Hausdorff spaces and T a weakly continuous linear mapping of E into F. Let E^* have the strong topology. Then T maps bounded sets into totally bounded sets if and only if T^* maps equicontinuous sets into totally bounded sets. (Use 16A applied to $\langle F, F^* \rangle$, with $\mathscr{A}$ the family of images of bounded sets and $\mathscr{B}$ the family of equicontinuous sets, and use the identity $\langle T(x), y^* \rangle = \langle x, T^*(y^*) \rangle$.)

The linear mapping T is called *totally bounded* if it maps bounded sets into totally bounded sets. If, in addition to the above hypotheses, F is evaluable and F^* has the strong topology, then T is totally bounded if and only if T^* is totally bounded. When E and F are Banach spaces, this becomes Schauder's theorem: T is completely continuous if and only if T^* is completely continuous.

E CLOSABLE MAPPINGS

Let E and F be locally convex Hausdorff spaces. This problem is concerned with linear mappings which are defined only on subspaces of E and take values in F. Let T be a linear mapping with domain $D \subset E$ and range in F; T is called *closed* if its graph G is a closed subspace of $E \times F$, and *closable* if it has some closed extension. This is the case if and only if G^- is the graph of a closed linear mapping which has the smallest domain of all the closed extensions of T. The mapping T is closed if and only if $y = T(x)$ whenever there is a net $\{x_\gamma : \gamma \in \Gamma\}$ in D with $x_\gamma \to x$ and $T(x_\gamma) \to y$; it is closable if and only if $y = 0$ whenever there is a net $\{x_\gamma : \gamma \in \Gamma\}$ in D with $x_\gamma \to 0$ and $T(x_\gamma) \to y$.

Let E^* and F^* be the adjoints of E and F. The mapping $x \to \langle T(x), g \rangle$ defines a linear functional on D. Let D_* be the subspace of F^* consisting of those elements g for which this linear functional is continuous on D. This sets up a mapping of D_* into E^*/D°. Now if (and only if) D is dense in E, there is a unique continuous linear functional f on E with $f(x) = \langle T(x), g \rangle$ for all $x \in D$; in this case we may write $f = T^*(g)$, and T^* is a linear mapping of D_* into E^*. Suppose then that D is dense in E.

(a) With the weak* topologies on the adjoint spaces, T^* is closed. In fact, the graph of $-T^*$ in $F^* \times E^*$ is the polar of the graph of T.

(b) The mapping T is closable if and only if D_* is dense in F^*. (For if D_* is dense in F^*, the process can be repeated to give a closed extension T^{**}. Conversely, if T is closable, let S be the closed extension of T with smallest domain, and let $\mathscr{U}$ be a local base of convex circled neighborhoods of 0 in this domain. Then $D_*{}^\circ = (T^{*-1}[E^*])^\circ = (S^{*-1}[E^*])^\circ = (\bigcup \{S^{*-1}[U^\circ] : U \in \mathscr{U}\})^\circ = (\bigcup \{(S[U])_\circ : U \in \mathscr{U}\})^\circ = \bigcap \{\overline{S[U]} : U \in \mathscr{U}\}$, and 11H leads to the required results.)

F   STONE-ČECH COMPACTIFICATION

Let X be a completely regular space and $B(X)$ the set of bounded continuous real-valued functions on X with the norm $\|f\| = \sup \{|f(x)| : x \in X\}$.

(a) The mapping e of X into the adjoint $M(X)$ of $B(X)$ with the weak* topology, defined by $(e(x))(f) = f(x)$, is a homeomorphism of X onto a subset of the surface of the unit sphere of $M(X)$. Let $\beta(X) = e[X]^-$; $\beta(X)$ is a compact Hausdorff space, the Stone-Čech compactification of X.

(b) Every $f \in B(X)$ has a unique continuous extension to $\beta(X)$ (suppose $e[X]$ and X identified).

(c) If t is a continuous mapping of X into any compact Hausdorff space Y, then t has a unique continuous extension F mapping $\beta(X)$ into Y. (Let T be the linear mapping of $B(Y)$ into $B(X)$ induced by t; obtain F by restricting T^* to $\beta(X)$.)

(d) Let H be the subset of $M(X)$ consisting of homomorphisms: that is, $H = \{\phi : \phi \in M(X), \phi(fg) = \phi(f)\phi(g) \text{ for all } f,g \in B(X)\}$. Then H is weak* closed and $H \sim \{0\}$ is a subset of the surface of the unit sphere. (Consider the mappings $\phi \to \phi(fg) - \phi(f)\phi(g)$.)

(e) $H = \beta(X) \cup \{0\}$. (Clearly $e[X] \subset H$. Suppose that X is compact, and show that if $\phi \in H$, but $\phi \neq 0$, then $\phi \in e[X]$. For this, show that there is some x for which $\phi(f) = 0$ implies $f(x) = 0$. For if not, a function g can be constructed (as a finite sum of squares) so that $\phi(g) = 0$ but $g(x) > 0$ for all $x \in X$. Thus $g^{-1} \in B(X)$ and it follows that $\phi = 0$. For general X, use (b).)

22   PSEUDO-METRIZABLE SPACES

This section is concerned with special properties of pseudo-metrizable spaces. Such spaces are always bound, and both the strong and the weak* topologies for the adjoint have localizability properties. However, the adjoint may still have a rather intricate structure, and we are able to discover very little about the structure of the second adjoint.

This section is devoted to a number of results which are peculiar to locally convex pseudo-metrizable spaces. The topology of such a space E is always bound, and the weak* topology for E^* has a noteworthy "localization" property. The strong topology for E^* may, however, exhibit pathological features, and we are able to establish relatively little concerning its structure.

We begin with a boundedness property of E.

22.1   Sᴇǫᴜᴇɴᴛɪᴀʟ Aʙsᴏʀᴘᴛɪᴏɴ Tʜᴇᴏʀᴇᴍ   *If E is a pseudo-metrizable linear topological space and A is a subset of E which absorbs each sequence converging to 0, then A is a neighborhood of 0.*

ᴘʀᴏᴏꜰ   Let $\{U_n\}$ be a local base for the topology of E, and assume that A is not a neighborhood of 0. Then, for each positive integer n, there is a point x_n in $\left(\dfrac{1}{n} U_n\right) \sim A$. Evidently the sequence $\{nx_n\}$ converges to 0, but $nx_n \notin nA$. It follows that A does not absorb the sequence $\{nx_n\}$, and hence each set which absorbs every sequence which converges to 0 is necessarily a neighborhood of 0.|||

22.2   Cᴏʀᴏʟʟᴀʀʏ   *A family F of linear transformations of a pseudo-metrizable linear topological space E into a linear topological space H is equicontinuous if and only if F is uniformly bounded on each sequence in E which converges to 0.*

ᴘʀᴏᴏꜰ   Each equicontinuous family is uniformly bounded on every bounded set (see 12.1), and in particular on each sequence which converges to 0. Conversely, let U be a circled neighborhood of 0 in H and let $A = \bigcap \{f^{-1}[U] : f \in F\}$. Then, the assumption that F is uniformly bounded on sequences converging to 0 implies that A absorbs each such sequence, hence A is a neighborhood of 0, and F is therefore equicontinuous.|||

Theorem 22.1 implies that every pseudo-metrizable locally convex space is bound. Proposition 19.5 then yields the properties of pseudo-metrizable locally convex spaces which are listed in the following corollary.

22.3   Cᴏʀᴏʟʟᴀʀʏ   *If E is a locally convex pseudo-metrizable space, then the topology of E is bound and hence also Mackey.*

Consequently, for such a space E, the adjoint E* is strongly complete, and the following conditions on weak* closed convex circled subsets A of E* are equivalent: A is equicontinuous, A is weak* compact, and A is strongly bounded.

The next results concern the weak* topology for the adjoint of a pseudo-metrizable linear topological space E. We are primarily concerned with "localizing" the weak* topology, in the sense that we infer (in certain cases) that a set is weak* closed if and only if its intersection with each equicontinuous subset A of E^* is weak* closed in A. The usefulness of the results lies in the fact that, in proving that a particular set B is weak* closed, we may restrict our attention to an equicontinuous (and hence to a weak* compact) subset of E^*. One theorem of this general type has already been proved: theorem 17.7 implies that a hyperplane in the adjoint of a complete locally convex space is weak* closed if and only if its intersection with each equicontinuous subset A of E^* is weak* closed in A.

The next lemma is the critical step in the identification of the strongest topology which agrees with the weak* topology on equicontinuous subsets of E^*.

22.4 LEMMA *Let E be a pseudo-metrizable linear topological space, and let W be a subset of E^* such that $0 \in W$ and $U^\circ \cap W$ is weak* open in U° for each neighborhood U of 0 in E.   Then there is a countable set S in E such that $S^\circ \subset W$ and the points of S can be arranged to form a sequence converging to 0.*

PROOF Let $\{U_n: n = 1, 2, \cdots\}$, with $U_1 = E$, be a decreasing sequence of sets forming a local base for E, and suppose that W satisfies the hypothesis of the lemma.   The proof of the lemma involves an induction, which in turn depends on the following pre-lemma: if U and V are neighborhoods of 0 in E, and A is a subset of E such that $A^\circ \cap U^\circ \subset W$, then there is a finite subset B of U such that $(A \cup B)^\circ \cap V^\circ \subset W$. This pre-lemma is established as follows.   The set U° is the intersection of the sets $\{x\}^\circ$, for x in U, by the definition of polar.   Since $A^\circ \cap U^\circ \subset W$, it follows that $V^\circ \cap A^\circ \cap U^\circ = \bigcap \{V^\circ \cap A^\circ \cap \{x\}^\circ : x \in U\} \subset V^\circ \cap W$. The set V° is weak* compact because V is a neighborhood of 0, and $V^\circ \cap W$ is weak* open in V° by hypothesis. We then have an intersection of closed subsets of a compact space contained in the open subset $V^\circ \cap W$, and by a well known compactness argument, some finite intersection is a subset of $V^\circ \cap W$.   That is, there is a finite subset B of U such that $\bigcap \{V^\circ \cap A^\circ \cap \{x\}^\circ : x \in B\} = V^\circ \cap A^\circ \cap B^\circ \subset V^\circ \cap W \subset W$, and the pre-lemma is established.

To prove the lemma, let A_1 be a finite subset of $E = U_1$ such that $A_1^\circ \cap U_2^\circ \subset W$.   Such A_1 exists because $W \cap U_2^\circ$ is weak* open in U_2° and contains 0.   Then the pre-lemma yields a finite subset A_2 of U_2 such that $(A_1 \cup A_2)^\circ \cap U_3^\circ \subset W$. Repeating this process inductively, one can choose a sequence $\{A_n\}$ of finite sets such that $A_n \subset U_n$ and $(A_1 \cup A_2 \cup \cdots \cup A_{n-1})^\circ \cap U_n^\circ \subset W$.   Let $S = \bigcup \{A_n: n = 1, 2, \cdots\}$; then clearly S is countable and can be arranged to form a sequence converging to 0.   To show that $S^\circ \subset W$, first observe that $S^\circ \cap U_n^\circ \subset W$ for each n.   Hence $S^\circ = \bigcup \{S^\circ \cap U_n^\circ : n = 1, 2, \cdots\} \subset W.|||$

If E is a pseudo-metrizable linear topological space, $\mathscr{T}_S$ is defined to be the locally convex topology for E^* which has for a local base the set of all polars S°, where S is the set of values of a sequence which converges to 0. This topology will be described, somewhat loosely, as the **topology of uniform convergence on sequences converging to** 0. The topology might also be described as the topology of uniform convergence on convergent sequences, for the uniform convergence on each sequence converging to 0 implies the uniform convergence on each convergent sequence. It is clear that the topology $\mathscr{T}_S$ is stronger than the weak* topology and weaker than that of uniform convergence on compact sets, or totally bounded sets. Among other facts, the next theorem states that the locally convex topology $\mathscr{T}_S$ is the strongest topology which agrees with the weak* topology on equicontinuous sets. It is not hard to see that this strongest topology which agrees with the weak* topology on equicontinuous sets consists of all sets U such that $U \cap A$ is weak* open in A for each equicontinuous subset A of E^*.

22.5 THE LOCALIZED WEAK* TOPOLOGY   *Let E be a pseudo-metrizable linear topological space. Then the following four topologies for E^* coincide.*

   (i) *the topology $\mathscr{T}_S$ of uniform convergence on sequences converging to* 0,
   (ii) *the topology of uniform convergence on compact sets,*
   (iii) *the topology of uniform convergence on totally bounded sets, and*
   (iv) *the strongest topology which agrees with the weak* topology on equicontinuous sets.*

PROOF   Clearly the topology (ii) is stronger than the topology (i); the topology (iii) is stronger than the topology (ii). The topology of pointwise convergence for an equicontinuous set is identical with that of uniform convergence on totally bounded sets (theorem 8.17), and hence the topology (iv) is stronger than the topology (iii). It remains to show that the topology (i) is stronger than the topology (iv). Let U be an open set relative to the topology (iv), and let $f_0 \in U$. Since, for each equicontinuous subset A of E^*, it is true that $U \cap A$ is weak* open in A, the same is true for $U - f_0$, that is, $(U - f_0) \cap A$ is weak* open in A whenever A is an equicontinuous subset. Hence by lemma 22.4, $U - f_0$ contains a $\mathscr{T}_S$ neighborhood of 0; that is, U is a $\mathscr{T}_S$ neighborhood of f_0. Therefore U is a $\mathscr{T}_S$-open set.|||

22.6 KREIN-ŠMULIAN THEOREM ON WEAK* CLOSED CONVEX SETS
A locally convex pseudo-metrizable space E is complete if and only if each

convex subset of E* is weak* closed whenever its intersection with every equicontinuous subset C of E* is weak* closed in C.

PROOF Assume that E is complete. Then since the closed convex circled extension of each totally bounded subset of E is compact and hence weakly compact, the topology $\mathcal{T}_S$ is weaker than the Mackey topology $m(E^*,E)$ and stronger than the weak* topology $w(E^*,E)$. Hence by 18.8, E represents $(E^*,\mathcal{T}_S)^*$, and by 17.1, each convex subset of E^* which is $\mathcal{T}_S$-closed is weak* closed; that is, if A is a convex subset of E^* such that $A \cap C$ is weak* closed in C whenever C is an equicontinuous subset, then A is weak* closed. The converse is obvious from theorem 17.7.|||

The foregoing theorem can be strengthened somewhat. If E is locally convex and pseudo-metrizable and $E^\wedge$ is a completion of E, then $(E^\wedge)^*$ is identical with (more precisely, isomorphic to) E^*. Each equicontinuous subset A of E^* is contained in a $w(E^*,E^\wedge)$-compact subset B, and since $w(E^*,E)$ is a weaker Hausdorff topology for B, the relativization to B of $w(E^*,E^\wedge)$ and $w(E^*,E)$ must coincide. This remark, in conjunction with the last theorem, yields the following proposition.

22.7 COROLLARY *Let E be a locally convex pseudo-metrizable space, and let $E^\wedge$ be a completion of E. Then a convex subset A of E* intersects each equicontinuous set B of E* in a set which is $w(E^*,E)$-closed in B if and only if A is $w(E^*,E^\wedge)$-closed.*

The rest of this section is devoted to a study of the strong topology for E^*, and to a few propositions about the second adjoint. We begin with a discussion of metrizability.

22.8 METRIZABILITY OF E^* *Let E be a locally convex pseudo-metrizable space which is not pseudo-normable, and let E^* have the strong topology. Then each bounded subset of E^* is nowhere dense, E^* is of the first category in itself, and E^* is not metrizable.*

PROOF Suppose that B is a strongly bounded, somwhere dense subset of E^*. Then B is equicontinuous, and hence there is a neighborhood U of 0 in E such that the polar U° contains B. The set U° is weak* closed, hence strongly closed, and 0 belongs to the (strong) interior of U° because this interior is convex, circled, and non-void. Consequently there is a bounded subset A of E such that $A^\circ \subset U^\circ$, and hence $U \subset U^\circ{}_\circ \subset A^\circ{}_\circ$. The set $A^\circ{}_\circ$ is bounded because E is locally convex, $U^\circ{}_\circ$ is therefore a bounded convex neighborhood of 0, and theorem 6.1 shows that E is pseudo-normable.   This contradiction establishes the first of the conclusions of the theorem.   If $\{U_n\}$ is a local base for E, then E^* is the union of the sets $U_n{}^\circ$, each of these is nowhere dense, and hence E^* is of the first category.   Finally, E^* is (strongly) complete, and if E^* were pseudo-metrizable, then the Baire theorem 9.4 would imply that E^* be of the second category.|||

The following theorem is fundamental for the rest of this section.

22.9   COUNTABLE INTERSECTION OF STRONG NEIGHBORHOODS   *Let E be a locally convex pseudo-metrizable space, and let $\{U_n\}$ be a sequence of convex circled strong neighborhoods of 0 in E^*.   Then the intersection $\bigcap \{U_n : n = 1, 2, \cdots\}$ is a strong neighborhood of 0 if and only if this intersection absorbs each strongly bounded set.*

PROOF   Let $U = \bigcap \{U_n : n = 1, 2, \cdots\}$.   If U is a strong neighborhood of 0, then clearly U absorbs strongly bounded sets.   Conversely, assume that U absorbs strongly bounded sets; then to show that U is a strong neighborhood of 0, it is sufficient to exhibit a weak* closed convex circled subset V which is radial at 0, since such a subset is the polar of a bounded subset of E.   Let $\{B_n\}$ be a co-base for the strongly bounded subsets of E^* such that each B_n is weak* compact convex circled (there is such a co-base by 22.3).   For each n, there is a positive number t_n such that $t_n B_n \subset \frac{1}{2} U$, and there is a weak* closed convex circled strong neighborhood W_n of 0 which is contained in $\frac{1}{2} U_n$.   Let $V_n = \langle t_1 B_1 \cup \cdots \cup t_n B_n \rangle + W_n$.   Then, since $\langle t_1 B_1 \cup \cdots \cup t_n B_n \rangle$ is weak* compact, V_n is a weak* closed convex circled strong neighborhood of 0 contained in U_n.   Let $V = \bigcap \{V_n : n = 1, 2, \cdots\}$; then it is clear that V is a weak* closed convex circled subset of U which absorbs each B_n, and consequently absorbs each point of E^*.|||

We obtain as a corollary to 22.9 the fact that E^*, which may fail to be evaluable, nevertheless satisfies a weakened form of the evaluability condition.

22.10   COROLLARY   *Let E be a locally convex pseudo-metrizable space.   If a strongly bounded subset B of E^{**} is the union of a sequence $\{B_n\}$ of strongly equicontinuous sets, then B is strongly equicontinuous.*

PROOF   Since $B = \bigcup \{B_n : n = 1, 2, \cdots\}$, taking polars in E^*, $B_0 = \bigcap \{(B_n)_0 : n = 1, 2, \cdots\}$.   Because $(B_n)_0$ is a strong neighborhood of 0 and B_0 absorbs strongly bounded sets, B_0 is a strong neighborhood of 0.|||

The next theorem states an extremal property of the strong topology for E^* which is similar to that proved for the topology of uniform convergence on compact sets in theorem 22.5. The result may be stated: The strong topology $\mathscr{S}$ for the adjoint E^* of a pseudo-metrizable locally convex space is the strongest of the locally convex topologies which agree with $\mathscr{S}$ on strongly bounded subsets of E^* (equivalently, on equicontinuous subsets of E^*).

22.11 LOCALIZATION OF STRONG TOPOLOGY *Let E be a locally convex pseudo-metrizable space. Then a convex circled subset A of E^* is a strong neighborhood of 0 if and only if $A \cap B$ is a neighborhood of 0 in B for each strongly bounded convex circled subset B of E^*.*

PROOF   If A is a strong neighborhood of 0, then $A \cap B$ is surely a neighborhood of 0 in B for each subset B of E^* which contains 0. Conversely, assume that A is a convex circled subset of E^* such that, for each strongly bounded convex circled subset B of E^*, $A \cap B$ is a neighborhood of 0 in B. Let $\{B_n\}$ be a co-base for strongly bounded sets in E^* such that each B_n is convex and circled and $B_n \subset B_{n+1}$. Then by the assumption, for each n, there is a convex circled strong neighborhood U_n of 0 in E^* such that $A \cap B_n \supset U_n \cap B_n$. Let $V_n = (A \cap B_n) + U_n$. Then $V_n \cap B_n \subset 3(A \cap B_n)$, for the following reason. If $h \in V_n \cap B_n$, then $h = f + g$, where $f \in A \cap B_n$ and $g \in U_n$, hence $g = h - f \in B_n + A \cap B_n \subset 2B_n$, and therefore $g \in 2(B_n \cap U_n) \subset 2(A \cap B_n)$ and $h = f + g \in 3(A \cap B_n)$. It follows that $\bigcap \{V_n : n = 1, 2, \cdots\} \subset 3A$, and by theorem 22.9, the proof reduces to showing that $\bigcap \{V_n : n = 1, 2, \cdots\}$ absorbs strongly bounded sets. For a fixed m, there is a positive number t such that $tB_m \subset U_m \cap B_m \subset A \cap B_m \subset A \cap B_n$ for $n \geq m$. Hence $tB_m \in \bigcap \{V_n : n \geq m\}$. Since each of $V_1, \cdots, V_{m-1}$ absorbs B_m, it is evident that B_m is absorbed by $\bigcap \{V_n : n = 1, 2, \cdots\}$.|||

22.12 COROLLARY *Let E be a locally convex pseudo-metrizable space. Then a linear transformation T of E^*, with the strong topology, into a locally convex space F is continuous if and only if T is continuous on each equicontinuous set (or equivalently, on each strongly bounded set) in E^*. Consequently, E^{**} with the strong topology is always a locally convex complete metrizable space.*

PROOF   The first half is evident. E^{**} with the strong topology is obviously locally convex and metrizable, and the completeness of E^{**} follows immediately from part (iii) of theorem 18.4.|||

Although the strong topology for the adjoint of pseudo-metrizable space may fail to be bound, there are boundedness properties which are of importance.   The following result is the key to these properties.

22.13   BOUND ABSORBING SUBSETS OF E^* *Let E be a pseudo-metrizable locally convex space, and let C be a convex circled subset of E^* which absorbs strongly bounded sets.   Then there is a strongly closed convex circled subset of C which absorbs strongly bounded sets.*

PROOF   Let $\{B_n\}$ be a co-base for the strongly bounded subsets of E^* such that each B_n is weak* compact convex and circled.   Then there is a sequence $\{t_n\}$ of positive numbers such that $\langle \bigcup \{t_n B_n : n = 1, 2, \cdots\} \rangle \subset C$.   Let $A_n = \langle t_1 B_1 \cup \cdots \cup t_n B_n \rangle$; then A_n is strongly closed (in fact, weak* compact) convex circled, $A_n \subset A_{n+1}$, and $A = \bigcup \{A_n : n = 1, 2, \cdots\} \subset C$.   If it is shown that $\frac{1}{2} A^- \subset A$, then the theorem is proved, because A^- absorbs strongly bounded sets.   Let $f_0 \notin A$; then $f_0 \notin A_n$ for each n, and since A_n is closed, there is a strong neighborhood V_n of 0 such that $(f_0 + V_n) \cap A_n$ is void.   Let $W_n = V_n + \frac{1}{2} A_n$.   Then $(f_0 + W_n) \cap \frac{1}{2} A_n$ is void, and the intersection $\bigcap \{W_n : n = 1, 2, \cdots\}$ absorbs each A_n and hence each B_n.   By theorem 22.9, $f_0 + \bigcap \{W_n : n = 1, 2, \cdots\}$ is a strong neighborhood of f_0, and this neighborhood is disjoint from $\frac{1}{2} A$. Hence $f \notin \frac{1}{2} A^-$.|||

Recall that if E is a linear topological space, then E^b is the space of all bounded linear functionals (that is, functionals which are bounded on each bounded subset of E).

22.14   COROLLARY   *If E is a locally convex pseudo-metrizable space, then each strongly bounded subset of E^{*b} is contained in the weak* closure (that is, $w(E^{*b}, E^*)$-closure) of a strongly bounded subset of E^{**}.*

PROOF   Let B be a strongly bounded subset of E^{*b}.   Then B_0 is a convex circled subset of E^* which absorbs strongly bounded sets.   Hence, in view of the foregoing theorem, there is a strongly closed convex circled subset C of B_0 which absorbs strongly bounded sets.

Let D be the polar of C in E^{**}; then D is strongly bounded, $C = D_o$, and $D_o \subset B_o$. It follows that the polar in E^{*b} of D_o, namely the weak* closure of D in E^{*b}, includes $B.|||$

22.15 STRONG TOPOLOGY FOR THE ADJOINT *If E is a locally convex pseudo-metrizable space, then the following three statements concerning the strong topology for E^* are equivalent.*

(i) *E^* with the strong topology is evaluable.*
(ii) *The strong topology for E^* is bound.*
(iii) *E^* with the strong topology is barrelled.*

PROOF   That (i) implies (ii) is clear from theorem 22.13.   Since E^* is complete relative to the strong topology (22.3), (ii) implies (iii) by virtue of theorem 19.5.   By theorem 20.4, (iii) implies (i).$|||$

Finally, there are two special cases in which the adjoint is a barrelled space (equivalently, the strong topology is bound).

22.16 REFLEXIVE METRIZABLE SPACES *If E is a reflexive locally convex metrizable space, then the strong topology on E^* is bound.*

PROOF   If E is reflexive, so is E^*, and hence E^* is evaluable (see 20.5).$|||$

22.17 SEPARABLE ADJOINT OF A PSEUDO-METRIZABLE SPACE *Let E be a locally convex pseudo-metrizable space such that E^* is separable relative to the strong topology.   Then the strong topology for E^* is bound.*

PROOF   In view of theorem 22.15, it is sufficient to show that E^* with the strong topology is evaluable, that is, each convex circled strongly closed subset A of E^* which absorbs strongly bounded sets is a strong neighborhood of 0.   Since E^* is separable and A is closed, there is a sequence $\{f_n\}$ which is dense in $E^* \sim A$.   For each n, let U_n be a convex circled strong neighborhood of 0 such that $f_n \notin U_n$ and $A \subset U_n$.   Then $U = \bigcap \{U_n : n = 1, 2, \cdots\} \supset A$ and, by theorem 22.9, U is a strong neighborhood of 0.   Since $E^* \sim U^i$ (U^i is the interior of U) is a closed set containing $\{f_n\}$, $E^* \sim A \subset E^* \sim U^i$ and $A \supset U^i$; hence, A is a strong neighborhood of 0.$|||$

PROBLEMS

A CONDITION FOR COMPLETENESS

A locally convex pseudo-metrizable space is complete if the closed convex circled extension of every sequence convergent to zero is weakly

compact.   In particular, it is sufficient if the closed convex circled extension of every compact, or totally bounded, set is weakly compact.

B EMBEDDING THEORIES

(a) Let R be the space of real numbers and $C(R)$ the space of continuous scalar-valued functions on R with the topology of compact
convergence.   Any locally convex metrizable separable space E is isomorphic to a subspace of $C(R)$.   (The theorem can be established by
proving the following

Lemma   There is a continuous mapping f of R onto the adjoint E^*
with the weak* topology, such that the family of sets $f[K]$, where K is a
compact subset of R, form a co-base for the equicontinuous subsets of E^*.

To prove this, choose a co-base $\{B_n: n = 0, 1, \cdots\}$ of the equicontinuous
subsets of E^* so that each B_n is convex, circled, weak* compact and weak*
metrizable, and so that $B_n \subset B_{n+1}$ for each n.   There is a continuous
function f_n on the Cantor set in $[0,1]$ onto B_n with the weak* topology: see,
for example, Kelley [5] Ch. V, O.   Define f by $f(2n + s) = f_n(s)$ whenever
s belongs to the Cantor set, and by continuity and linearity elsewhere.)

(b) A separable normed space is isometrically isomorphic to a subspace
of the space $C([0,1])$ of continuous scalar-valued functions on $[0,1]$ with
the topology of uniform convergence.

C INDUCTIVE LIMITS V (see 1I, 16C, 17G, 19A)

Let $(E,\mathcal{T})$ be the inductive limit of an increasing sequence of Fréchet
spaces $(E_n,\mathcal{T}_n)$.   Then E is called a *generalized LF space*.   If each E_n is a
closed subspace of $(E_{n+1},\mathcal{T}_{n+1})$, the inductive limit is called an *LF space*.
In this case, the open mapping theorem shows that the injection of E_n into
E_{n+1} is a homeomorphism; thus E is a strict inductive limit (see 17G(b)).
It follows in particular that E is a Hausdorff space and is therefore complete
(17G(a)).

(a) A generalized LF space is a barrelled and a bound space, and so also a
Mackey space and evaluable.

(b) If $\{E_n\}$ is a strictly increasing sequence and E is an LF space, then E
cannot be of the second category and so cannot be metrizable.   An LF
space has a countable co-base for bounded sets if and only if it is the inductive limit of a sequence of Banach spaces; equivalently, the adjoint
with its strong topology is metrizable.

(c) The adjoint of a reflexive Fréchet space with its strong topology is a
generalized LF space.   The adjoint of a reflexive Fréchet space with its
strong topology is a complete, barrelled, and bound space.

(d) A closed subspace, or a quotient by a closed subspace, of even a
reflexive LF space may fail to be an LF space.   The Krein-Šmulian
theorem 22.6 does not extend even to reflexive LF spaces; further, an LF
space is not necessarily fully complete.   Not all generalized LF spaces
are complete.   (See 20D.)

(e) A linear mapping of one Hausdorff generalized LF space into
another is continuous if its graph is closed; a linear mapping of one onto
another is open if its graph is closed.   (See 19B.)

(f) Let X be a topological space in which there is a countable co-base for the compact sets (for example, a space which is locally compact and is the union of a sequence of compact sets), and let $K(X)$ be the space of continuous functions on X of compact support with the inductive limit topology (see 8I, 16C(j)). Then $K(X)$ is an LF space; if X is not compact, $K(X)$ cannot be metrizable (because of (b)).

D SPACES OF ANALYTIC FUNCTIONS II (see 20H, 22J)

Let C be a closed proper subset of the Riemann sphere, and consider the family $\mathcal{U}$ of all open proper subsets containing C, directed by putting $U \geq V$ iff $U \subset V$. For each $U \in \mathcal{U}$, let $A(U)$ be the space of analytic functions on U with the topology of compact convergence (see 20H). For $U \geq V$, let I_{UV} be the natural injection of $A(V)$ into $A(U)$; then $\{A(U): U \in \mathcal{U}, \geq, I_{UV}\}$ is an inductive system. Let $F(C) = \lim \text{ind} \{A(U): U \in \mathcal{U}\}$ with the inductive limit topology (see 16C). The space $F(C)$ can be identified with the set of equivalence classes of functions analytic on (some open set containing) C, calling f and g equivalent iff there is some $U \in \mathcal{U}$ on which f and g agree.

(a) Each of the mappings I_{UV} is continuous. Since $\mathcal{U}$ contains a countable cofinal subset $\{U_n: n = 1, 2, \cdots\}$, $F(C) = \lim \text{ind} \{A(U_n): n = 1, 2, \cdots\}$ and the inductive limit topology defined by the Fréchet spaces $A(U_n)$ is the topology of $F(C)$. Thus $F(C)$ is a generalized LF space. In fact, $F(C)$ is the inductive limit of a sequence of Banach spaces. (Let B_n be the space of functions analytic on U_n, continuous on U_n^-, and vanishing at ∞ in case $\infty \in U_n$, with the topology of uniform convergence on U_n^-.)

(b) The spaces $A(U)$ and $F(\sim U)$ may be paired, in the following way. If $g' \in F(\sim U)$, there is a member g of g' defined on an open set $V \supset \sim U$ (and we may suppose $\infty \notin U \cap V$); there is an open set W with $\sim V \subset W \subset W^- \subset U$ and whose oriented boundary ∂W is rectifiable. For each $f \in A(U)$, put $\langle f, g' \rangle = (1/2\pi i) \int_{\partial W} f(z) g(z) dz$. The value of the integral is independent of the member of g' chosen and of the particular choice of V and W satisfying the above requirements. This pairing is separated.

(c) The adjoint of $A(U)$ is (algebraically) isomorphic to $F(\sim U)$. (If $\phi \in A(U)^*$, there is a compact subset K of $U \sim \{\infty\}$ and a constant k with $|\phi(f)| \leq k \sup \{|f(z)|: z \in K\}$. There is a natural mapping of $A(U)$ into $F(K)$ and so ϕ may be regarded as a linear functional on the image of this mapping; extend ϕ to $F(K)$, so that if $h' \in F(K)$, $|\phi(h')| \leq k \sup \{|h(z)|: z \in K\}$. For each $\zeta \notin K$, let $h_\zeta(z) = \dfrac{1}{\zeta - z}$; then $h_\zeta' \in F(K)$. Put $g(\zeta) = \phi(h_\zeta')$. Then $g \in F(\sim U)$; show that, for some suitable open set W,

$$\langle f, g' \rangle = \frac{1}{2\pi i} \int_{\partial W} f(\zeta) \phi(h_\zeta') d\zeta = \phi(f_0') \quad \text{where } f_0(z) = \frac{1}{2\pi i} \int_{\partial W} \frac{f(\zeta)}{\zeta - z} d\zeta,$$

using, for example, approximating sums or vector integration. Now $f_0(z) = f(z)$ in some open set containing K; thus $\phi(f_0') = \phi(f)$.)

(d) With its strong topology, $A(U)^*$ is isomorphic to $F(\sim U)$. (Since both are generalized LF spaces, it is sufficient, by 22C(e), to show that the algebraic isomorphism of $F(\sim U)$ onto $A(U)^*$ is continuous. This can be achieved by proving that, for each open set V containing $\sim U$, the mapping of $A(V)$ into $A(U)^*$ is continuous.)

E SPACES $l^p(\omega)$

A non-negative real valued function ω defined on a set A defines a measure μ_ω on the σ-ring of all subsets of A by the formula $\mu_\omega(B) = \sum \{\omega(t); t \in B\}$; ω is called the *weight function* for the measure μ_ω. The space $L^p(A, \mu_\omega)$ bears a close resemblance to the l^p spaces and it will therefore be denoted by $l^p(\omega)$. Thus $l^p(\omega)$ is the space of (equivalence classes of) real or complex valued functions x with

$$\|x\|_p = (\int |x(t)|^p d\mu_\omega(t))^{1/p} = (\sum \{|x(t)|^p \omega(t): t \in A\})^{1/p} < \infty,$$

two functions x and y being regarded as equivalent iff $x(t) = y(t)$ whenever $\omega(t) \neq 0$. The adjoint of $l^p(\omega)$ for $1 \leq p < \infty$ is $l^q(\omega)$, where q is the index conjugate to p ($q = \infty$ if $p = 1$ and $q = p/(p-1)$ if $p > 1$).

(a) If y is a real or complex valued function on A such that $xy \in l^1(\omega)$ for all $x \in l^p(\omega)$, then $y \in l^q(\omega)$.

(b) If $\omega(t) > 0$ for all $t \in A$, then $l^p(\omega)$ is reflexive for $1 < p < \infty$, and $l^1(\omega)$ and $l^\infty(\omega)$ are reflexive if and only if A is a finite set.

(c) In $l^1(\omega)$ each sequence which is Cauchy in the weak topology is convergent in the norm topology, and each set which is compact in the weak topology is compact in the norm topology. (The second part is a consequence of the first, which can be proved in the same way as the special case of l^1 in 16F.)

F KÖTHE SPACES

These spaces are projective and inductive limits of spaces of the type $l^p(\omega)$, defined in the previous problem, and their adjoints. For each weight function ω on A, denote by $l^p(\omega)^+$ the space of real or complex valued functions y on A such that $y(t) = 0$ whenever $\omega(t) = 0$ and $\sum \{|x(t)y(t)|: t \in A\} < \infty$ for all $x \in l^p(\omega)$.

(a) The spaces $l^p(\omega)$ and $l^p(\omega)^+$ are paired by the bilinear functional $\langle x, y \rangle = \sum \{x(t)y(t): t \in A\}$; $y \in l^p(\omega)^+$ if and only if there is a $z \in (l^p(\omega))^*$ with $y = \omega z$, and the natural norm associated with the pairing is $\|y\| = \|z\|_q$, where q is the index conjugate to p. This pairing does not depend explicitly on the function ω and is convenient when a family of weight functions comes simultaneously under consideration.

(b) There is a natural partial ordering among weight functions in which $\omega \geq \sigma$ means that $\omega(t) \geq \sigma(t)$ for all t in A. Let Ω be a family of weight functions directed by this partial ordering. If $\omega \geq \sigma$, $l^p(\omega) \subset l^p(\sigma)$ and so, with injections for the canonical mappings, the family $\{l^p(\omega): \omega \in \Omega\}$ forms a projective system. The projective limit of this system is algebraically isomorphic to $E^p(\Omega) = \bigcap \{l^p(\omega): \omega \in \Omega\}$. The space $E^p(\Omega)$ is dense in each $l^p(\omega)$. We suppose that $E^p(\Omega)$ has the projective limit topology; the intersections of spheres in $l^p(\omega)$ with $E^p(\Omega)$ form a local base for this topology.

(c) The space $E^p(\Omega)$ is Hausdorff iff to each $t \in A$ corresponds an $\omega \in \Omega$ with $\omega(t) > 0$. For each p with $1 \leqq p \leqq \infty$, $E^p(\Omega)$ is complete. The adjoint of $E^p(\Omega)$ is algebraically isomorphic to $E^p(\Omega)^+ = \bigcup \{l^p(\omega)^+ : \omega \in \Omega\}$ with the canonical pairing $\langle x, y \rangle = \sum \{x(t)y(t) : t \in A\}$ for $x \in E^p(\Omega)$ and $y \in E^p(\Omega)^+$.

(d) In $E^p(\Omega)$ each sequence which is Cauchy in the weak topology is convergent in the projective limit topology.

(e) If $\omega \geqq \sigma$, $l^p(\sigma)^+ \subset l^p(\omega)^+$ and so, with injections for the canonical mappings, the family $\{l^p(\omega)^+ : \omega \in \Omega\}$ forms an inductive system, for which $E^p(\Omega)^+$ is the inductive limit. The inductive limit topology on $E^p(\Omega)^+$ is stronger than the strong topology, but if $1 < p < \infty$ they coincide and $E^p(\Omega)$ is semi-reflexive. (Prove that any linear functional continuous relative to the inductive limit topology on $E^p(\Omega)^+$ is the evaluation at a member of $E^p(\Omega)$.)

The case of l^1 itself shows that this last result may fail for $p = 1$.

G   COUNTER-EXAMPLE ON FRÉCHET SPACES

The Fréchet space E constructed in this example has the property that the strong topology of its adjoint E^* is not bound. This is equivalent, by 22.15, to saying that E^* is not evaluable or that it is not barrelled. Thus the example shows that none of these three properties carries over to the adjoint with its strong topology. Next, the Fréchet space E may be embedded as a closed subspace of a countable product F of Banach spaces; then F^* with its strong topology is a countable direct sum of Banach spaces, and E^* is algebraically isomorphic to F^*/E°. The quotient of $s(F^*, F)$ is bound; thus the strong topology $s(F^*/E^\circ, E)$ is not the quotient of $s(F^*, F)$. Finally, this provides an example of a strong adjoint of a projective limit which is not the inductive limit of the corresponding strong adjoints.

Let I be the set of positive integers, let $A = I \times I$, and let Ω be a sequence $\{\omega_n\}$ with the following properties: if $n > m$, then $\omega_n \geqq \omega_m$; for each $(i, j) \in A$ and each n, $\omega_n(i, j) > \frac{1}{2}$; for $i \leqq n$, $\lim_j \omega_n(i, j) = \infty$; for $i > n$, $\lim \sup_j \omega_n(i, j) < 1$.

Let $\mathscr{B}$ be the family of subsets B of A of the following type: there are a positive integer i_0 and a sequence $\{k_i\}$ of positive integers such that $B = \{(i, j) : i \geqq i_0 \text{ and } j > k_i\}$. If for each $B \in \mathscr{B}$, x_B denotes the characteristic function of the singleton $\{(i_0, k_{i_0})\}$, then $\{x_B, B \in \mathscr{B}, \subset\}$ is a net in $E = E^1(\Omega)$; regard it as a net in the algebraic dual of $E^* = E^1(\Omega)^+$.

(a) A universal subnet of $\{x_B\}$ converges relative to $w((E^*)', E^*)$ to a linear functional ϕ on E^* which is continuous relative to the inductive limit topology, that is, which is bounded on E^*.

(b) Each strong neighborhood of 0 in E^* contains a neighborhood of the form $V = \langle \bigcup_n e_n S_n^\circ \rangle^-$, where the closure is taken with respect to the $w(E^*, E)$-topology, S_n is the unit sphere in $l^1(\omega_n)$, and $e_n > 0$.

(c) Given V of the form in (b), choose the sequence $\{k_n\}$ so that, if $j \geqq k_n$, then $1 \leqq (e_n/2^n)\omega_n(n, j)$, and let $y(i, j) = 1$ if $j \geqq k_i$, $y(i, j) = 0$ otherwise. Then $y \in V$, and $\phi(y) = 1$. (Let $y_n(n, j) = 1$ if $j \geqq k_n$, and $y_n(i, j) = 0$ otherwise. Then $y_n \in (e_n/2^n)S_n^\circ$, while $\sum \{y_n : n = 1, 2, \cdots\}$ converges to y with respect to $w(E^*, E)$.)

(d) E is a Fréchet space. The strong topology of its adjoint is not bound.

H   An adjoint with a bound topology

Let E be a locally convex pseudo-metrizable space and let s^b be the bound extension of $s = s(E^*,E)$.

(a) If A is a strongly separable subset of E^*, then s and s^b coincide on A. In particular, a sequence is s-convergent if and only if it is s^b-convergent.

(b) If every strongly bounded subset of E^* is strongly metrizable, then E^* is a bound space.

(For (a), use a method similar to 22.17; for (b), use 22.11 and (a).)

I   Example on Montel spaces

For each p, $1 \leqq p < \infty$, there is a Fréchet Montel space E_p which contains a closed subspace F_p such that E_p/F_p is topologically isomorphic to l_p.

With the definitions of 22F, let $E_p = E^p(\Omega)$, where I is the set of positive integers, $A = I \times I$, and Ω is the sequence $\{\omega_n : n = 1, 2, \cdots\}$ with $\omega_n(i,j) = j^n$ if $i \leqq n$ and $\omega_n(i,j) = i^n$ if $i > n$.

(a) For each p, $1 \leqq p < \infty$, E_p is a Fréchet Montel space. (To show that a closed bounded set B is compact, show that for each $e > 0$ and each n there is a bounded finite dimensional set B' such that every point of B is within distance e of B' with respect to the norm in $l^p(\omega_n)$.)

(b) The transformation T defined by

$$T(x)(j) = \sum \{x(i,j) : i = 1, 2, \cdots\}$$

is a continuous and open linear transformation from E_p onto l^p. (Prove that the transformation T^* defined by

$$T^*(y)(i,j) = y(j)$$

is a transformation from l^q (where q is the conjugate of p) into $E_p^* = E^p(\Omega)^+$, that T^* is the adjoint of T, that T^* is one-to-one, and that the range of T^* is $w(E_p^*,E_p)$-closed.)

J   Spaces of analytic functions III (see 20H, 22D)

This problem gives a Köthe space representation of the space $A(U)$ of functions analytic in the interior $\{z : |z| < 1\}$ of the unit circle in the complex plane. Take any (fixed) sequence $\{r_k\}$ of positive numbers increasing to 1. Then the topology of $A(U)$ is determined by the sequence $\{p_k : k = 1, 2, \cdots\}$ of pseudo-norms, where $p_k(f) = \sup\{|f(z)| : |z| \leqq r_k\}$.

(a) For each k, let ω_k be the weight-function defined on the set I of nonnegative integers by putting $\omega_k(i) = (r_k)^i$, and let $\Omega = \{\omega_k : k = 1, 2, \cdots\}$. Then for each $f \in A(U)$, the sequence $\{a_n : n = 0, 1, 2, \cdots\}$ of coefficients in the Taylor expansion of f about the origin lies in $E^1(\Omega)$ and the mapping J defined by $J(f) = \{a_n\}$ is a topological isomorphism of $A(U)$ onto $E^1(\Omega)$.

(b) For each sequence $\{b_n : n = 0, 1, 2, \cdots\}$ in $E^1(\Omega)^+$, put $J^*(\{b_n\}) = f$, where $f(z) = \sum\{b_n z^{-n-1} : n = 0, 1, 2, \cdots\}$. Then J^* is a topological isomorphism of $E^1(\Omega)^+$ onto the space $F(\sim U)$, which is the adjoint of $A(U)$ under its strong topology. (Use 22D(d).) Also J^* is the adjoint of J.

(c) Let T be a continuous linear mapping of $A(U)$ into itself. Then there exists a unique kernel B such that $T(f) = g$ where, for $|z| < 1$,

$$g(z) = \frac{1}{2\pi i} \int_{|\zeta| = r(z)} f(\zeta) B(\zeta, z) d\zeta.$$

Here the contour integral is taken round a circle of radius $r(z) < 1$, which may depend on z, and the kernel B has the following properties:

(i) the mapping $(\zeta, z) \to B(\zeta, z)$ is continuous in the two variables, for $|\zeta| \geq 1$ and $|z| < 1$;

(ii) to each $r < 1$ corresponds an $r' < 1$ such that, whenever $|z| < r$, the mapping $\zeta \to B(\zeta, z)$ is analytic on $\{\zeta : |\zeta| > r'\}$ and vanishes at infinity;

(iii) for each ζ with $|\zeta| > 1$ the mapping $z \to B(\zeta, z)$ is analytic on U.

Conversely, if B is a kernel satisfying the conditions (i), (ii), (iii), it defines a unique continuous linear mapping of $A(U)$ into itself. (Let $v_z(\zeta) = (\zeta - z)^{-1}$; if T is given, put $B(\zeta, z) = u_z(\zeta)$ where $u_z = T^*(v_z)$. In the converse, the main difficulty is to show that the function $T(f)$ defined by such a kernel is an analytic function in U. This can be done using (ii) and Hartogs' lemma—see, for example, Bochner and Martin [2].)

Appendix

ORDERED LINEAR SPACES

This appendix is a brief introduction to the theory of ordered linear spaces. Many linear topological spaces have a natural order, and the relation between order and topology is of considerable interest. For example, we may ask whether positive linear functionals are necessarily continuous, or whether each continuous linear functional is the difference of two positive functionals. The first section of this chapter is devoted to introductory material and to a few results on problems of this type. The second section contains two theorems of Kakutani which characterize two "extremal" types of Banach lattice.

23 ORDERED LINEAR SPACES

Each continuous linear functional on an ordered normed linear space E is the difference of positive functionals if and only if there is an equivalent monotonic norm; in particular, this is the case if the positive cone E is normal in the sense of Krein. A linear functional on a vector lattice is the difference of positive functionals if and only if it is bounded on order bounded sets. Hence the Mackey topology $m(E,E^*)$, where E^* is the order dual of E, is bound. The class of lattice pseudo-norms for E is a base for the class of $m(E,E^*)$-continuous pseudo-norms, and the lattice operations are continuous relative to $m(E,E^*)$. The evaluation map of E into E^{**} is a lattice homomorphism.

It is assumed throughout the section that E is a real linear space; no topology is assumed, *ab initio*. It has already been noted that there is a natural correspondence between cones in E (convex subsets, closed under multiplication by non-negative scalars) and vector orderings $\geq$; that is, between cones and partial orderings $\geq$ such that: $x \geq y$ if and only if $x - y \geq 0$; if $x \geq 0$ and t is a non-negative

224

scalar, then $tx \geqq 0$; and if $x \geqq 0$ and $y \geqq 0$, then $x + y \geqq 0$. An **ordered linear space (partially ordered linear space)** is a real linear space E with a vector ordering. In view of the correspondence which has just been noted, the ordering $\geqq$ is completely determined by the cone $\{x \colon x \geqq 0\}$. This cone is called the **positive cone** of the ordering and members of the cone are called **positive**.

If C is a cone in E, then $C - C$ is clearly a linear subspace F of E. In the most interesting cases, the subspace F is identical with E; in particular, if the radial kernel of C is non-void, then $C - C = E$. If $C - C$ is the entire space E, then the cone **generates** E (the cone C is **reproducing**). This condition can be interpreted easily in terms of the order $\geqq$ determined by the cone. If C generates E, then for each x in E there are u and v in C such that $x = u - v$, and consequently there is a positive element u such that $u \geqq x$. Conversely, if for each x in E there is a positive element u, such that $u \geqq x$, then $x = u - (u - x)$ and it follows that $E = C - C$. Consequently, C generates E if and only if for each x there is u such that $u \geqq 0$ and $u \geqq x$. An equivalent form of this requirement can be obtained by a simple translation argument: the positive cone C generates E if and only if the ordering $\geqq$ directs E; that is, if and only if for members x and y of E there is z in C such that $z \geqq x$ and $z \geqq y$.

An ordering $\geqq$ of a linear space is studied by means of the linear functionals which are order preserving; that is, by means of the linear functionals which are non-negative on the positive cone C. The set $C^\star$ of all such linear functionals on E is the **dual cone**, and the set $C^\star - C^\star$ of all differences of members of $C^\star$ is the **order dual** $E^\star$ of E. Of course, the dual cone may consist of the single element 0, as is certainly the case if $C = E$. The cone $C^\star$ defines an order on $E^\star$ which is called the **dual ordering**. If C generates E, then the dual ordering is anti-symmetric, in the sense that, if $f \geqq g$ and $g \geqq f$, then $g = f$, for in this case $f - g$ is zero on each member of C and hence on $E = C - C$.

If a cone C is closed relative to a locally convex topology for E and $x \notin C$, then by the separation theorem 14.4, there is a linear functional f such that $f(x) < \inf \{f(y) \colon y \in C\}$. Since C is a cone $f(x) < 0$, and f is non-negative on C; that is, $f \in C^\star$. In particular, if C is closed relative to the strongest locally convex topology $\mathscr{T}$, which has the family of all convex circled radial sets at 0 as a local base, then each point which does not belong to C can be separated from C by a member of $C^\star$. The preceding discussion can be summarized:

23.1   THEOREM   *If C is a cone which is closed in the strongest locally convex topology for E, then for each x in E $\sim$ C there is f in the dual cone C^* such that $f(x) < 0$.   Consequently, in this case, $x \in C$ if and only if $f(x) \geq 0$ for each f in C^*.*

If a linear space has both a topology and an order, it is of some importance to know under what conditions each continuous linear functional is the difference of continuous positive functionals.   This problem will be investigated now, deferring until later certain questions of a purely order theoretic nature.   The first result concerns ordered pseudo-normed spaces.

23.2   THEOREM   *Let C be a cone in a pseudo-normed space E and let B be the set of all positive functionals which are of norm at most one. Then each member of E^* is the difference of bounded positive linear functionals if and only if $B - B$ is a neighborhood of 0 in E^*.*

PROOF   If $B - B$ is a neighborhood of 0 in E^*, then clearly each member of E^* is the difference of continuous positive functionals, and only the converse requires proof.   First observe that B is weak* compact, because it is a weak* closed subset of the unit sphere in E^*, and hence $B - B$ is weak* compact.   It follows that $B - B$ is weak* closed, convex, and circled.   If $C^* \cap E^*$ generates E^*, then for each f in E^*, some non-zero scalar multiple of f belongs to $B - B$: that is, $B - B$ is radial at 0.   Finally, since the unit sphere in E^* is weak* compact, the absorption theorem 10.2 asserts the existence of a scalar a, such that $a(B - B)$ contains the unit sphere, and hence $B - B$ is a neighborhood of 0.|||

Observe that if each continuous linear functional is the difference of two continuous positive functionals, then there is a certain uniformity present, in the following sense.   If a sphere of radius r is contained in $B - B$, then any linear functional of norm at most one is the difference of two positive functionals of norm at most $1/r$.

There is a useful corollary to the foregoing theorem: if E is a pseudo-normed space with a cone C such that each member of E^* is the difference of positive members then the set B of all continuous positive functionals of norm at most one may be used to construct a new pseudo-norm of E, as follows: let $p(x) = \sup \{|f(x)| : f \in B\}$. Clearly $p(x) \leq \|x\|$, because B is contained in the unit sphere, and since $B - B$ contains a multiple, say $1/r$, of the unit sphere, $\|x\| \leq 2rp(x)$.   Hence p is equivalent to the original pseudo-norm.   But notice that p is **monotonic** on E; that is, if x and y are members of C and $x \geqq y$, then $p(x) \geqq p(y)$.

23.3 COROLLARY *If each continuous linear functional on an ordered, pseudo-normed space E is the difference of continuous positive functionals, then there is an equivalent monotonic pseudo-norm for E.*

It is true that if p is a pseudo-norm which is monotonic on a cone C, then each continuous linear functional is the difference of continuous positive functionals; however, a slightly stronger statement can be proved. A cone C in a pseudo-normed space E is **normal** relative to the pseudo-norm if and only if there is a positive number e such that if x and y are positive, $\|x\| \geqq 1$, and $\|y\| \geqq 1$, then $\|x + y\| \geqq e$. Geometrically, this amounts to requiring that the angle between positive vectors, as computed in terms of distance, is bounded away from π. This condition can be rephrased in a convenient way.

23.4 LEMMA *A cone C in a pseudo-normed linear space is normal if and only if $(S + C) \cap (S - C)$ is bounded, where S is the unit sphere.*

PROOF If C is normal, there is a constant k (the reciprocal of the constant e of the definition) such that, if x and y are positive and $\|x + y\| < 1$, then either $\|x\|$ or $\|y\|$ is less than k. If z is a member of $(S + C) \cap (S - C)$, then $z = u + x = v - y$ for some x and y in C and u and v in S. Then $\|(x + y)/2\| \leqq 1$, and hence either $\|x/2\| < k$ or $\|y/2\| < k$. In either case, since both u and v belong to S, $\|z\| < 2k + 1$, and half of the statement of the lemma is proved. To prove the converse, suppose that $\|z\|$, for z in $(S + C) \cap (S - C)$, is bounded by k. If x and y are positive elements and $\|x + y\| < 1$ then, since $y = (x + y) - x$, it is true that $y \in (S + C) \cap (S - C)$. It follows that $\|y\| \leqq k$, and it is clear that the cone C is normal.|||

23.5 THEOREM *If a cone C is normal in a pseudo-normed space, or if, equivalently, $(S - C) \cap (S + C)$ is bounded, where S is the unit sphere about 0, then each bounded linear functional is the difference of two bounded positive linear functionals.*

PROOF Let B be the set of positive functionals of norm at most one. In view of theorem 23.2 above, it must be shown that $B - B$ is a neighborhood of 0 in the adjoint E. Since $B - B$ is weak* compact and therefore closed, and is convex and circled, $B - B$ is the polar of the subset $(B - B)_0$ of E, and the problem reduces to showing that $(B - B)_0$ is bounded; it will, in fact, be shown that B_0, which contains $(B - B)_0$, is bounded. To this end, notice that if x is a member of E such that $x/2$ does not belong to the convex set $S + C$, which is radial at 0, then there is a non-zero linear functional f on E such that $f(x/2) \leq \inf \{f(y): y \in S + C\}$ by theorem 14.2. Since f is not identically zero on S, $f(x/2)$ is negative, and it may be supposed that $f(x/2) = -1$; since C is a cone, f must be a positive functional, and since f is bounded below by -1 on S, $\|f\| \leq 1$. Then $f \in B$, and it follows that x is not a member of B_0. This proves that $B_0 \subset 2(S + C)$, and since B_0 is circled, $B_0 \subset 2(-S - C) = 2(S - C)$. It follows that B_0 is a subset of $2[(S + C) \cap (S - C)]$, and is hence bounded, and the proof is complete.|||

The following is an immediate consequence of 23.3 and 23.5.

23.6 THEOREM *Each continuous linear functional on an ordered, pseudo-normed space is the difference of positive continuous linear functionals if and only if there is an equivalent pseudo-norm which is monotonic relative to the order.*

The existence of an order in a linear space leads to a natural definition of order bounded set. A subset A of an ordered linear space is **order bounded** if and only if there are members x and y of the space which, respectively, precede and follow each member of A in the ordering $\leq$; that is, such that $A \subset \{z: x \leq z \leq y\}$. Each linear functional which is the difference of positive linear functionals is necessarily bounded on order-bounded sets; under certain circumstances the converse is true.

23.7 THEOREM *If C is a normal cone in a pseudo-normed space E, then each order bounded set is bounded. If C is a cone containing an interior point x of E then C is normal if and only if the set $\{y: -x \leq y \leq x\}$ is bounded.*

PROOF If C is a normal cone then each continuous linear functional is the difference of positive functionals and is hence bounded on order bounded sets. That is, each order bounded set is weakly bounded, and is therefore bounded. To prove the second statement, notice that if C is normal, then surely the order bounded set $\{y: -x \leq y \leq x\}$ is bounded, by the preceding remarks. Finally, if x is an interior point of C, then for some positive r, $x + rS \subset C$, where S

is the unit sphere.   Then $x + rS + C \subset C$, hence $\{y: -x \leqq y \leqq x\} =$ $(-x + C) \cap (x - C) \supset (rS + C) \cap (-rS - C)$ and, if $\{y: -x \leqq y \leqq x\}$ is bounded, then $(rS + C) \cap (-rS - C)$ is bounded.   It follows that $(S + C) \cap (S - C)$ is bounded.|||

The theory of ordered spaces is simplified a great deal if further assumptions are made about the ordering.   An ordered space $(E, \geqq)$ is called a **vector lattice** if and only if for each x and y in E there is a unique supremum of x and y in E; that is, there is a unique member z of E with the property that z is greater than each of x and y, and is less than every member of E which has this property.   The supremum of x and y is called the **join** and is denoted $x \vee y$.   A lattice ordering is necessarily anti-symmetric (if $x \geqq y$ and $y \geqq x$, then $x = y$) because the supremum of two elements is supposed to be unique.   Geometrically, the ordering is a lattice ordering if and only if the intersection of two translates, $x + C$ and $y + C$, of the positive cone is a translate, $x \vee y + C$, of the cone.   As an example one may consider two-dimensional real Euclidean space with the positive cone $C_1 = \{(r,s): r \geqq 0 \text{ and } s \geqq 0\}$ or the cone $C_2 = \{(r,s): r > 0, \text{ or } r = 0 \text{ and } s \geqq 0\}$.   The orderings corresponding to these two cones are intrinsically quite different.   For the first, C_1, it is true that if x and y are positive elements and $tx \leqq y$ for every positive scalar t, then $x = 0$.   Such a lattice is called **Archimedean**; it is sometimes said, in this case, that there is no positive element which is "infinitely small" relative to another positive element.   The ordering defined by C_2 is non-Archimedean, since $t(0,1) \leqq (1,1)$ for all positive t.   Geometrically, the ordering is Archimedean if and only if, for each positive x, the set $C \cap (x - C)$ contains no half line.   It is known, although it will not be proved here, that any finite dimensional vector lattice can be constructed from the usual ordering of the real numbers by taking products, assigning either the product ordering (as for C_1) or the lexicographic ordering (as for C_2).

If E is any vector lattice, then scalar multiplication by -1 is order inverting.   It follows that there is then a greatest element which is less than or equal to each of two given elements; that is, there is an infimum of each pair of elements.   The **meet** of x and y, denoted $x \wedge y$, is defined to be the infimum of x and y.   Using the fact that multiplication by -1 is order inverting, it is evident that $x \wedge y = -[(-x) \vee (-y)]$.   Either of the operations, meet or join, determine the ordering entirely, for $x \geqq 0$ if and only if $0 = x \wedge 0$, and $x \leqq 0$ if and only if $0 = x \vee 0$.

A few simple computation rules are necessary.   First, because the ordering is invariant under translation by a member of E, the smallest element of E which is greater than both $x + z$ and $y + z$ is $x \vee y + z$. It follows that $x \vee y + z = (x + z) \vee (y + z)$, and a similar relation holds for the meet. If, in the preceding equation, z is replaced by $-x - y$, then, rearranging terms and using the definition of the meet of x and y, one obtains the useful identity: $x + y = x \vee y + x \wedge y$. In particular, $x = x \vee 0 + x \wedge 0$. The first of the two latter terms, $x \vee 0$, is called the **positive part** of x and is denoted x^+. The element $-(x \wedge 0) = (-x) \vee 0$ is called the **negative part** of x and is denoted x^-. Clearly $x^+ \geq 0$, $x^- \geq 0$, and $x = x^+ - x^-$. A particular consequence of this equality is that the positive cone generates E.

There is another way of describing the positive part of an element x of a vector lattice. Two positive elements, x and y are **disjoint** if and only if $x \wedge y = 0$. Observe that x^+ and x^- are always disjoint, for $(x \vee 0) \wedge [-(x \wedge 0)]$ may be written, using the translation invariance, as $[(x \vee 0 + x \wedge 0) \wedge 0] - x \wedge 0 = x \wedge 0 - x \wedge 0 = 0$. On the other hand, if x and y are arbitrary disjoint positive elements, then $(x - y)^+ = (x - y) \vee 0 = x \vee y - y = x - x \wedge y = x$; a similar calculation shows that $(x - y)^- = y$. The positive and negative part of an element z of a vector lattice can then be described as the unique disjoint positive elements whose difference is z. There is another simple result on disjointness which will be useful. If x and y are arbitrary elements, then $x - x \wedge y$ and $y - x \wedge y$ are disjoint, as may easily be verified using the translation invariance of the ordering.

The **absolute value** of a member x of a vector lattice is denoted $|x|$ and is defined to be $x^+ + x^-$. It is easy to see that $|tx| = |t| \cdot |x|$, for each scalar t.

The following result will be essential.

23.8 DECOMPOSITION LEMMA *Let E be a vector lattice and let x, y, and z be positive elements such that $x \leq y + z$. If $u = x \wedge y$ and $v = x - x \wedge y$, then u and v are positive, $x = u + v$, $u \leq y$ and $v \leq z$.*

PROOF Clearly $0 \leq u \leq x$, and it must be proved that $0 \leq v \leq z$. Since $x \wedge y$ is less than x, $v = x - x \wedge y$ is positive. To show that $z \geq x - x \wedge y$ observe that $z \geq x - y$, from the hypothesis, and since $z \geq 0$, it follows that $z \geq (x - y) \vee 0$. Applying a translation by $-x$, it follows that $z - x \geq (x - y - x) \vee (-x) = -x \wedge y$, which is the desired inequality.$|||$

The order dual of a vector lattice can be characterized very nicely. It has already been remarked that each member of the order dual is bounded on order bounded sets, and the converse is correct if the space is a lattice.

23.9 ORDER DUAL OF A VECTOR LATTICE   *A linear functional f on a vector lattice E belongs to the order dual E⋆ if and only if f is bounded on order bounded sets.   Under the dual ordering E⋆ is a vector lattice, and, for an f in E⋆, $f^+ = f \vee 0$ is given by $f^+(x) = \sup\{f(u): 0 \leqq u \leqq x\}$ for each positive element x in E.*

PROOF   Let f be a linear functional on E which is bounded on order bounded sets, and for each positive element x let $f^+(x)$ be the value as defined in the statement of the theorem.   Clearly $f^+(x)$ is finite and non-negative.   If x and y are positive elements, then $0 \leqq u \leqq x$ and $0 \leqq v \leqq y$ imply that $0 \leqq u + v \leqq x + y$; hence, $f(x) + f(y) = f(x + y) \leqq f^+(x + y)$.   It follows that $f^+(x) + f^+(y) \leqq f^+(x + y)$. On the other hand, if $0 \leqq w \leqq x + y$, then by the decomposition lemma there are elements u and v such that $w = u + v$, $0 \leqq u \leqq x$, and $0 \leqq v \leqq y$.   Hence, $f(w) = f(u + v) = f(u) + f(v) \leqq f^+(x) + f^+(y)$ from which it follows that $f^+(x + y) \leqq f^+(x) + f^+(y)$. Therefore, we have $f^+(x + y) = f^+(x) + f^+(y)$ for positive elements x and y.   Obviously, $f^+(tx) = tf^+(x)$ for a non-negative real number t.   It is not hard to see that f^+ defined on the positive cone can be extended to a linear functional (positive) on E.   (The extension will also be denoted by f^+.)   Since $f \leqq f^+$, $f^+ - f$ is positive, and $f = f^+ - (f^+ - f) \in E^\star$.   In view of the remark preceding the theorem, the first assertion is proved.   The functional f^+ defined above is an upper bound of f and 0, and, to see that it is the least such, it is enough to notice that if $g \geqq 0$ and $g \geqq f$, then for $0 \leqq u \leqq x$, $f(u) \leqq g(u) \leqq g(x)$.   By the usual translation argument, we see that for each f and g in $E^\star$, $f \vee g$ exists.|||

Let f and g be members of the order dual of a vector lattice E, and let x be a positive element of E.   Then it is easy to compute $(f \vee g)(x)$, $(f \wedge g)(x)$ and $|f|(x)$.   In fact, $(f \vee g)(x) = (f - g)^+(x) + g(x) = \sup\{f(u) + g(x - u): 0 \leqq u \leqq x\}$, or $(f \vee g)(x) = \sup\{f(u) + g(v): 0 \leqq u, v \leqq x, u + v = x\}$.   Similarly, $(f \wedge g)(x) = \inf\{f(u) + g(v): 0 \leqq u, v \leqq x, u + v = x\}$. $|f|(x) = f^+(x) + (-f)^+(x) = \sup\{f(u - v): 0 \leqq u, v \leqq x\}$, but an element y is of the form $y = u - v$ for some u and v such that $0 \leqq u, v \leqq x$ if and only if $|y| \leqq x$; hence, $|f|(x) = \sup\{f(y): |y| \leqq x\}$.

There are other more or less immediate consequences of the preceding theorem. Let E be a vector lattice and let $\mathscr{A}$ be the family of all sets which, for some positive y, are of the form $\{x: -y \leqq x \leqq y\}$. Then $\mathscr{A}$ is a co-base for the order bounded sets of E, and each member of $\mathscr{A}$ is convex and circled. Let $\mathscr{A}^{\sim}$ be the family of all convex circled sets which absorb each member of $\mathscr{A}$. Then in view of lemma 19.1, $\mathscr{A}^{\sim}$ is a local base for a bound locally convex topology $\mathscr{T}$ for E, and $\mathscr{T}$ is the strongest locally convex topology relative to which each member of $\mathscr{A}$ is bounded. Each $\mathscr{T}$-continuous linear functional is bounded on order bounded sets and conversely, if a linear functional f is bounded on order bounded sets, then $\{x: |f(x)| \leqq 1\}$ is a member of $\mathscr{A}^{\sim}$ and f is therefore $\mathscr{T}$-continuous. Hence the adjoint of $(E,\mathscr{T})$ is simply E^*. Recall (19.4) that a bound topology is always the Mackey topology of the pairing of E and E^*, and hence $\mathscr{T}$ must be the Mackey topology $m(E,E^*)$. Consequently the family of all convex circled sets which absorb each order bounded set is a local base for $m(E,E^*)$. The following theorem is then clear.

23.10 The Topology $m(E,E^*)$. *Let E be a vector lattice, let E be its order dual, and let $\mathscr{T}$ be the Mackey topology $m(E,E^*)$. Then:*

(i) *A linear functional on E is $\mathscr{T}$-continuous if and only if it is bounded on order bounded sets.*

(ii) *The topology $\mathscr{T}$ is bound and is the strongest locally convex topology with the property that each order bounded set is bounded.*

(iii) *A pseudo-norm p on E is $\mathscr{T}$-continuous if and only if each order bounded set is p-bounded.*

The next theorem describes the pseudo-norms which are continuous relative to the Mackey topology $m(E,E^*)$. A set P of pseudo-norms for a linear topological space E is called a **base** for the family of continuous pseudo-norms if and only if each member of P is continuous and every continuous pseudo-norm is dominated by some member of P. This is entirely equivalent to requiring that the family of all unit spheres about 0, constructed from the members of P, be a local base for the topology. It is natural to suspect that the set of all monotonic pseudo-norms for E is a base for the $m(E,E^*)$-continuous pseudo-norms; actually, an even stronger result can be proved. A pseudo-norm p is a **lattice pseudo-norm** if and only if $p(x) \geqq p(y)$ whenever $|x| \geqq |y|$. It is easy to see that p is a lattice pseudo-norm if and only if p is monotonic (if $x \geqq y \geqq 0$, then $p(x) \geqq p(y)$ and $p(x + y) = p(x - y)$ for disjoint positive elements x

and y. The latter condition is evidently equivalent to the requirement: $p(x) = p(|x|)$ for all x.

23.11 ORDER DUAL AND LATTICE PSEUDO-NORMS *If E is a vector lattice, then the set of all lattice pseudo-norms is a base for the family of pseudo-norms which are ·continuous relative to the Mackey topology $m(E,E^\star)$.*

PROOF   In view of the preceding theorem a pseudo-norm p is continuous relative to $m(E,E^\star)$ if and only if order bounded sets are p-bounded.   If p is a lattice pseudo-norm, then an order bounded set $\{z: x \leq z \leq y\}$ is evidently p-bounded, and consequently such a pseudo-norm is $m(E,E^\star)$-continuous.   Suppose now that p is an arbitrary $m(E,E^\star)$-continuous pseudo-norm, and let $q(x) = \sup\{p(u): |u| \leq |x|\}$.   This supremum is finite because the order bounded set $\{u: -|x| \leq u \leq |x|\}$ is supposed to be p-bounded.   Clearly $q \geq p$, $q(tx) = |t|q(x)$, and if $|x| \geq |y|$, then $q(x) \geq q(y)$.   It remains to be proved that $q(x + y) \leq q(x) + q(y)$, and, since $|x| + |y| \geq |x + y|$, it suffices to consider only the case where x and y are positive.   In view of the definition of q, the problem then reduces to showing that if x and y are positive and $|z| \leq x + y$, then there are elements u and v such that $|u| \leq x$, $|v| \leq y$ and $z = u + v$.   Since $z^+ \leq x + y$ and $z^- \leq x + y$, there are, in view of the decomposition lemma 23.8, positive elements u_1, v_1, u_2, and v_2 such that $z^+ = u_1 + v_1$, $u_1 \leq x$, $v_1 \leq y$, $z^- = u_2 + v_2$, $u_2 \leq x$, and $v_2 \leq y$.   Since z^+ and z^- are disjoint, so are u_1 and u_2 and v_1 and v_2, respectively.   Hence $|u_1 - u_2| = u_1 + u_2 = u_1 \vee u_2 + u_1 \wedge u_2 = u_1 \vee u_2 \leq x$, and similarly $|v_1 - v_2| \leq y$.   But $z = u_1 - u_2 + v_1 - v_2$, and the proof is complete.|||

We have seen that the order dual $E^\star$ of a vector lattice is a vector lattice, and it follows that the second order dual $E^{\star\star}$ is again a vector lattice.   Let e denote the evaluation map of E into its second order dual $E^{\star\star}$; then it is easy to see that e takes a positive element into a positive element.   The fact that e preserves lattice operations (that is, $e(x \vee y) = e(x) \vee e(y)$ and $e(x \wedge y) = e(x) \wedge e(y)$) is less obvious.   In order to prove this fact we require a lemma which is useful in other connections too.

23.12 LEMMA *Let E be a vector lattice, let u be a positive element of E, and let f be a positive linear functional on E.   Then there is a positive linear functional g with the properties: (a) $g \leq f$, (b) $g(u) = f(u)$, and (c) $g(x) = 0$ for each positive element x such that $x \wedge u = 0$.*

PROOF   For each positive element x, define $g(x) = \sup \{f(y):$ $0 \leq y \leq x$ and $y \leq tu$ for some real number $t \geq 0\}$. It will be shown that g is additive on the positive cone.   Let x and y be arbitrary positive elements, and assume that z is an element such that $0 \leq z \leq$ $x + y$ and $z \leq tu$ for some $t \geq 0$.   Then by the decomposition lemma 23.8, there are elements a and b such that $z = a + b$, $0 \leq a \leq x$, and $0 \leq b \leq y$.   Clearly $a \leq tu$ and $b \leq tu$.   Hence $f(z) = f(a) +$ $f(b) \leq g(x) + g(y)$, and consequently $g(x + y) \leq g(x) + g(y)$.   On the other hand, if a and b are elements such that $0 \leq a \leq x$, $0 \leq$ $b \leq y$, $a \leq tu$ and $b \leq su$ for some t and s, then $0 \leq a + b \leq x + y$ and $a + b \leq (t + s)u$; therefore $f(a) + f(b) = f(a + b) \leq g(x + y)$. It follows that $g(x) + g(y) \leq g(x + y)$.   Hence $g(x + y) = g(x) +$ $g(y)$.   It is easy to see that $g(tx) = tg(x)$ for a positive x and a non-negative real number t.   Therefore, g can be extended to a linear functional on E which will be again denoted by g.   That g satisfies (a) and (b) is clear.   In order to see (c), let x be a positive element such that $x \wedge u = 0$, and let y be an element such that $0 \leq y \leq x$ and $y \leq tu$ for some real number $t \geq 0$.   Then for some positive integer n, $y \leq nu$; hence $0 \leq y \leq x \wedge nu \leq n(x \wedge u) = 0$, or $y = 0$, and (c) is proved.|||

23.13   THE EVALUATION INTO THE SECOND ORDER DUAL   *Let E be a vector lattice.   Then the evaluation map e on E into the second order dual E^{**} preserves the lattice operations.   (See the remarks before the lemma.)*

PROOF   By the usual argument it suffices to show that $e(x)^{+} = e(x^{+})$ or, equivalently, to show that $e(x)^{+}(f) = f(x^{+})$ for each element x of E and each positive linear functional f on E.   In view of theorem 23.9, this last equation becomes $f(x^{+}) = \sup \{h(x): 0 \leq h \leq f\}$, which can be established as follows.   For each linear functional h which satisfies $0 \leq h \leq f$, $h(x) \leq h(x^{+}) \leq f(x^{+})$; hence, $\sup \{h(x):$ $0 \leq h \leq f\} \leq f(x^{+})$.   Next, by applying the lemma to f and x^{+}, we see that there is a linear functional g on E such that $0 \leq g \leq f$, $g(x^{+}) = f(x^{+})$, and $g(y) = 0$ for each positive y such that $y \wedge x^{+} =$ 0.   Since $x^{+} \wedge x^{-} = 0$, it follows that $g(x) = g(x^{+}) - g(x^{-}) =$ $g(x^{+}) = f(x^{+})$.   Hence, $f(x^{+}) = \sup \{h(x): 0 \leq h \leq f\}$ which proves the theorem.|||

The evaluation map of a vector lattice E into its second order dual is not necessarily one-to-one, for, if this were the case, every vector lattice would be Archimedean since the order dual of a vector lattice is always Archimedean.   However, being Archimedean alone does not ensure that the evaluation map be one-to-one. Consider, for instance, the space of all real valued measurable functions on the interval [0,1] modulo null functions (functions which are zero almost everywhere) under the usual ordering. This space is an Archimedean vector lattice, but it can be proved that the order dual is zero dimensional. In order to give some necessary and sufficient conditions for the evaluation map of a vector lattice to be one-to-one, we require some preliminary facts.

For arbitrary u and v in a vector lattice E, obviously $(u + v)^+ \leqq u^+ + v^+$ holds. Hence for any x and y, $x^+ \leqq (x - y)^+ + y^+$ or $x^+ - y^+ \leqq (x - y)^+$. It follows that $(x^+ - y^+)^+ \leqq (x - y)^+$. By interchanging x and y, we see that $(x^+ - y^+)^- \leqq (x - y)^-$. By adding the last two inequalities, we obtain $|x^+ - y^+| \leqq |x - y|$.

23.14 CONTINUITY OF THE LATTICE OPERATIONS *Let E be a vector lattice and let $\mathscr{T}$ be a locally convex topology for E such that the set of all continuous pseudo-norms admits a base consisting of lattice pseudo-norms. Then the lattice operations on E are continuous relative to the topology $\mathscr{T}$, that is, the functions M and J defined by $M(x,y) = x \wedge y$ and $J(x,y) = x \vee y$ are continuous on $E \times E$ into E. In particular the lattice operations are continuous relative to the Mackey topology $m(E,E^\star)$.*

PROOF It can be seen easily that the functions J and M are continuous if and only if the function which takes x to x^+ is continuous. If p is a lattice pseudo-norm, then from the remark preceding the theorem one obtains $p(x^+ - y^+) \leqq p(x - y)$ for all x and y; hence the lattice operations are continuous relative to the pseudo-norm topology determined by p. The theorem follows.|||

23.15 ONE-TO-ONENESS OF THE EVALUATION *Let E be a vector lattice. Then the following statements are all equivalent.*

(a) *The evaluation map on E into the second order dual is one-to-one.*
(b) *There are enough positive linear functionals to distinguish points of E.*
(c) *The topology $w(E,E^\star)$ is Hausdorff.*
(d) *The topology $m(E,E^\star)$ is Hausdorff.*
(e) *For some locally convex Hausdorff topology the map which takes x to x^+ is continuous.*
(f) *The positive cone is closed for some locally convex topology on E.*

PROOF (a) $\Leftrightarrow$ (b) $\Rightarrow$ (c) $\Rightarrow$ (d) is obvious. (d) implies (e) because of the previous theorem. (e) implies (f) because the positive cone is the set $\{x: x = x^+\}$. That (f) implies (b) is a consequence of theorem 23.1.|||

24 L AND M SPACES

We prove two theorems of Kakutani which characterize Banach lattices of (roughly speaking) continuous function type, and Banach lattices of L^1 type. The adjoint of an M space is an L space, and dually, the adjoint of an L space is an M space.

In this section two especially important vector lattices will be examined. The principal theorems concern vector lattices with a lattice norm—that is, a norm such that, if $|x| \geq |y|$, then $\|x\| \geq \|y\|$. Such spaces are called **normed lattices** and, if the space is complete relative to the norm, they are **Banach lattices.**

Many, if not most, of the real Banach spaces which are commonly considered are Banach lattices under some natural ordering. For example, if m is a measure on a σ-ring $\mathscr{S}$ of subsets of a set X and $p \geq 1$, then the space $L^p(m)$ of all $\mathscr{S}$-measurable real functions f such that $|f|^p$ is m-integrable is a Banach lattice, relative to the usual ordering of the real functions and with the norm: $\|f\| = (\int |f|^p dm)^{1/p}$. (Measure theoretic terminology is that of Halmos [4].) The space $L^\infty(m)$ of all m-essentially bounded $\mathscr{S}$-measurable real functions is also a Banach lattice; it has the noteworthy features that $\|f \vee g\| = \max[\|f\|, \|g\|]$ for all $f \geq 0$ and $g \geq 0$, and the function which is constantly one is interior to the positive cone. The space of all continuous real valued functions on a compact Hausdorff space is also a Banach lattice, if the norm is the usual supremum norm, and this space shares the two special properties of $L^\infty(m)$ noted above. The space $L^1(m)$ also possesses a special property: if f and g are positive members of $L^1(m)$, then $\|f + g\| = \|f\| + \|g\|$. If X is σ-finite relative to the measure m, then there is another description of (an isomorph) of $L^1(m)$ which is useful. Each countably additive real function on $\mathscr{S}$ which is of bounded total variation and is absolutely continuous with respect to m is, by the classical Radon-Nikodym theorem, the indefinite integral of a member of $L^1(m)$, and the space of all such countably additive functions, with variation for norm, is isomorphic and isometric to $L^1(m)$. It will be shown that a Banach lattice is essentially of this sort if it is true that $\|f + g\| = \|f\| + \|g\|$ for all positive elements f and g.

It is convenient, before proceeding, to establish a connection between ordering and extreme points, preparatory to an application of the theorem 15.1 on the existence of extreme points.   The following lemma is given in a form somewhat more general than is necessary here.

24.1   LEMMA   *Let E be an ordered linear topological space, let f be a positive linear functional on E, and let A be a subset of the positive cone of E such that:*

(i) *the set A is compact and convex, and $f(x) \leq 1$ for each x in A.*
(ii) *if $x \in A$ and $x \geq y \geq 0$, then $y \in A$, and*
(iii) *if $x \in A$ and $x \neq 0$ then $f(x) \neq 0$ and $x/f(x)$ is a member of A.*

Then a member x of A is an extreme point of A if and only if for each y such that $x \geq y \geq 0$ it is true that $y = f(y)x$.

PROOF   Suppose that x is an extreme point of A and that $x \geq y \geq 0$. If $x = 0$, then $f(x) = f(y) = 0$; by (iii) it follows that $y = 0$, and hence $y = f(y)x$.   If $x \neq 0$ then, since $f(x) \leq 1$ and both 0 and $x/f(x)$ belong to A, $f(x) = 1$; otherwise x could not be an extreme point.   If $x \neq 0$ and $x - y = 0$, then again $y = f(y)x$.   If neither y nor $x - y$ is 0, then $f(y) + f(x - y) = 1$, and writing $x = f(y)[y/f(y)] + f(x - y)[(x - y)/f(x - y)]$, the fact that x is extreme implies that $y/f(y) = x$.   To prove the converse, suppose that x is a member of A such that $y = f(y)x$ for each y such that $x \geq y \geq 0$. If $x = 0$, then since 0 is the unique member of A which is a zero of f, and since f is positive, x is an extreme point.   If $x \neq 0$ then $f(x) = 1$ because $x = f(x)x$.   In this case, if $x = ty + (1 - t)z$ where $0 < t < 1$ and y and z belong to A, then clearly $x \geq y \geq 0$; hence $y = f(y)x$ and similarly $z = f(z)x$.   Finally, since $f(x) = 1 = tf(y) + (1 - t)f(z)$, and since $f(y)$ and $f(z)$ are each at most one, it follows that $f(y) = f(z) = 1$, and hence $y = f(y)x = x$ and $z = x$.   It follows that x is an extreme point.|||

A linear function f on a vector lattice E to another vector lattice F is a **lattice homomorphism** if and only if $f(x \vee y) = f(x) \vee f(y)$ for all x and y in E.   A one-to-one lattice homomorphism is a **lattice isomorphism**.   A **real lattice homomorphism** is a homomorphism whose range is contained in the lattice of real numbers, with the usual ordering.   It follows easily from the relation between meet and join that a lattice homomorphism also preserves meets, and that a meet preserving linear function is necessarily a homomorphism. It is clear that a real lattice homomorphism is a positive functional, for, if x is a positive element of E, then $x = x \vee 0$ and $f(x) = f(x) \vee f(0) \geq 0$.   The following theorem identifies the lattice homomorphisms geometrically.

24.2   Characterization of Lattice Homomorphism   *If E is a vector lattice and f is a linear functional on E, then the following three conditions are equivalent:*

    (i) *the functional f is positive and, if x and y are disjoint positive elements of E, then $f(x) \wedge f(y) = 0$;*

    (ii) *the functional f is a lattice homomorphism; and*

    (iii) *the functional f is positive and, if g is a linear functional such that $0 \leq g \leq f$, then g is a scalar multiple of f.*

PROOF   It is first shown that (i) implies (ii). Assuming (i), suppose that x and y are arbitrary positive members of E. Then $x - x \wedge y$ and $y - x \wedge y$ are disjoint positive elements and consequently $0 = (f(x) - f(x \wedge y)) \wedge (f(y) - f(x \wedge y)) = f(x) \wedge f(y) - f(x \wedge y)$, the latter equality being a consequence of the translation formula for meets: $(a + b) \wedge (c + b) = a \wedge c + b$. It follows that f preserves meets of positive elements, and, using the translation formula again, it is easy to see that meets of arbitrary elements are preserved.

To show that (ii) implies (iii) assume that f is a lattice homomorphism, that $0 \leq g \leq f$, and that x is a point of E at which f vanishes. It will be shown that g also vanishes at x, and hence, since the null space of g includes the null space of f, it will follow that g is a scalar multiple of f. Write $x = x^+ - x$ and notice that, since x^+ and x^- are disjoint, either $f(x^+)$ or $f(x^-)$ is 0. Since $f(x) = 0$ also, $f(x^+) = f(x^-) = 0$, and since $0 \leq g \leq f$, it follows that $g(x^+) = g(x^-) = 0$ and therefore $g(x) = 0$.

Finally, it must be proved that (iii) implies (i). Suppose that x and y are disjoint positive elements of E and that $f(x) \neq 0$. The application of lemma 23.12 to f and x ensures the existence of a linear functional g such that $0 \leq g \leq f$, $g(x) = f(x)$ and $g(y) = 0$. The first two properties of g and (iii) imply that $f = g$; hence, $f(y) = g(y) = 0$. It follows that $f(x) \wedge f(y) = 0$ whether $f(x) = 0$ or not.|||

The two types of normed lattices which are the concern of this section satisfy especial conditions on the norm and the ordering. A normed lattice is of **type M** (respectively of **type L**) if and only if for every pair x and y of positive elements it is true that $\|x \vee y\| = \|x\| \vee \|y\|$ (respectively, $\|x + y\| = \|x\| + \|y\|$). A Banach lattice of type M (of type L) is an **M space** (respectively, an **L space**). These two sorts of normed lattice are, in a sense, dual to each other. Each is studied by means of its adjoint space, which turns out to be a space of the dual sort, as noted in the following theorem.

Before turning to the theorem it should be stated that for any normed lattices each continuous linear functional is the difference of continuous positive functionals, by 23.6; it is not true that, for an arbitrary normed lattice, each positive functional is norm bounded. For Banach lattices, each positive linear functional is bounded; however this fact is not needed for the present discussion.

The largest element in the unit sphere of a normed lattice, if such exists, is called the **unit**.

24.3 ADJOINTS OF SPACES OF TYPE *L* AND *M*   *The adjoint of a normed lattice is, with the dual ordering, a Banach lattice. The adjoint of a space of type M (respectively, of type L) is an L space (an M space with unit).*

PROOF   If E is a normed lattice, then surely E^* is a Banach space. If f is a continuous positive linear functional on E, then $\|f\| = \sup \{f(x): x \geq 0$ *and* $\|x\| \leq 1\}$ because $\|x\| = \|\, |x|\, \| \geq \|x^+\| \vee \|x^-\|$ and one of the two numbers $f(x^+)$ and $f(x^-)$ is necessarily as great as $|f(x)|$. Consequently, if f and g are bounded positive functionals, then $f \wedge g$ is also bounded, and it follows that E^* is a lattice. Moreover, since the meet of f and g in the order dual E^* is identical with the meet in the adjoint E^*, it follows from the calculation made after 23.9 that, for a positive element x, $|f|(x) = \sup \{f(y): |y| \leq x\}$. Hence, $\|\, |f|\, \| = \sup \{|f|(x): x \geq 0$ *and* $\|x\| \leq 1\} = \sup \{f(y): \|y\| \leq 1\} = \|f\|$. Since the norm of E^* is clearly monotone, the adjoint E^* is a normed lattice.

Finally, it must be shown that the adjoint of a space of type M (type L) is an L space (an M space with unit). If E is of type M and f and g are positive members of E^*, then there are positive members x and y of E of norm 1 such that $f(x)$ and $g(y)$ are, respectively, approximately $\|f\|$ and $\|g\|$. Then $x \vee y$ is positive and of norm 1, and $(f + g)(x \vee y) \geq f(x) + g(y)$. Hence $\|f + g\| \geq \|f\| + \|g\|$, and it follows that the adjoint is an L space. If E is of type L, then the norm is linear on the positive cone of E, and consequently there is a linear functional u such that $u(x) = \|x\|$ for $x \geq 0$. For x positive and f in E^* it is then true that $f(x) \leq \|f\| \, \|x\| = \|f\| \, u(x)$, and hence $f \leq \|f\|u$. It follows that u is a unit, and, moreover, for positive elements f and g of E^*, because of the inequality $f \vee g \leq (\|f\| \vee \|g\|)u$, it is true that $\|f \vee g\| \leq (\|f\| \vee \|g\|)\|u\| = \|f\| \vee \|g\|$. Therefore E^* is an M space with unit.‖|

If E is a normed lattice, then E^{**} is a Banach lattice. Let e be the evaluation map on E into E^{**}; then, for each x in E and positive element f of E^*, $e(x)^+(f) = \sup \{g(x): 0 \leq g \leq f$ and $g \in E^*\} =$ $\sup \{g(x): 0 \leq g \leq f$ and $g \in E^*\} = f(x^+) = e(x^+)(f)$ (the third equality follows from 23.13), or $e(x)^+ = e(x^+)$. Hence, we have proved the following:

24.4 THE EVALUATION MAP *Let E be a normed lattice: then the evaluation map on E into its second adjoint E^{**} is a lattice isomorphism.*

If the image of the evaluation map on a normed lattice E into E^{**} is again denoted by E, then the closure E^- in E^{**} of E relative to the norm topology has the property that, whenever x and y belong to E^-, $x \wedge y$ and $x \vee y$ also belong to E^-. This fact is a consequence of the continuity of lattice operations in E^{**} (see 23.14). Hence E^- is a Banach lattice in which the given normed lattice E is densely embedded.

The structure of an M space with unit will now be described completely. If E is a normed space of type M which has a unit u, the **spectrum** X of E is defined to be the set of all real homomorphisms of E which are of norm one, with the weak* topology. It is evident that the set of all real lattice homomorphisms is weak* closed and, since X is precisely the set of those homomorphism h such that $h(u) = 1$, it follows that X is a weak* closed subset of the unit sphere of E^*. (A homomorphism h is positive, and hence its norm is the supremum of values at positive elements of norm at most one, and hence $\|h\| = h(u)$.) The spectrum is then a compact Hausdorff space. The normed lattice E may be mapped into the lattice $C(X)$ of continuous real valued functions on X by means of the evaluation map ϕ. Explicitly, for x in E and h in X, $\phi(x)(h) = h(x)$. For x and y in E and h in X it is true that $\phi(x \wedge y)(h) = h(x \wedge y) = h(x) \wedge h(y) = (\phi(x) \wedge \phi(y))(h)$, because h is a lattice homomorphism. Hence ϕ is a lattice homomorphism of E into $C(X)$. It will now be shown that ϕ is an isomorphism, and an isometry, and that the range of ϕ is dense in $C(X)$.

24.5 REPRESENTATION OF SPACES OF TYPE M *Let E be a normed lattice of type M with unit, and let X be the spectrum of E. Then the evaluation map ϕ of E into the function space $C(X)$ is an isometric lattice isomorphism of E onto a dense subspace of $C(X)$.*

PROOF In view of the remarks preceding the theorem there remain just two facts to be demonstrated: that ϕ is an isometry, and that the range of ϕ is dense in $C(X)$. In order to show that ϕ is an isometry it is necessary only to show that $\|x\| = \|\phi(x)\|$ for positive members x of E, since both E and $C(X)$ are normed lattices. Since each member of X is of norm one, $\sup\{|\phi(x)(h)|: h \in X\} = \sup\{|h(x)|: h \in X\} \leq \|x\|$. On the other hand, given a positive member x, let $A = \{g: g \geq 0 \text{ and } \|g\| \leq 1\}$ and $B = \{g: g \in A \text{ and } g(x) = \|x\|\}$; then A is weak* compact and B is non-void and, since B is the set of members of A where the evaluation at x is a maximum, B is a support of A (that is, if B contains an interior point of a line segment in A, then B contains the entire segment). It follows that each extreme point of B is an extreme point of A, and, in view of theorem 15.1, B has extreme points. Consequently there is an extreme point h of A such that $h(x) = \|x\|$. Finally, lemma 24.1 may be applied (the requisite linear functional on A is evaluation at the unit of E), and 24.2 then shows that h is a lattice homomorphism. It follows that $\|x\| = \|\phi(x)\|$.

It remains to prove that the range of ϕ is dense in $C(X)$. For this purpose the following prelemma, which is of interest in itself, is useful.

PRELEMMA   *Let X be a compact space and let A be a subset of $C(X)$ which is closed under the lattice operations (that is, f, $g \in A$ implies $f \vee g \in A$ and $f \wedge g \in A$). Then a function h in $C(X)$ belongs to the uniform closure of A if for each $e > 0$ and for each pair of points x and y of X there is an f in A such that $|f(x) - h(x)| < e$ and $|f(y) - h(y)| < e$.*

Suppose for a moment that the prelemma is established. Notice that, if h and g are distinct members of X, then for some x in E, $h(x) \neq g(x)$, and hence $\phi(x)$, assumes different values at the members g and h of X. It is also true that the image under ϕ of the unit u of E is the function which is constantly one on X. Therefore, given a pair of real numbers a and b, it is possible to choose a linear combination y of x and u such that $\phi(y)(g) = a$ and $\phi(y)(h) = b$. Also the range of ϕ is closed under the lattice operations; hence, in view of the prelemma, the range of ϕ is dense in $C(X)$.

PROOF OF THE PRELEMMA   Let a positive number e be given, and let $f_{x,y}$ be a function in A such that $|f_{x,y}(x) - h(x)| < e$ and $|f_{x,y}(y) - h(y)| < e$. If $U_{x,y} = \{z: |f_{x,y}(z) - h(z)| < e\}$, then clearly $U_{x,y}$ is an open set containing x and y. Since X is compact, for a fixed x, there are points $y_1, \cdots, y_n$ such that $X = \bigcup \{U_{x,y_i}: i = 1, \cdots, n\}$. Let $g_x = f_{x,y_1} \wedge \cdots \wedge f_{x,y_n}$; then $g_x(y) \leq h(y) + e$ for each y in X and

$g_x(y) \geqq h(y) - e$ for each y in $V_x = \bigcap \{U_{x, y_i} : i = 1, \cdots, n\}$. Since V_x is an open set containing x and X is compact, there are $x_1, \cdots, x_m$ such that $X = \bigcup \{V_{x_i} : i = 1, \cdots, m\}$. Let $g = g_{x_1} \vee \cdots \vee g_{x_m}$. Then clearly $g \in A$ and $h(y) - e \leqq g(y) \leqq h(y) + e$ for each y in X.|||

24.6 CHARACTERIZATION OF M SPACES *Each M space with unit is isomorphic and isometric, under evaluation, to the space of all continuous real valued functions on its spectrum.*

The preceding discussion of spaces of type M gives insight into the structure of an Archimedean vector lattice E whose positive cone is radial at some point. For if the cone is radial at a point u, then one may construct a norm for E such that E, with this norm, is a space of type M with unit u. In fact:

24.7 THEOREM *If E is an Archimedean vector lattice and the positive cone is radial at a point u, then E with the norm $\|x\| = \inf \{t : t > 0, |x| \leqq tu\}$ is a space of type M with unit u.*

PROOF First observe that the infimum in the definition of the norm is assumed, because, if one lets $s = \|x\|$, then, for any positive number t, $|x| - su \leqq tu$ or $(|x| - su)^+ \leqq tu$ and it follows from the fact that the ordering is Archimedean that $(|x| - su)^+ = 0$ or $|x| \leqq su$. From this remark all the assertions in the theorem can be seen easily once it is shown that $\| \ \|$ is a pseudo-norm. The function $\| \ \|$ is a pseudo-norm because it is precisely the Minkowski functional of the set $(u - C) \cap (C - u)$ which is convex, circled, and radial at 0.|||

Spaces of type L will now be studied, utilizing the known structure of M spaces. This mode of attack, while perhaps not the most direct, yields a rather concrete representation theorem and has the advantage that elementary measure theory is available for some of the proofs. The procedure is the following. If E is a space of type L, then E^* is an M space with unit, according to 24.5, and hence is isomorphic to the space $C(X)$ of continuous real functions on its spectrum X. The evaluation map of E into E^{**} preserves the norm and the lattice operations. But a simple concrete representation for the adjoint of $C(X)$ is given by the Riesz representation theorem (14J), for, to an isomorphism, $C(X)^*$ is the space of signed regular Borel measures on X, with variation for norm. Explicitly, the Borel σ-ring $\mathscr{B}$ of X is the smallest σ-ring which contains each compact set. A signed regular Borel measure m is a countably additive real valued function on $\mathscr{B}$ such that for each Borel set B and a positive number e

there is a compact subset of C of B with the property $|m(B) - m(C)| < e$, and the variation $\|m\|$ of m is sup $\{m(A) - m(B): A \text{ and } B \text{ disjoint}$ *members of $\mathscr{B}$}. The map which carries m into the functional on $C(X)$ whose value at f is $\int f dm$ is an isometry, and order preserving. (See Halmos [4] for details.) The following theorem summarizes the foregoing remarks.

24.8 REPRESENTATION OF SPACES OF TYPE *L* *Let E be a space of type L, let E^* be its adjoint, and let X be the spectrum of E. Then E is isometric and lattice isomorphic to a sublattice of the space of regular Borel measures on X.*

The characterization of L spaces will now be completed by describing precisely the class of regular Borel measures on X which are the images of members of E. The proof of the following theorem depends on a sequence of lemmas, which are given after the statement of the theorem.

24.9 CHARACTERIZATION OF *L* SPACES *Let E be an L space, and let X be the spectrum of its adjoint E^*. Then E is isometric and lattice isomorphic to the space of all those signed regular Borel measures on X which vanish on each Borel set of the first category.*

Throughout the following E will be a fixed L space, E^* its conjugate, and X will be the spectrum of E^*. To avoid notation, no distinction will be made between E^* and the space $C(X)$ of real continuous functions on X. If x is a member of E, then m_x will denote the corresponding signed measure on X.

24.10 LEMMA *A subset of E or of E^* which has an upper bound has a least upper bound. Each monotonically increasing net in E (respectively E^*) which is bounded above converges relative to the norm topology (the w^*-topology) to its least upper bound.*

PROOF The second assertion will be proved first. If $\{f_\alpha, \alpha \in D\}$ is a monotonically increasing net in E^* which is bounded above by g, then for each positive x in E the net $\{f_\alpha(x), \alpha \in D\}$ is a bounded monotonically increasing net of real numbers, hence converges, and there is therefore a linear functional f such that $\{f_\alpha(x), \alpha \in D\}$ converges to $f(x)$ for all positive x in E; hence $f_\alpha \to f$ relative to the topology w^*. For each positive x, $f_\alpha(x) \leq f(x) \leq g(x)$, from which it follows that f is bounded, and it is evident that f is the supremum of the functionals f_α. To prove that a bounded monotonic net $\{x_\alpha, \alpha \in D\}$ in E converges relative to the norm topology it is first shown that the net is a

Cauchy net. Observe that for α fixed, $\|x_\beta - x_\alpha\|$ is monotonic for $\beta \geq \alpha$, and bounded, and hence converges. Since the norm is linear on the positive cone, for $\gamma \geq \beta \geq \alpha$ it is true that $\|x_\gamma - x_\beta\| = \|x_\gamma - x_\alpha\| - \|x_\beta - x_\alpha\|$. From this it follows easily that the net is Cauchy and hence converges to a member x of E. It is not hard to verify that x is the supremum of the members x_α, by making use of the fact that the set $\{z : z \geq y\}$ is always closed. Finally, to prove the first statement of the lemma, if B is a subset of E (or of E^*) with an upper bound, then let $\mathscr{A}$ be the family of finite subsets of B directed by $\supset$, and for A in $\mathscr{A}$ let x_A be the supremum of the set A. Then the net $\{x_A, A \in \mathscr{A}\}$ is monotonically increasing and bounded, and its limit is the supremum of B.|||

The preceding lemma has important consequences which concern the topological structure of the spectrum X and the nature of the measures m_x.

24.11 LEMMA *The closure of each open set in X is both open and closed. For each x in E, the measure m_x vanishes on Borel sets of the first category.*

PROOF Let U be an open set in X, let B be the family of all non-negative continuous real functions which are 0 on $X \sim U$ and are bounded by 1, and let f be the supremum of the set B. Then f is a continuous function which is 1 on U, because for each s in U there is a member of B which assumes the value 1 at s. On the other hand, if $s \notin U^-$ then there is a function g which is an upper bound for B such that $g(s) = 0$. It follows that f must be the characteristic function of U^-, and hence U^- is both open and closed. To prove the second statement of the lemma it is necessary only to show that $m_x(S)$ vanishes for each nowhere dense Borel subset S of X, and it may be assumed that x is a positive member of E. Let B be the family of all continuous real functions on X which are bounded by 1, are non-negative, and are zero on S. Then B is directed by $\geq$, and, since S is nowhere dense, it is easy to see that the supremum of B is the function which is identically one. The net $\{f, f \in B\}$ converges w^* to 1, and hence $\int f \, dm_x \to \int 1 \, dm_x$. For $e > 0$ there is then f in B such that $e > \int (1 - f) \, dm_x \geq 0$, and since $(1 - f)(s) = 1$ for s in S, $m_x(S) < e$.|||

The second statement of the preceding lemma shows that the L space E maps into a sublattice of the space of all signed regular Borel measures which vanish on first category Borel subsets of X. It remains to show that E maps onto the latter class. The following method of attack is used. Suppose n is a positive regular Borel measure which vanishes on sets of the first category. Let B be the set of measures of the form m_x such that $0 \leqq m_x \leqq n$, and let p be the supremum of B. This supremum exists, in view of lemma 24.10 applied to the L space E^{**}, and the suprema of finite subsets of B converge to p relative to the norm topology. Since E is isometric to a closed subspace of E^{**} it follows that $p = m_y$ for some y in E. If $p = n$ then n belongs to the image of E; otherwise $n - p$ is a positive regular Borel measure, zero on sets of the first category, such that no non-zero positive m_x is less than $n - p$. The following lemma then completes the proof.

24.12 LEMMA *If n is a positive regular Borel measure vanishing on Borel sets of the first category and $\|n\| \neq 0$, then for some positive non-zero x in E it is true that $m_x < n$.*

PROOF As a preliminary, it is to be observed that, if A is a Borel subset of X, then there is an open and closed subset A' of X such that the symmetric difference $(A \sim A') \cup (A' \sim A)$ is of category one. This may be proved by showing that the class $\mathscr{A}$ of all sets A such that, for some open set C, the symmetric difference of A and C is of first category, is closed under countable union and complementation, and contains all compact sets and hence all Borel sets (this is a well-known lemma of set theory). Since each open set C differs from the open and closed set C^- by a nowhere dense set, the stated result follows.

Turning to the proof of the lemma, it is first shown that if p is a positive measure belonging to E (more precisely, if $p = m_x$ for some positive member of E) and if A is a Borel subset of X, then the measure p_A belongs to E, where $p_A(B) = p(A \cap B)$ for each B. Since p vanishes on first category sets there is, in view of the remark above, no loss in generality in assuming A is open and closed. Let f and g be the characteristic functions of A and $X \sim A$, respectively. Then, in view of the remark after 23.9, $0 = (f \wedge g)(p) = \inf\{f(p - q) + g(q): 0 \leqq q \leqq p,\ q \in E\}$. Hence, for a positive e, there is q in E such that $0 \leqq q \leqq p$, $(p - q)(A) < e$, and $q(X \sim A) < e$. Thus $q_A \leqq p_A$ and $\|q - p_A\| = \|q_A - p_A + q_{X \sim A}\| < 2e$. Since E is closed and e arbitrary, $p_A \in E$.

Suppose now that n is a positive regular Borel measure, not identically zero, which vanishes on first category Borel subsets of X. Let U be the union of all open subsets of X with n-measure zero. Since

n vanishes on each compact subset of U, $n(U) = 0$. It follows that $n(U^-) = 0$. Let A be the complement of U^- in X; then A is open and closed, and it has the properties: $n(X \sim A) = 0$, and if B is a non-void open subset of A, then $n(B) > 0$. From this fact it follows that a Borel subset B of A is of category one if and only if $n(B) = 0$, for given such a set B there is an open set B' such that $(B \sim B') \cup (B' \sim B)$ is of category one, hence $n(B) = n(B') = 0$, therefore B' is disjoint from A, and consequently B is a subset of the first category set $B \sim B'$. It follows that a Borel set B in X is of n-measure 0 if and only if $B \cap A$ is of the first category.

Finally, let f be the characteristic function of A, and choose a positive measure p in E such that $\int f dp \neq 0$. Then, applying the result of the second paragraph, $p_A \in E$, and clearly $p_A \neq 0$. In view of the characterization of the null sets of n, p_A is absolutely continuous with respect to n, and by virtue of the Radon-Nikodym theorem there is a non-negative Borel function g such that $p_A(B) = \int_B g \, dn$ for any Borel set B in X. For some positive integer r the set $C = \{x : g(x) \leqq r\}$ has a positive p_A-measure. Then $p_{A \cap C}$ is a non-zero element of E and is dominated by rn; the lemma is proved.|||

24.13 NOTES A few general remarks on methodology may clarify the representation problem. The obvious method of representation, by means of real lattice homomorphisms, is essentially equivalent to embedding the lattice in a lattice of real functions, by virtue of the argument given in this section. This method is completely successful for M spaces. However, there are lattices, such as L^1 relative to Lebesgue measure on the unit interval, for which there are no real homomorphisms. A possible mode of attack is to seek lattice homomorphisms into the extended reals. Another possible attack is suggested by the fact that many lattices have enough order-continuous positive linear functionals to distinguish points (a functional is order-continuous if the supremum of the values on a monotone increasing sequence is the value at the supremum—a suggestive statement of this requirement: the functional satisfies the Lebesgue bounded convergence theorem).

The fact that the spectrum of the adjoint of an L space is totally disconnected suggests a strong connection with the theory of Boolean algebras, and such a connection does, in fact, exist.

The Boolean algebra of open and closed subsets of the spectrum of the adjoint of an L space is actually of very special kind; it is always isomorphic to the Boolean algebra of measurable sets modulo sets of measure zero for a suitably chosen measure. The adjoint of an *L* space is isomorphic to the space of linear operators which preserve absolute continuity and it also permits a representation as the essentially bounded functions relative to some measure.

Much of the motivation for the study of *L* spaces was derived from the applications to the theory of Hermitian operators on Hilbert space. In the terminology of this section the essential content of the spectral theorem may be stated: the smallest norm (respectively, strongly) closed real algebra of operators containing a given Hermitian operator is, under the natural order, an *M* space whose spectrum is homeomorphic to the spectrum of the operator (respectively, is the adjoint of an *L* space).

BIBLIOGRAPHY

[1] Banach, S.,
Théorie des opérations lineaires, Monografje matematyczne, Warszawa, 1932.

[2] Bochner, S. and Martin, W. T.,
Several Complex Variables, Princeton University Press, Princeton, N. J., 1948.

[3] Bourbaki, N.,
Espaces vectoriels topologiques, Actualités Scientifiques et Industrielles 1189 and 1229, Hermann, Paris, 1953 and 1955.

[4] Halmos, P. R.,
Measure Theory, D. Van Nostrand Co., Inc., Princeton, N. J., 1950.

[5] Kelley, J. L.,
General Topology, D. Van Nostrand Co., Inc., Princeton, N. J., 1955.

LIST OF SYMBOLS

(Page number indicates where symbol first appears)

249

INDEX

251